# COMPLEX ANALYSIS

# COMPLEX ANALYSIS

## The Argument Principle in Analysis and Topology

Alan F. Beardon

Dover Publications
Garden City, New York

This Dover edition, first published in 2020, is a republication of the work originally printed by John Wiley & Sons, Ltd., New York, in 1979. The author has provided an update for Theorem 5.4.1 on page 62.

*Library of Congress Cataloging-in-Publication Data*

Names: Beardon, Alan F., author.
Title: Complex analysis : the argument principle in analysis and topology / Alan F. Beardon.
Description: Dover edition | Garden City : Dover Publications, 2020. | Originally published: New York : Wiley, 1979; with an update for theorem 5.4.1, provided by author. | Includes bibliographical references and index. | Summary: "Intended for advanced undergraduates and graduate students in mathematics, this book represented a new approach to the teaching of complex analysis with emphasis on the argument principle in analysis and topology. The presentation, developed in three main parts—Angles, Basic Complex Analysis, and Interactions with Plane Topology—focuses on the concepts of angle and winding numbers and their role in complex analysis, providing a geometrical insight into the subject"—Provided by publisher.
Identifiers: LCCN 2019032473 | ISBN 9780486837185 (trade paperback)
Subjects: LCSH: Functions of complex variables. | Numbers, Complex.
Classification: LCC QA331 .B34 2020 | DDC 515/.9—dc23
LC record available at https://lccn.loc.gov/2019032473

Manufactured in the United States of America
83718102 2023
www.doverpublications.com

*To*
*my parents*

# *Contents*

*PART I ANGLES*

## *PART II BASIC COMPLEX ANALYSIS*



## *PART III INTERACTIONS WITH PLANE TOPOLOGY*

# *Preface*

> It is impossible for an expositor not to write too little for some and too much for others. He can only judge what is necessary by his own experience; and, how long soever he may deliberate, will at last explain many lines which the learned will think impossible to be mistaken and omit many for which the ignorant will want his help. These are censures merely relative and must be quietly endured.
>
> DR. JOHNSON'S *Preface to his edition of Shakespeare, 1765*

Complex analysis is the study of complex numbers and of functions of complex numbers. It is, at the same time, a fertile area of pure mathematics exhibiting many beautiful and surprising results, and a source of powerful techniques which are widely applied in the sciences. Regardless of one's preference, there is one inescapable fact: the motivation and understanding of complex analysis depends in an essential way on the concept and use of an angle. This text is a study of the angle from school geometry (revisited) through complex analysis to plane topology.

In the beginning, the angle is used to motivate (and sometimes define) the product of complex numbers, the $n$th roots of a complex number and the complex logarithm. When considering elementary functions as mappings, the conformal (angle preserving) nature of the functions is emphasized and this plays a vital role in, for example, the applications to electrostatics and to hydrodynamics. Then there is Cauchy's beautiful theory of line integrals and the subsequent methods of evaluating real integrals and infinite series. This whole edifice depends on line integrals of the functions $(z - a)^{-1}$ and these, in turn, depend on the notion of an angle. Finally, the deeper questions concerning plane curves also depend (though less obviously so) on the angle.

It would seem, then, that as a prerequisite of (or perhaps as part of) complex analysis one must look critically at the concept of an angle. This is rarely done, however, in the context of the rigour of post-school mathematics. Part I of this text consists of just this and has, as its ultimate goal, a proof of the fact that *as a non-zero complex number varies continuously so one may select a continuously changing value of its argument* (*or polar angle*). The tools are those of analysis: the role of geometry and of school mathematics here is to motivate and illuminate and they are rarely adequate for a proof. For example, I do not define angle at first: it is easier to develop the trigonometric functions first (as functions of a complex variable) and then to return (in Chapter 5) to the notion of an angle. Even then, it is the argument of a complex number (rather than an angle) which is defined.

One further comment about Part I is necessary. Part I is not intended to provide a thorough treatment of, for example, infinite series and continuous functions: it provides (as far as I can see) only the most elementary and minimal material which is necessary to reach the goal described above. The useful results in calculus (for example, the Mean Value Theorem) require compactness and I have avoided this: thus trigonometry is developed without the use of calculus (an exercise not without interest in itself). In any case, it is doubtful whether a self-contained treatment using calculus would be any shorter.

Having completed Part I, the argument and its continuous variation are available as a tool in further studies. In Part II, I observe first that the integrals of the functions $(z - a)^{-1}$, or winding numbers as they are often called, arise as a straightforward extension of the ideas in Part I and they are then studied (in Chapter 7) for their own sake. There are two major deviations from the usual development of complex analysis. First, I develop the theory and properties of winding numbers without reference to line integrals (the benefits of this occur only in Part III of the text) and second, this development precedes the study of derivatives and integrals of complex functions. In this way I use the winding numbers to obtain (in Chapter 8) the purely topological and geometrical aspects of the theory without reference to calculus. I believe that this leads to a better understanding of the essential nature of winding numbers and moreover, when they subsequently appear in Cauchy's Theorem they do so in a very natural way. Cauchy's Theorem occurs later in the text than is usually the case. It is not needed for the analysis of the local properties of analytic functions and I have preferred to base the evaluation of the elementary integrals on the easier Fundamental Theorem of Calculus (the integral of a derivative) and partial fractions.

It is perhaps worth stating quite explicitly that Part II contains only the results usually given in a first course on complex analysis: it is the approach which differs from that usually given. Throughout, I have adopted the philosophy that geometry and plane topology are necessary and vital parts of complex analysis and, as such, must be treated with the same respect and scrutiny as the analysis itself.

In reading Part II it becomes clear that complex analysis and plane topology can be used to assist each other and Part III consists of material which explores this link to the mutual benefit of both subjects. This material builds on that given in Part II and it is for this reason that winding numbers were developed without reference to integration. The material in Part III is written for the complex analyst and it is only intended to provide a base from which the interested reader may proceed.

Initially, we adopt the point of view that a function is analytic if it has a power series expansion about every point in its domain of definition. In Chapter 9 we prove (using Cauchy's Theorem in the usual way) that a differentiable function is analytic. It is not necessary, however, to appeal to integration theory to prove this (the result itself does not involve integration) and I have added an Appendix in which I indicate how integration may be avoided.

In writing a text on complex analysis, an author must decide whether to include or omit a general discussion on metric spaces. I have omitted such a discussion because it seems to me that the omission will give less trouble to the more knowledgeable student than the inclusion would give to the less knowledgeable student. Nevertheless, I have included the chordal metric (in Chapter 10) for only in this context can Möbius transformations be properly discussed.

From a teaching point of view, it is possible to cover the material in Parts I and II in a single course. If this is to be done, the material can be shortened considerably by defining line integrals immediately after Chapter 6 and then developing the material in Chapter 7 with winding numbers defined as line integrals. This makes Chapter 7 shorter and easier and, in effect, means taking Theorem 9.5.2 as a definition in Chapter 7. The material can be shortened still further by the simple expedience of omitting most of Part I and basing the initial material on school work. Although this is against the very spirit of this text, I believe that even then, there is still much to be gained from the view adopted in Chapter 8.

I wish to thank Mr. C. J. Burrows and Dr. S. J. Patterson who read the manuscript and who provided me with corrections and many useful comments. I also acknowledge the inspiration gained from the texts by L. V. Ahlfors, S. Saks and A. Zygmund and, of course, G. T. Whyburn.

Alan F. Beardon

# PART I
## *Angles*

# Chapter 1

## 1.1 SETS

We begin by accepting the concepts of a *set* of objects (or 'points') and of *belongs to*. If $X$ is a set and if $x$ belongs to $X$ we write $x \in X$: if not we write $x \notin X$. The phrases '*x is in X*' and '*x is a member (or element) of X*' are also in common use and are synonomous with '*x belongs to X*'.

Let $X$ and $Y$ be any two sets. Then *X is a subset of Y* (or *X is contained in Y*) if and only if each element of $X$ is also an element of $Y$: if $X$ is a subset of $Y$ we write $X \subset Y$. The two sets $X$ and $Y$ are equal if and only if they have the same elements, that is if and only if $X \subset Y$ and $Y \subset X$.

Let $X$ be any set and let $P(x)$ be any statement which for each $x$ in $X$ is either true or false. We then use

$$\{x \in X : P(x)\}$$

to denote the set of elements in $X$ for which $P(x)$ is true. We denote by $\{x_1, \ldots, x_n\}$ the set whose only elements are $x_1, \ldots, x_n$ and we use $\emptyset$ to designate the empty set, that is the set with no elements. We say that $X$ is *non-empty* if $X \neq \emptyset$.

To illustrate these ideas let $\mathbf{R}$ be the set of real numbers. Then

$$\{x \in \mathbf{R} : x^2 = 4\} = \{-2, 2\}$$

and

$$\{x \in \mathbf{R} : x^2 = -1\} = \emptyset.$$

An *ordered pair* $(x, y)$ is the set $\{\{x\}, \{x, y\}\}$ with elements $\{x\}$ and $\{x, y\}$. With this definition we can see that $(x, y) = (x', y')$ if and only if $x = x'$ and $y = y'$ and this is the characteristic property of ordered pairs (Exercise 1.1.2). Note that this definition avoids the use of the idea of a 'first coordinate' (for this we would need the notion of a function and this, in turn, depends on ordered pairs).

If $X$ and $Y$ are any sets then the *Cartesian product set* $X \times Y$ is, by definition, the set of ordered pairs $(x, y)$ for which $x \in X$ and $y \in Y$. As a familiar example, the set of points in the Euclidean plane is simply the Cartesian product $\mathbf{R} \times \mathbf{R}$, that is, the set of ordered pairs of real numbers.

If $X$ and $Y$ are any non-empty sets then a function $f$ from $X$ to $Y$ is a 'rule' which assigns to each $x$ in $X$ a unique $y$ in $Y$. The formal definition of a function simply gives the elements of $X$ and their assigned elements in $Y$ as ordered pairs.

Precisely, a *function* $f$ from $X$ to $Y$ is a subset of $X \times Y$ such that for each $x$ in $X$ there is a unique ordered pair $(x, y)$ in $f$. For example, the function $f(x) = x^3$ on $\mathbf{R}$ is defined by

$$f = \{(x, y) \in \mathbf{R} \times \mathbf{R} : y = x^3\}.$$

This definition of a function is important because it shows that the concept of a function depends only on the notion of a set. It is, however, cumbersome and we revert to the more natural view of a function as a rule. If $f$ is a function from $X$ to $Y$ we say that *$f$ maps $X$ into $Y$* and write $f : X \to Y$. We shall normally (but not always) use $f(x)$ to denote the unique element of $Y$ assigned to $x$ by $f$.

We may regard $f$ as acting on the subsets of $X$ in the following way. For each subset $E$ of $X$ we define the set $f(E)$ by

$$f(E) = \{y \in Y : \text{for some } x \text{ in } E, y = f(x)\}:$$

this is the subset of $Y$ consisting of all points in $Y$ of the form $f(x)$, $x \in E$. Similarly for each subset $K$ of $Y$ we define

$$f^{-1}(K) = \{x \in X : f(x) \in K\},$$

and this is the set of points in $X$ which $f$ maps into $K$.

If $f : X \to Y$ and if $f(X) = Y$ then we say that $f$ maps $X$ *onto* $Y$: each element of $Y$ is then of the form $f(x)$. If $f(x) \neq f(x')$ whenever $x$ and $x'$ are distinct points in $X$, then $f$ is said to be *one-to-one* (written 1–1) on $X$. If $f$ is 1–1 and maps $X$ into $Y$ then it has a unique *inverse*: this is a function $g$ from $f(X)$ to $X$ such that for all $x$ in $X$ and all $y$ in $f(X)$,

$$g(f(x)) = x, \qquad f(g(y)) = y.$$

The existence of $g$ follows directly from the definitions (see Exercise 1.1.5) with

$$g = \{(y, x) : (x, y) \in f\}. \tag{1.1.1}$$

When the inverse function exists, it will be denoted by $f^{-1}$.

Two sets $X$ and $Y$ are said to be in *one-to-one correspondence* with each other if and only if there is a 1–1 function which maps $X$ onto $Y$. A set $X$ is said to be *countable* if and only if it is in 1–1 correspondence with a subset of the positive integers (thus finite sets are countable): otherwise $X$ is said to be *uncountable.* Note that $X$ is countable if and only if there is a 1–1 function which maps $X$ into the set of positive integers.

A collection or family $\mathscr{F}$ of sets is said to be *indexed* or *labelled* by the set $A$ if and only if there is a 1–1 correspondence between $A$ and $\mathscr{F}$. If this is so we denote by $X_a$ the set in $\mathscr{F}$ which corresponds to the element $a$ in $A$. Every set in $\mathscr{F}$ arises in this way and $\mathscr{F}$ is then simply the collection

$$\{X_a : a \in A\}. \tag{1.1.2}$$

We define the *union* of the family (1.1.2) as the set consisting of those elements which are in at least one $X_a$: this union is denoted by

$$\bigcup_{a \in A} X_a.$$

In a similar way we define the *intersection* of the family (1.1.2) as the set consisting of those elements which are in $X_a$ for every $a$ in $A$: this is denoted by

$$\bigcap_{a \in A} X_a.$$

Observe that for each $b$ in $A$,

$$\bigcap_{a \in A} X_a \subset X_b \subset \bigcup_{a \in A} X_a.$$

For any two sets $X$ and $Y$ we define $X - Y$ as the set of those $x$ in $X$ which are not in $Y$; thus

$$X - Y = \{x \subset X : x \not\subset Y\}.$$

Given the family (1.1.2) and any set $X$, we have

$$X - (\bigcap_{a \in A} X_a) = \bigcup_{a \in A} (X - X_a) \tag{1.1.3}$$

and

$$X - (\bigcup_{a \in A} X_a) = \bigcap_{a \in A} (X - X_a). \tag{1.1.4}$$

The proofs of these results are easy (see Exercise 1.1.8.).

For a finite family $\{X_1, X_2, \ldots, X_n\}$ of sets we usually write the union and intersection as

$$X_1 \cup X_2 \cup \cdots \cup X_n, \qquad X_1 \cap X_2 \cap \cdots \cap X_n$$

respectively. Two sets $X$ and $Y$ are said to be *disjoint* if $X \cap Y = \emptyset$: we often say that $X$ *meets* $Y$ if $X \cap Y \neq \emptyset$.

**Exercise 1.1**

1. Let $X$ be any set. Prove that $\emptyset$ and $X$ are subsets of $X$. [*Hint*: suppose that $\emptyset$ is not a subset of $X$.]
2. Prove that
$$\{\{x\}, \{x, y\}\} = \{\{x'\}, \{x', y'\}\}$$
if and only if $x = x'$ and $y = y'$.
3. Prove that $X \times Y = \emptyset$ if and only if $X = \emptyset$ or $Y = \emptyset$ (equivalently: $X \times Y \neq \emptyset$ if and only if $X \neq \emptyset$ and $Y \neq \emptyset$).
4. Prove that $(U \times V) \cap (X \times Y) = (U \cap X) \times (V \cap Y)$.
5. Let $f : X \to Y$ be any function and let $g$ be the *set* defined by (1.1.1). Prove that $g$ is a function if and only if $f$ is a 1–1 map of $X$ into $Y$ ($g$ is then defined on $f(X)$).
6. Let $f : W \to X$ and $g : X \to Y$ be functions. Show that
$$\{(w, y) : \text{for some } x \text{ in } X, (w, x) \in f \text{ and } (x, y) \in g\}$$
is a function (this is the function $y = g(f(w))$ ).

7. Let $f : X \to Y$ be any function. Prove that $f(\emptyset) = \emptyset, f^{-1}(\emptyset) = \emptyset$, and $f^{-1}(Y) = X$. [*Note*: we need not have $f(X) = Y$.]
8. Prove (1.1.3) and (1.1.4).
9. Prove that the set **Z** of integers is countable. [*Hint*: consider

$$f(n) = \begin{cases} 2n + 2 & \text{if } n \geqslant 0, \\ -(2n + 1) & \text{if } n < 0.] \end{cases}$$

10. Prove that any subset of a countable set is countable.
11. Prove that if $X$ and $Y$ are countable, then so is $X \times Y$. [*Hint*: if $f$ and $g$ are 1–1 maps of $X$ and $Y$ into the set of positive integers, consider $F(x, y) = 2^{f(x)}3^{g(y)}$.]
12. Prove that if $f : X \to Y$ is 1–1 and if $Y$ is countable, then $X$ is also countable.
13. Let $f : X \to Y$ be any function and let $\{Y_a : a \in A\}$ be a family of subsets of $Y$. Prove that

$$f^{-1}\left(\bigcap_{a\in A} Y_a\right) = \bigcap_{a\in A} (f^{-1}(Y_a)),$$

$$f^{-1}\left(\bigcup_{a\in A} Y_a\right) = \bigcup_{a\in A} (f^{-1}(Y_a)).$$

If $\{X_b : b \in B\}$ is a family of subsets of $X$, what can be said of

$$f\left(\bigcup_{b\in B} X_b\right), \quad \left(f \bigcap_{b\in B} X_b\right)?$$

## 1.2 COMPLEX NUMBERS

We shall assume familiarity with the usual algebraic operations on the set **R** of real numbers. Briefly, to each real number $x$ there is an associated real number $-x$ and, if $x \neq 0$, a real number $x^{-1}$ such that for all real numbers $x, y$ and $z$:

(1) $x + y \in \mathbf{R}, \ x \cdot y \in \mathbf{R}$;
(2) $x + y = y + x, \ x \cdot y = y \cdot x$;
(3) $x + (y + z) = (x + y) + z, \ (x \cdot y) \cdot z = x \cdot (y \cdot z)$;
(4) $(x + y) \cdot z = (x \cdot z) + (y \cdot z)$;
(5) $0 + x = x, \ 1 \cdot x = x$;
(6) $x + (-x) = 0$;
(7) if $x \neq 0$ then $x \cdot x^{-1} = 1$.

We shall use $xy$ for the product $x \cdot y$, and for each positive integer $n$, $x^n$ denotes the product of $x$ with itself $n$ times. Further we use $x/y$ for $x \cdot y^{-1}$ so, for example, $1/2 = 2^{-1}$. If $x \neq 0$, we put $x^0 = 1$.

More generally, we say that a set $X$ is a *field* if it has elements 0 and 1 and an addition and a multiplication defined on it which satisfy (1)–(7) (with **R** replaced by $X$ in (1)). For example, the set **Q** of rational numbers is a field. To avoid the trivial field $\{0\}$ we shall assume that $1 \neq 0$.

Another important feature of **R** is the existence of the subset $\mathbf{R}^+$ of *positive* numbers of **R** with the properties

(8) if $x$ and $y$ are positive then so are $x + y$ and $xy$ and
(9) for each real number $x$ exactly one of the following is true: (a) $x$ is positive, (b) $x = 0$, (c) $-x$ is positive.

We shall see later that 1 is positive. As usual, we write $x - y$ for $x + (-y)$, $x > y$ for $x - y \in \mathbf{R}^+$ and $x \geqslant y$ for $x > y$ or $x = y$. Thus $x > 0$ means $x \in \mathbf{R}^+$ ($x$ is positive); $x$ is *negative* if $-x > 0$, equivalently $x < 0$.

The set **R** is deficient in the sense that it does not contain a square root of any negative number. It follows from (8) and (9) that for each real $x$, either $x^2$ is positive, $x^2 = 0$ or $(-x)^2$ is positive. However $(-x)^2 = x^2$ (Exercise 1.2.1) and so in all cases $x^2 \geqslant 0$. It is now clear that if $k$ is negative then $x^2 \neq k$ and there is no square root of $k$ in **R**.

In order to guarantee the existence of square roots we seek to extend **R** and its algebraic structure to a larger set. The set **C** of complex numbers is the appropriate extension and we shall now outline the construction of **C** from **R**. In Section 1.4 we shall verify that every complex number has at least one square root in **C** and our immediate objective will be attained.

A naïve approach to the construction of **C** is to denote by $i$ some 'quantity' whose square is $-1$ and then to think of complex numbers as being of the form $x + iy$, where $x$ and $y$ are real numbers. Of course, it is the quantity $i$ whose very existence is in doubt and we circumvent this difficulty by using the ordered pair $(x, y)$ (which does exist) instead of $x + iy$.

*Definition 1.2.1 A complex number is an ordered pair of real numbers.*

We define addition and multiplication of complex numbers by

$$(x, y) \oplus (x', y') = (x + x', y + y')$$

and

$$(x, y) \otimes (x', y') = (xx' - yy', xy' + x'y)$$

respectively and we have taken care to distinguish between, for example, the addition $\oplus$ of complex numbers and the addition + of real numbers. These definitions are, of course, motivated by the suggested form $x + iy$.

The alert reader will have observed that **C** is precisely **R** x **R** and consequently **R** is not a subset of **C**. In order to view it as such a subset we make the natural identification of the real number $x$ with the complex number $(x, 0)$ and we henceforth regard **R** as a subset of **C** (we omit the formal treatment).

This identification implies that

$$\begin{aligned} x \oplus x' &= (x, 0) \oplus (x', 0) \\ &= (x + x', 0) \\ &= x + x' \end{aligned}$$

and similarly

$$x \otimes x' = xx':$$

thus we may also identify $\oplus$ and $\otimes$ with the operations of addition and multiplication of real numbers. In conclusion, we regard a real number as a complex number and we write $z + w$ and $zw$ for the sum and product respectively of the complex numbers $z$ and $w$. Observe that **C** is a field (Exercise 1.2.4).

Henceforth we shall use $i$ to denote the complex number (0, 1). Then for all real $x$ and $y$,

$$\begin{aligned} x + iy &= (x, 0) + [(0, 1) \cdot (y, 0)] \\ &= (x, y) \end{aligned}$$

and we may revert to the familiar notation $x + iy$ for complex numbers. Observe that for real $x, y, u$ and $v$, $x + iy = u + iv$ if and only if $x = u$ and $y = v$. Moreover,

$$\begin{aligned} i^2 &= (0, 1) \cdot (0, 1) \\ &= (-1, 0) \\ &= -1, \end{aligned}$$

and so we can at least find a square root of $-1$ in **C**, namely $i$.

We complete this section with a few elementary remarks about the algebra of **C**. First, if $z_1, \ldots, z_n$ are any complex numbers then there is a uniquely defined sum $z_1 + \cdots + z_n$ and the additions may be carried out in any order (Exercise 1.2.8). We emphasize that no meaning is yet attached to an infinite sum $z_1 + z_2 + z_3 + \cdots$.

There are two algebraic identities which are particularly useful. The first is the Binomial Theorem: for each positive integer $n$,

$$(z + w)^n = \sum_{k=0}^{n} \binom{n}{k} z^k w^{n-k} \tag{1.2.1}$$

where

$$\binom{n}{0} = \binom{n}{n} = 1, \qquad \binom{n}{k} = \frac{n!}{k!(n-k)!}, \qquad (1 \leqslant k \leqslant n - 1).$$

This identity is easily proved by induction. The second identity is

$$z^n - w^n = (z - w) \sum_{k=0}^{n-1} z^k w^{n-1-k} \tag{1.2.2}$$

and this is easily proved as each expression is equal to

$$(z^n + z^{n-1}w + \cdots + zw^{n-1}) - (z^{n-1}w + \cdots + zw^{n-1} + w^n).$$

**Exercise 1.2**

1. Prove the following results directly from the field properties (2)–(7):
   (a) $-0 = 0$, $1^{-1} = 1$;
   (b) if for some $x$, $x + 0^* = x$, then $0^* = 0$;
   (c) if for some $x$, $x \neq 0$ and $x \cdot 1^* = x$, then $1 = 1^*$;

(d) if $x + x_1 = x + x_2$, then $x_1 = x_2$;
(e) if $x \neq 0$ and $x \cdot x_1 = x \cdot x_2$, then $x_1 = x_2$;
(f) $x \cdot y = 0$ if and only if $x = 0$ or $y = 0$;
(g) $-(-x) = x$;
(h) if $x \neq 0$, then $(x^{-1})^{-1} = x$;
(i) $(-x)(-y) = xy$;
(j) $(-x)(-x) = x^2$;
(k) $(-y) = (-1) \cdot y$.

2. Prove that $\{a + ib : a, b \in \mathbf{Q}\}$ is a field.
3. Prove the following:
(a) if $x > y$, then $x + t > y + t$;
(b) if $t > 0$ and $x < y$, then $xt < yt$;
(c) if $t < 0$ and $x < y$, then $xt > yt$;
(d) if $x < y$, then $-y < -x$;
(e) if $w < x$ and $x < y$ then $w < y$.
4. Prove that **C** is a field with the definitions

$$-(x, y) = (-x, -y)$$

and

$$(x, y)^{-1} = \left(\frac{x}{x^2 + y^2}, \frac{-y}{x^2 + y^2}\right), \qquad (x, y) \neq (0, 0).$$

5. For each complex number $z$ and each positive integer $n$ define $z^n$ inductively by

$$z^0 = 1, \qquad z^{n+1} = z \cdot z^n.$$

If $n$ is a negative integer define $z^n = 1/z^{|n|}$. Prove that for all integers $m$ and $n$, $z^{m+n} = z^m \cdot z^n$.

6. Prove (1.2.1).
7. Prove (1.2.2).
8. Let $z_1, z_2, \ldots, z_n$ $(n \geqslant 2)$ be complex numbers and define

$$s_n = z_1 + \cdots + z_n$$

inductively by $s_2 = z_1 + z_2$ and

$$s_{j+1} = s_j + z_{j+1}, \qquad j = 2, \ldots, n + 1.$$

Prove (for example, by induction) that if $\sigma$ is any permutation of $\{1, \ldots, n\}$, so

$$\{1, \ldots, n\} = \{\sigma(1), \sigma(2), \ldots, \sigma(n)\},$$

then

$$z + \cdots + z_n = z_{\sigma(1)} + \cdots + z_{\sigma(n)}.$$

## 1.3 UPPER BOUNDS

We now turn our attention to the remaining fundmental property of **R**. A real number $u$ is said to be an *upper bound* of a subset $E$ of **R** if and only if $x \leqslant u$ for all

$x$ in $E$. Observe that if $u$ is an upper bound of $E$ and if $t \geqslant u$ then $t$ is also an upper bound of $E$. We are interested in the possible existence of a smallest upper bound of $E$: accordingly we define $u$ to be a *least upper bound* of $E$ if

(a) $u$ is an upper bound of $E$ and
(b) $u \leqslant t$ for every upper bound $t$ of $E$.

If $u$ and $v$ are both least upper bounds of $E$ then by (b) (with $t = v$), $u \leqslant v$. As we may interchange $u$ and $v$, we also obtain $v \leqslant u$ and so $u = v$. Thus *if* the least upper bound of $E$ exists, it is uniquely determined by $E$. The existence cannot, however, be established from the earlier properties of **R** and so we are led to our final assumption concerning **R**.

*The Least Upper Bound Axiom* *If a non-empty subset of* **R** *has an upper bound, then it has a least upper bound.*

We shall say that $E$ is *bounded above* if $E$ has an upper bound. Thus if $E$ is non-empty and bounded above, then there is a unique least upper bound of $E$, and we denote this by sup ($E$) (the supremum of $E$). Note that sup ($E$) may or may not be in $E$.

Let us illustrate the use of this axiom by deriving the following simple, but essential, result.

*Theorem 1.3.1* *The set* **Z** *of integers is not bounded above.*

*Proof* We assume that **Z** is bounded above; then, by the above axiom, sup (**Z**) exists. For every integer $n$, $n + 1$ is also an integer and so $n + 1 \leqslant \sup(\mathbf{Z})$. Thus $n \leqslant \sup(\mathbf{Z}) - 1$ for every integer $n$ and this violates the assertion that sup (**Z**) is the smallest upper bound of **Z**. The original assumption is therefore false and the proof is complete.

There is a parallel development in terms of lower bounds. A real number $v$ is a *lower bound* of $E$ if $x \geqslant v$ for all $x$ in $E$: if such a $v$ exists then $E$ is said to be *bounded below.* We say that $v$ is the *greatest lower bound* of $E$ if $v$ is a lower bound of $E$ which is not less than any other lower bound of $E$. When this exists it is unique and is denoted by inf ($E$) (the infimum of $E$).

*Theorem 1.3.2* *If a non-empty subset of* **R** *has a lower bound, then it has a greatest lower bound.*

*Proof* Let $E$ be a non-empty subset of **R** and let $v$ be a lower bound of $E$. We denote by $L$ the set of all lower bounds of $E$. As $v \in L$, $L$ is not empty. As $E$ contains an element $e$, $x \leqslant e$ for all $x$ in $L$ and so $e$ is an upper bound of $L$. We conclude that sup ($L$) exists and satisfies $\sup(L) \leqslant e$ for any $e$ in $E$. This shows that sup ($L$) is a lower bound of $E$.

In fact, sup ($L$) is the greatest lower bound of $E$ for if $y$ is any lower bound of $E$, then $y \in L$ and so $y \leqslant \sup(L)$.

A direct consequence of Theorems 1.3.1 and 1.3.2 is that

$$\inf\{n^{-1} : n \in \mathbf{Z}, n > 0\} = 0. \tag{1.3.1}$$

Indeed, if $n > 0$ then $n^{-1} > 0$ and so 0 is a lower bound of

$$E = \{n^{-1} : n \in \mathbf{Z}, n > 0\}.$$

Now let $\alpha = \inf(E)$ and note that $\alpha \geqslant 0$. If $\alpha > 0$ then $n \leqslant \alpha^{-1}$ for all positive integers $n$ and hence for all integers $n$. This contradicts Theorem 1.3.1 and so $\alpha = 0$.

Finally, we say that $E$ is *bounded* if it is bounded both above and below: otherwise, $E$ is *unbounded*. If $E$ is not empty and not bounded above, we write

$$\sup(E) = +\infty:$$

this is purely notational and does not assert the existence of a number $+\infty$.

**Exercise 1.3**

1. Let $E = \{x \in \mathbf{R} : x^2 < 1\}$. Prove that $\sup(E) = 1$ and that $\sup(E) \notin E$. [This shows that $\sup(E)$ is *not* the largest element of $E$.]
2. Prove Theorem 1.3.2 as follows. Define $K = \{x : -x \in E\}$: show that $\sup(K)$ exists and that $-\sup(K) = \inf(E)$.
3. Show that $\sup\{x^4 - 2x^2 + 1 : -5 \leqslant 4x \leqslant 5\} = 1$.
4. Prove that $\{2^n : n = 1, 2, \ldots\}$ is not bounded above.
5. Prove that any non-empty *finite* set $K$ of real numbers has a largest element. The largest element is denoted by $\max(K)$.

   Prove that if any non-empty set $E$ of real numbers has a largest element, say $\max(E)$, then $\max(E) = \sup(E)$. [See Question 1.]
6. Prove that

   $$\inf\{x \in \mathbf{R} : x > 0\} = 0.$$

   Deduce that if a real number $x$ satisfies $x \leqslant \epsilon$ for every positive number $\epsilon$, then $x \leqslant 0$.
7. Let $A$ and $B$ be non-empty bounded sets of real numbers and define

   $$AB = \{ab : a \in A, b \in B\}.$$

   Prove that

   $$\sup(A)\sup(B) \leqslant \sup(AB).$$

   Give an example in which the strict inequality occurs.
8. Prove (by induction) that if $n \in \mathbf{Z}$ and $n > 0$, then $n \geqslant 1$. Thus

   $$\inf\{n \in \mathbf{Z} : n > 0\} = 1$$

   (so 1 is the smallest positive integer). Deduce that if $m$ and $n$ are integers and if $m > n$, then $m \geqslant n + 1$.
9. Prove that if $a$ and $b$ are real numbers with $a \neq 0$, then there is an integer $n$ with $an > b$.

## 1.4 SQUARE ROOTS

A complex number $z$ is a *square root* of $w$ if $z^2 = w$. As the reader well knows, our ability to solve quadratic equations depends on our ability to compute square roots. A more urgent reason to study square roots, however, is that they are needed in order to define the distance between points in the plane. This section is devoted to a study of square roots.

*Theorem 1.4.1* (a) *There is no solution $x$ in* **Q** *of $x^2 = 2$.*
(b) *If $k > 0$ there is a positive solution $x$ in* **R** *of $x^2 = k$: there is no solution in* **R** *if $k < 0$.*
(c) *If $w \in$* **C** *there is a solution $z$ in* **C** *of $z^2 = w$.*

Observe that this shows that we can distinguish between the fields **Q**, **R** and **C** simply by a knowledge of the square roots of, say, 2 and $-1$.

*Proof* The proof of (a) is easy. The set **Q** of rational numbers consists of all numbers of the form $p/q$, where $p$ and $q$ are integers. We may cancel common factors of $p$ and $q$, hence for a given $p/q$ we may assume that not both $p$ and $q$ are even. If $(p/q)^2 = 2$, then $p^2 = 2q^2$ and so $p^2$ and hence $p$ is even. We now write $p = 2m$ and a similar argument shows that $q$ is also even. This is contrary to our assumption and so there is no solution of $x^2 = 2$ in **Q**.

To prove (b) we first recall that we have already considered the case $k < 0$. We now select any positive $k$ and define

$$K = \{x \in \mathbf{R} : x^2 \leqslant k, x \geqslant 0\}.$$

As $0 \in K$, $K$ is not empty. If $x \in K$, then $x \leqslant 1 + k$ (otherwise $x^2 > k$) and so $K$ is bounded above. We deduce that sup $(K)$ exists and, for brevity, we let $t = \sup(K)$. By Theorem 1.3.1 there is a positive integer $n$ satisfying $n > k$ and hence

$$(k/n)^2 \leqslant k^2/n < k.$$

Thus $k/n \in K$ and so $t > 0$.

We first show that $t^2 \geqslant k$. Let $n$ be any positive integer, then $t + 1/n > t$ and so $t + 1/n \notin K$: thus $(t + 1/n)^2 > k$.
This shows that

$$t^2 + \frac{2t+1}{n} \geqslant \left(t + \frac{1}{n}\right)^2 > k$$

and so

$$\frac{k - t^2}{2t+1} \leqslant \frac{1}{n}, \qquad n = 1, 2, \ldots .$$

By (1.3.1), this lower bound is negative or zero, hence $t^2 \geqslant k$.

We now show that $t^2 \leqslant k$. As $t > 0$ we can find a positive integer $n$ with $1/n < t$. Then

$$0 < t - 1/n < t$$

and so there is an $x$ in $K$ with $t - 1/n < x$. Thus

$$t^2 - \frac{2t}{n} < \left(t - \frac{1}{n}\right)^2 < x^2 \leqslant k$$

and (as above) we find that $t^2 \leqslant k$. We conclude that $t^2 = k$.

It remains to prove (c). First, if $w = 0$, then zero is a square root of $w$. Otherwise we write $w = u + iv$ with $u$ and $v$ real and not both zero. In particular, $u^2 + v^2$ is positive and has a positive square root $s$, say.

For the moment, assume that $z$, $z = x + iy$, satisfies

$$(x + iy)^2 = u + iv.$$

Then

$$x^2 - y^2 = u, \; 2xy = v$$

and elimination of $y$ yields

$$(x^2 - u/2)^2 = s^2/4.$$

As $s + u > 0$ we can now *define* $x$ to be the positive square root of $\frac{1}{2}(s + u)$ and $y$ to be $v/2x$: a simple computation then gives $z^2 = w$.

Observe that (1.2.2) shows that

$$z^2 - z_0^2 = (z - z_0)(z + z_0)$$

and so if $z$ and $z_0$ are distinct square roots of, say, $w$, then $z = -z_0$ (see Exercise 1.2.1(f)). In particular, if $k \geqslant 0$, there is a *unique* non-negative square root $t$ of $k$ and we denote this by $\sqrt{k}$.

Some readers will have met the 'polar' form of complex numbers (that is $w = re^{i\theta}$, where $r = |w|$ and $\theta = \arg(w)$) and will be tempted to use this to establish the existence of a square root of $w$, namely $\sqrt{r}e^{i\theta/2}$. This is correct but premature for we have not yet established the existence (less still the properties) of the argument of a complex number. Unless we dispense with rigour, this requires substantial effort and it will be carried out in Chapter 5.

**Exercise 1.4**

1. Show that there is no solution $x$ in **Q** of $x^2 = 3$.
2. Find all solutions of $z^3 = 1$, $z \in \mathbf{C}$.
3. Find all square roots of $i$ and of $-i$.
4. Find all solutions of $az^2 + bz + c = 0$, $a \neq 0$.
5. Find all solutions (in the form $x + iy$) of $z^4 - z^2 + 1 = 0$.

## 1.5 DISTANCE

Let $z$ and $w$ be two complex numbers, say $z = x + iy$ and $w = u + iv$ (with $x$, $y$, $u$ and $v$ real). We define $|z - w|$ by

$$|z - w| = \sqrt{[(x - u)^2 + (y - v)^2]} \geqslant 0$$

and observe that this always exists. This is the *distance* between $z$ and $w$ and is, of course, the same as the Euclidean distance between the points $(x, y)$ and $(u, v)$ in the Euclidean plane.

The two basic properties of this distance are

$$|zw| = |z| \cdot |w| \tag{1.5.1}$$

and

$$|z| - |w| \leqslant |z + w| \leqslant |z| + |w|. \tag{1.5.2}$$

The identity (1.5.1) is easily proved by expanding the product $zw$ in terms of $x, y, u$ and $v$ and computing $|zw|^2$. The inequality (1.5.2) is best proved by the use of other elementary functions of $z$ and we shall introduce these first.

We define the *conjugate* $\bar{z}$ of $z$ by

$$\bar{z} = x - iy$$

and the *real* and *imaginary parts* of $z$, denoted by Re $[z]$ and Im $[z]$ respectively, by

$$\text{Re}\,[z] = x, \qquad \text{Im}\,[z] = y.$$

The following identities and inequalities are trivial and are left to the reader to verify:

$$z = \text{Re}\,[z] + i\,\text{Im}\,[z]; \qquad z\bar{z} = |z|^2;$$

$$z + \bar{z} = 2\,\text{Re}\,[z]; \qquad z - \bar{z} = 2i\,\text{Im}\,[z];$$

$$\overline{z_1 + z_2} = \bar{z}_1 + \bar{z}_2; \qquad \overline{z_1 z_2} = \bar{z}_1 \cdot \bar{z}_2;$$

$$\overline{(z^{-1})} = (\bar{z})^{-1}; \qquad \overline{z_1 - z_2} = \bar{z}_1 - \bar{z}_2;$$

$$\bar{\bar{z}} = z; \qquad |\bar{z}| = |z|;$$

$$|\text{Re}\,[z]| \leqslant |z|; \qquad |\text{Im}\,[z]| \leqslant |z|.$$

Let us now establish the inequalities (1.5.2). For each $z$ and $y$ we have

$$\begin{aligned} |z + w|^2 &= (z + w)\overline{(z + w)} \\ &= (z + w)(\bar{z} + \bar{w}) \\ &= |z|^2 + |w|^2 + (z\bar{w} + \bar{z}w) \\ &= |z|^2 + 2\,\text{Re}\,[z\bar{w}] + |w|^2 \\ &\leqslant |z|^2 + 2|z| \cdot |\bar{w}| + |w|^2 \\ &= |z|^2 + 2|z| \cdot |w| + |w|^2 \\ &= (|z| + |w|)^2. \end{aligned}$$

We conclude that

$$|z + w| \leqslant |z| + |w|.$$

This shows that

$$|z| = |(z + w) - w| \leqslant |z + w| + |w|$$

and the proof of (1.5.2) is complete.

Obviously the real relations (1.5.1) and (1.5.2) extend (by induction) to finite sums and products:

$$|z_1 \cdots z_n| = |z_1| \cdots |z_n|$$

and

$$|z_1 + \cdots + z_n| \leqslant |z_1| + \cdots + |z_n|. \tag{1.5.3}$$

There is no doubt of the importance of the geometric interpretation of complex numbers and of the insight and understanding that this brings. We are adopting the attitude that a geometric argument is no substitute for a formal proof and it is precisely this attitude which prevents us (for the moment) from using the concepts of angle and of the argument of a complex number. Nevertheless, we must not ignore the geometry.

The geometric interpretation is presumably familiar to the reader. The set **C** is the Euclidean plane **R** x **R**. Addition in **C** is precisely the addition of vectors in **R** x **R** and may be expressed by the 'parallelogram law' as represented in Diagram 1.5.1.

The conjugate $\bar{z}$ of $z$ is simply the image of $z$ after reflection in the real axis. In terms of complex numbers, the 'cosine rule' reads

$$|z - w|^2 = |z|^2 + |w|^2 - 2|z||w| \cos \theta.$$

We have proved that

$$|z - w|^2 = |z|^2 + |w|^2 - 2\mathrm{Re}\,[z\bar{w}]$$

and later we shall be able to identify these two equations.

Finally, we have the 'triangle inequality':

$$\begin{aligned} |z_1 - z_2| &= |(z_1 - z_3) + (z_3 - z_2)| \\ &\leqslant |z_1 - z_3| + |z_3 - z_2|. \end{aligned}$$

In Euclidean terms this simply says that if the triangle with vertices $z_1, z_2, z_3$ has sides of lengths $L_1, L_2$ and $L_3$, then $L_3 \leqslant L_1 + L_2$.

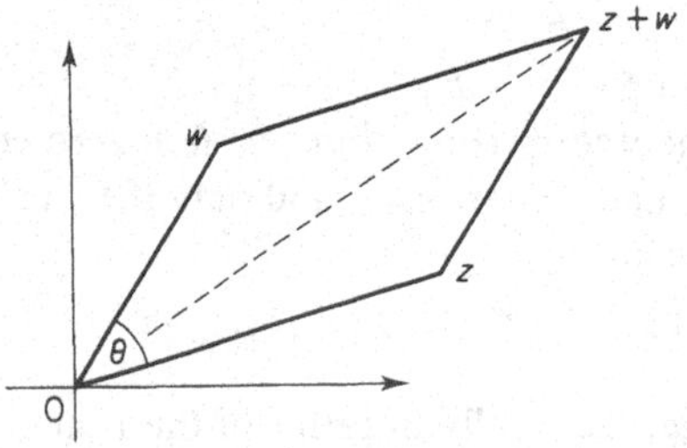


Diagram 1.5.1

**Exercise 1.5**

1. Establish the Cauchy–Schwarz inequality : for complex numbers $z_1, \ldots, z_n$, $w_1, \ldots, w_n$,

$$\left| \sum_{j=1}^{n} z_j w_j \right|^2 \leqslant \left( \sum_{j=1}^{n} |z_j|^2 \right) \left( \sum_{j=1}^{n} |w_j|^2 \right).$$

[*Hint*: consider, as a function of real $t$, the non-negative quadratic function $\Sigma(|z_j| - t|w_j|)^2$.]

2. Find the real and imaginary parts of

$$\frac{1+z}{1-z}, \qquad \left(\frac{1}{\bar{z}}\right)^2,$$

where $z = x + iy$.

3. Let $z = x + iy$, where $x$ and $y$ are real. Prove that

$$|z| \leqslant |x| + |y| \leqslant \sqrt{2}|z|.$$

Prove also that, in general, $\sqrt{2}$ cannot be replaced by a smaller constant.

4. Suppose that $t$ is positive and that

$$z_j = x_j + iy_j, \qquad j = 1, \ldots, n,$$

satisfies $|y_j| \leqslant tx_j$. Prove that

$$|z_1 + \cdots + z_n| \geqslant c(|z_1| + \cdots + |z_n|)$$

where $c\sqrt{(1+t^2)} = 1$. Interpret this in geometric terms (put $t = \tan \alpha$).

5. Find $\sup \{ |(1+i)z^2 + iz| : |z| < 1 \}$.

6. Given $r_1$ and $r_2$ satisfying $0 < r_1 < r_2$, prove that

$$\inf \{ |z - w| : |z| = r_1, |w| = r_2 \} = r_2 - r_1.$$

7. Let $n$ be a positive integer and let

$$p(z) = a_0 + a_1 z + \cdots + a_{n-1} z^{n-1} + z^n.$$

(a) Suppose that each $a_j$ is real. Prove that $p(z) = 0$ if and only if $p(\bar{z}) = 0$.
(b) Suppose that for all $z$, either both or neither of $p(z)$ and $p(\bar{z})$ are zero. Making (reasonable) assumptions about $p$ try and prove that each $a_j$ is real.

8. Let $k$ be positive. Identify (in geometric terms) each of the sets
(a) $\{z : |z - a| = k|z - b|\}$,
(b) $\{z : |z - a| + |z - b| = k\}$,
(c) $\{z : |z - a| - |z - b| = k\}$.

9. Using co-ordinate geometry show that two non-zero complex numbers $z$ and $w$ represent perpendicular directions if and only if $\operatorname{Re}[z\bar{w}] = 0$.

10. Prove that for all $z$ and $w$,

$$|z - w|^2 + |z + w|^2 = 2|z|^2 + 2|w|^2$$

and interpret this geometrically in terms of the triangle with vertices $0$, $z + w$, $z - w$.

# Chapter 2

## 2.1 INFINITE SERIES

We have already seen that given $n$ complex numbers $z_1, \ldots, z_n$ there is a unique complex number $z_1 + \cdots + z_n$ found by performing these additions in any order. Our next task is to define (where possible) the value of an infinite sum (or series): such a sum will be denoted, for example, by the expressions

$$z_0 + z_1 + \cdots, \qquad \sum_{n=0}^{\infty} z_n. \tag{2.1.1}$$

Before giving a formal definition we must be clear what is meant by our notation $z_n$. We are suggesting that to each integer $n$ under consideration there is assigned in a unique way a complex number $z_n$: this is precisely what is meant by a function from the set $\{0, 1, 2, \ldots\}$ to **C**. We shall explore this idea later; for the present we merely wish to clarify the notation.

*Definition 2.1.1* *The infinite sum* (2.1.1) *is said to converge to (or equal) $w$ if and only if for each positive number $\epsilon$ there is an integer $N$ such that*

$$\left| w - \sum_{k=0}^{n} z_k \right| < \epsilon$$

*whenever $n \geqslant N$. The infinite sum* (2.1.1) *is said to diverge if it does not converge.*

It is important to realize that the series (2.1.1) can equal at most one $w$. If the series is equal to both $w_1$ and $w_2$ then for each positive $\epsilon$ there are integers $N_1$ and $N_2$ such that

$$\left| w_j - \sum_{k=0}^{n} z_k \right| < \epsilon$$

whenever $n > N_j$ $(j = 1, 2)$. Thus if $n = N_1 + N_2$, then

$$|w_1 - w_2| \leqslant \left| w_1 - \sum_{k=0}^{n} z_k \right| + \left| w_2 - \sum_{k=0}^{n} z_k \right| < 2\epsilon$$

and so $w_1 = w_2$ (see Exercise 1.3.6).

The idea underlying Definition 2.1.1 is this: we are assigning the value $w$ to the infinite sum if we can guarantee as good an approximation to $w$ as we wish simply by adding a large but *finite* number of the $z_n$ together. The distinction must be clearly understood: any *finite* sum (of however many terms) is defined by the algebra of **C**, an *infinite* sum is not.

*Example 2.1.1* We shall prove that the infinite series

$$\sum_{k=0}^{\infty} z^k$$

(with $z^0 = 1$) converges if $|z| < 1$ and in this case its value is $1/(1-z)$.

First, the identity (1.2.2) shows that for all $z$,

$$1 - z^{n+1} = (1-z)(1 + z + z^2 + \cdots + z^n).$$

If $|z| < 1$ we can write

$$|z| = (1+\delta)^{-1}$$

where $\delta > 0$: then

$$\begin{aligned}\left|\frac{1}{1-z} - \sum_{k=0}^{n} z^k\right| &= \frac{|z|^{n+1}}{|1-z|}\\ &\leqslant \frac{|z|^{n+1}}{1-|z|}\\ &= \frac{1}{\delta(1+\delta)^n}\\ &< 1/n\delta^2,\end{aligned}$$

the last inequality being a consequence of the Binomial Theorem. Given any positive $\epsilon$ we select an integer $N$ satisfying $N > 1/\epsilon\delta^2$. Then for all $n$ greater than $N$, $1/n\delta^2 < \epsilon$ and the desired result follows,

We now collect together for future reference three elementary but useful results.

*Theorem 2.1.1* (a) *The infinite series* $\Sigma_{n=0}^{\infty} z_n$ *converges to* $w$ *if and only if the series* $\Sigma_{n=0}^{\infty} \bar{z}_n$ *converges to* $\bar{w}$.

(b) *If the series* $\Sigma_{n=0}^{\infty} z_n$ *and* $\Sigma_{n=0}^{\infty} z_n^*$ *converge to* $w$ *and* $w^*$ *respectively, then for all complex numbers* $s$ *and* $t$ *the series* $\Sigma_{n=0}^{\infty}(sz_n + tz_n^*)$ *converges to* $sw + tw^*$.

(c) *The series* $\Sigma_{n=0}^{\infty} z_n$ *converges to* $w$ *if and only if the series* $\Sigma_{n=0}^{\infty} \mathrm{Re}[z_n]$ *and* $\Sigma_{n=0}^{\infty} \mathrm{Im}[z_n]$ *converge to* $\mathrm{Re}[w]$ *and* $\mathrm{Im}[w]$ *respectively.*

*Proof* The proof of (a) is trivial as

$$\left|w - \sum_{k=0}^{n} z_k\right| = \left|\bar{w} - \sum_{k=0}^{n} \bar{z}_k\right|.$$

To prove (b) we consider any positive number $\epsilon$ and select a positive number $\delta$ such that $(|s| + |t|)\delta < \epsilon$. There are integers $N$ and $N^*$ such that

$$\left| w - \sum_{k=0}^{n} z_k \right| < \delta, \qquad \left| w^* - \sum_{k=0}^{n} z_k^* \right| < \delta$$

whenever $n$ is greater than both $N$ and $N^*$. For such $n$,

$$\left| (sw + tw^*) - \sum_{k=0}^{n} (sz_k + tz_k^*) \right| \leqslant |s| \left| w - \sum_{k=0}^{n} z_k \right| + |t| \left| w^* - \sum_{k=0}^{n} z_k^* \right|$$
$$< \epsilon$$

and (b) is proved.

The proof of (c) consists of several straightforward applications of (a) and (b) and is left to the reader.

**Exercise 2.1**

1. Suppose that $\Sigma_{n=0}^{\infty} z_n$ converges to $w$. Show that for any positive $\epsilon$ there is an integer $n$ such that $|z_m| < \epsilon$ whenever $m \geqslant n$. Deduce that the series in Example 2.1.1 does not converge if $|z| \geqslant 1$.
2. Prove that if $\Sigma_{n=0}^{\infty} x_n$ converges to $w$ and if each $x_n$ is real then $w$ is also real.
3. Describe geometrically the values of $z$ for which the series

$$\sum_{n=0}^{\infty} \left( \frac{z}{1+z} \right)^n$$

   converges.
4. Prove that

$$\sum_{n=0}^{\infty} \frac{z^n}{1 + z^{2n}}$$

   converges if $|z| \neq 1$.
5. Prove that

$$\sum_{n=0}^{\infty} (1 - z^{2^n})$$

   converges if and only if at least one of its terms is zero.
   [*Hint*: $1 - z^{2^{n+1}} = (1 - z^{2^n})(1 + z^{2^n})$.]
6. Let $a_n$ be defined by $a_0 = 0, a_1 = 1$ and $a_n = a_{n-1} + a_{n-2}$ when $n \geqslant 2$. For which values of $z$ does the series $\Sigma_{n=0}^{\infty} a_n z^n$ converge and what is its value?
7. Express $(1 + z)^{-1}(3 - z)^{-1}$ as a series in each of the regions (a) $|z| < 1$, (b) $1 < |z| < 3$, (c) $|z| > 3$.

## 2.2 TESTS FOR CONVERGENCE

There is a clear need to establish tests to decide whether or not a given infinite series actually converges. One method (which is not really a test) is to guess the answer and to attempt to verify it from first principles: we did this in the previous example.

In the case of an infinite series of real positive terms there is a simple and adequate test.

*Theorem 2.2.1* *Let $x_0, x_1, \ldots$ be real non-negative numbers. Then the series $\sum_{n=0}^{\infty} x_n$ convergences if and only if the set*

$$K = \left\{ \sum_{k=0}^{n} x_k : n = 0, 1, \ldots \right\}$$

*is bounded above. If the series does converge then it is equal to* sup $(K)$.

*Proof* Let us suppose that $K$ is bounded above and write $k$ for sup $(K)$. If $\epsilon$ is any positive number then $k - \epsilon$ is not an upper bound of $K$ and so there is some $N$ with

$$k - \epsilon < x_0 + x_1 + \cdots + x_N \leqslant k$$

(the latter inequality holds as $k$ is an upper bound of $K$). As each $x_n$ is positive we have

$$k - \epsilon < x_0 + x_1 + \cdots + x_n \leqslant k$$

and therefore

$$| k - (x_0 + \cdots + x_n) | < \epsilon$$

for all $n$ satisfying $n \geqslant N$, and so the series converges to $k$.

Suppose now that the series converges to $w$. Then there is an integer $N$ such that for all $n$ satisfying $n \geqslant N$,

$$| w - (x_0 + \cdots + x_n) | < 1.$$

For these $n$.

$$x_0 + \cdots + x_n \leqslant | x_0 + \cdots + x_n | \leqslant 1 + | w |$$

and, as each $x_n$ is non-negative, this must hold for all $n$. Thus $K$ is bounded above and by the preceding paragraph, $w = k$.

As an immediate corollary of this result we have the Comparison Test.

*The Comparison Test* *Suppose that the real numbers $x_n$, $a_n$ and $M$ satisfy*
(a) $0 \leqslant a_n \leqslant M x_n$ $(n = 0, 1, \ldots)$ *and*
(b) $\sum_{n=0}^{\infty} x_n$ *is convergent.*
*Then $\sum_{n=0}^{\infty} a_n$ is convergent.*

This is clear as obviously

$$a_0 + \cdots + a_n \leqslant M(x_0 + \cdots + x_n).$$

In an attempt to derive a result of this nature for complex series we are led to the notion of absolute convergence.

*Definition 2.2.1* *The series $\Sigma_{n=0}^{\infty} z_n$ is said to be absolutely convergent if and only if the series $\Sigma_{n=0}^{\infty} |z_n|$ is convergent.*

We are concerned with the question of convergence rather than the actual value of the infinite sum. The importance of this concept can be seen from the following result in conjunction with Theorem 2.2.1.

*Theorem 2.2.2* *If the series $\Sigma_{n=0}^{\infty} z_n$ is absolutely convergent, then it is also convergent.*

*Proof* We write $z_n = x_n + iy_n$ (with $x_n$ and $y_n$ real) and assume that $\Sigma_{n=0}^{\infty} |z_n|$ converges. The proof consists only of applications of the Comparison Test and Theorem 2.1.1 and we shall omit explicit reference to these results.

As

$$|x_n| \leqslant \sqrt{(x_n^2 + y_n^2)} = |z_n|$$

we see that $\Sigma_{n=0}^{\infty} |x_n|$ converges. Next,

$$0 \leqslant |x_n| + x_n \leqslant 2|x_n|, \qquad 0 \leqslant |x_n| - x_n \leqslant 2|x_n|$$

and so the series

$$\sum_{n=0}^{\infty} (|x_n| + x_n), \qquad \sum_{n=0}^{\infty} (|x_n| - x_n)$$

both convergence. This shows that

$$\tfrac{1}{2} \sum_{n=0}^{\infty} [(|x_n| + x_n) - (|x_n| - x_n)]$$

(which is $\Sigma_{n=0}^{\infty} x_n$) converges. A similar argument shows that $\Sigma_{n=0}^{\infty} y_n$ converges and hence

$$\sum_{n=0}^{\infty} (x_n + iy_n)$$

converges.

Theorem 2.2.2 enables us to extend an earlier inequality to infinite series.

*Theorem 2.2.3* *If the series $\Sigma_{n=0}^{\infty} z_n$ is absolutely convergent, then*

$$\left| \sum_{n=0}^{\infty} z_n \right| \leqslant \sum_{n=0}^{\infty} |z_n|.$$

*Proof* As the given series is absolutely convergent, the two terms in this inequality are meaningful. We define $\epsilon$ by

$$\epsilon = \left| \sum_{k=0}^{\infty} z_k \right| - \sum_{k=0}^{\infty} |z_k|$$

and suppose that $\epsilon > 0$. Then there is an integer $N$ such that

$$\left| \sum_{k=0}^{n} z_k - \sum_{k=0}^{\infty} z_k \right| < \epsilon$$

whenever $n \geqslant N$ and this implies that

$$\begin{aligned} \left| \sum_{k=0}^{\infty} z_k \right| &\leqslant \left| \sum_{k=0}^{\infty} z_k - \sum_{k=0}^{n} z_k \right| + \left| \sum_{k=0}^{n} z_k \right| \\ &< \left| \sum_{k=0}^{\infty} z_k \right| - \sum_{k=0}^{\infty} |z_k| + \sum_{k=0}^{n} |z_k| \end{aligned}$$

which is clearly false. Thus $\epsilon \leqslant 0$ and this is the desired inequality.

We end this discussion with three examples.

*Example 2.2.1* We show that

$$\sum_{n=1}^{\infty} \frac{1}{n^2} \leqslant 2.$$

Given any positive integer $n$ we choose $k$ satisfying $2^k > n$. Then

$$\begin{aligned} 1 + \frac{1}{2^2} + \cdots + \frac{1}{n^2} &< 1 + \left(\frac{1}{2^2} + \frac{1}{3^2}\right) + \left(\frac{1}{4^2} + \cdots + \frac{1}{7^2}\right) + \cdots \\ &\quad + \left(\frac{1}{(2^k)^2} + \cdots + \frac{1}{(2^{k+1}-1)^2}\right) \\ &\leqslant 1 + 2\left(\frac{1}{2^2}\right) + 4\left(\frac{1}{4^2}\right) + \cdots + 2^k\left(\frac{1}{2^{2k}}\right) \\ &= 1 + \frac{1}{2} + \frac{1}{2^2} + \frac{1}{2^3} + \cdots + \frac{1}{2^k} \\ &< 2 \end{aligned}$$

and the result follows from Theorem 2.2.1.

We have considered infinite sums in which the parameter $n$ takes the values $0, 1, 2, \ldots$. There is no particular reason why the first value of $n$ should be zero: the theory is equally applicable when $n$ takes the values, say, $N, N+1, \ldots$ (in the previous example, $N = 1$) and we leave the reader to make the appropriate modifications. This simple observation has one useful corollary: if the series $\sum_{n=N}^{\infty} z_n$ converges for one value of $N$ it necessarily converges for all values of $N$.

*Example 2.2.2* We shall show that the series

$$\exp(z) = \sum_{n=0}^{\infty} \frac{z^n}{n!}$$

is absolutely convergent regardless of the values of $z$.

Given any $z$ we select an integer $N$ satisfying $N > 2 \mid z \mid$. Then for $n \geqslant N$,

$$\left| \frac{z^{n+1}}{(n+1)!} \right| \leqslant \frac{1}{2} \left| \frac{z^n}{n!} \right|$$

and so for all positive integers $k$,

$$\left| \frac{z^{N+k}}{(N+k)!} \right| \leqslant \frac{1}{2^k} \frac{\mid z \mid^N}{N!} .$$

We conclude that for all $n \geqslant N$,

$$\sum_{k=N}^{n} \left| \frac{z^k}{k!} \right| \leqslant \frac{\mid z \mid^N}{N!} \left(1 + \frac{1}{2} + \cdots + \frac{1}{2^{n-N}}\right) \leqslant 2 \mid z \mid^N / N!$$

and so the given series is absolutely convergent.

*Example 2.2.3* We shall show that the series

$$\sum_{n=1}^{\infty} \frac{1}{n}$$

does not converge. If it does converge, say to $y$, then

$$y = \sup \left\{1 + \frac{1}{2} + \cdots + \frac{1}{n} : n = 1, 2, \ldots\right\}$$

and so there is some $n$ with

$$y - \frac{1}{2} < 1 + \frac{1}{2} + \cdots + \frac{1}{n} .$$

Then

$$\begin{aligned} y &\geqslant 1 + \frac{1}{2} + \cdots + \frac{1}{4n} \\ &= \left(1 + \cdots + \frac{1}{n}\right) + \left(\frac{1}{n+1} + \cdots + \frac{1}{4n}\right) \\ &> y - \frac{1}{2} + 3n \left(\frac{1}{4n}\right) \\ &= y + \frac{1}{4} \end{aligned}$$

and this is false.

**Exercise 2.2**

1. Does the series

$$\sum_{n=1}^{\infty} \frac{3 - 5n}{n^2 + 2n + 2}$$

converge or diverge? Justify your answer.

2. The Ratio Test. Suppose that each $z_n$ is non-zero.
   (a) Prove that $\Sigma_{n=0}^{\infty} z_n$ is absolutely convergent if for all $n$ and some constant $t$, $|z_{n+1}/z_n| \leqslant t < 1$.
   (b) Prove that the above series diverges if for all $n$ and some constant $t$, $|z_{n+1}/z_n| \geqslant t \geqslant 1$.
3. For which integers $p$ does $\Sigma_{n=1}^{\infty} p^n n^p$ converge?
4. Suppose that each $a_n$ is real and non-negative and suppose that $a_n \leqslant a_{n+1}$, $n = 0, 1, \ldots$. Prove that if the first of the series

$$\sum_{n=0}^{\infty} 2^n a_{2^n}, \qquad \sum_{n=0}^{\infty} a_n$$

   converges, then so does the second. Give an example to show that the converse is false.
5. (a) Prove that $\Sigma_{n=1}^{\infty} n^{-3/2}$ converges.
   (b) Does $\Sigma_{n=1}^{\infty} [\sqrt{(n+1)} - \sqrt{n}]$ converge?
   (c) Does $\displaystyle\sum_{n=1}^{\infty} \left[\frac{1}{\sqrt{n}} - \frac{1}{\sqrt{(n+1)}}\right]$ converge?
6. Let $x$ be real. Prove that $\Sigma_{n=1}^{\infty}(x^2 + n^2)^{-1}$ converges and that $\Sigma_{n=1}^{\infty}(x^2 + n)^{-1}$ diverges.
7. Prove that $\Sigma_{n=1}^{\infty}(-1)^n/n$ is convergent but not absolutely convergent. [*Hint*: first show that

$$\sum_{n=1}^{\infty} \left(\frac{1}{2n} - \frac{1}{2n-1}\right)$$

   converges; this is *not* the given series.]
8. Prove that if $\Sigma_{n=0}^{\infty} a_n z_0^n$ is absolutely convergent then
   (a) there is a constant $M$ such that for all $n$, $|a_n| \leqslant M/|z_0|^n$ and
   (b) $\Sigma_{n=0}^{\infty} a_n z^n$ is absolutely convergent when $|z| \leqslant |z_0|$.

## 2.3 THE CAUCHY PRODUCT

In the final section of this chapter we prove one result which has many applications. Given the convergent infinite series

$$\sum_{n=0}^{\infty} a_n, \qquad \sum_{n=0}^{\infty} b_n \tag{2.3.1}$$

we may attempt to evaluate their product as follows:

$$\left(\sum_{n=0}^{\infty} a_n\right)\left(\sum_{n=0}^{\infty} b_n\right) = (a_0 + a_1 + a_2 + \cdots)(b_0 + b_1 + b_2 + \cdots)$$
$$= (a_0 b_0) + (a_1 b_0 + a_0 b_1) + (a_2 b_0 + a_1 b_1 + a_0 b_2) + \cdots$$

This of course requires justification and we give this in the simplest case which is sufficient for our needs.

*Theorem 2.3.1* *Let the series in (2.3.1) be absolutely convergent to a and b respectively and define $c_n$ by*

$$c_n = \sum_{j=0}^{n} a_j b_{n-j}\,.$$

*Then $\Sigma_{n=0}^{\infty} c_n$ is absolutely convergent to ab.*

This result is fundamental to our development of the subject and the reader should become thoroughly familiar with it. Although it is possible to give a direct proof we shall first give the proof when each $a_n$ and $b_n$ are non-negative. In this way the reader should be able to appreciate fully the ideas involved.

Before giving the proof we remark that given any finite set $E$, say $E = \{e_1, \ldots, e_n\}$, and any function $f : E \to \mathbf{C}$, we can form the sum of $f(e_j)$ from $j = 1$ to $j = n$. This sum does not depend on the chosen order of the $e_j$ and so we can denote its value by

$$\sum_{x \in E} f(x),$$

a symbol which does not refer to the order. If $E_1$ and $E_2$ satisfy $E_1 \cap E_2 = \emptyset$, $E_1 \cup E_2 = E$, then

$$\sum_{x \in E} f(x) = \sum_{x \in E_1} f(x) + \sum_{x \in E_2} f(x).$$

*Proof of Theorem 2.3.1* We shall be interested in the function $f : E \to \mathbf{C}$ where

$$E = \{(i, j) : i, j \in \mathbf{Z}, i \geqslant 0, j \geqslant 0\}$$

and $f(i, j) = a_i b_j$. Indeed,

$$(a_0 + \cdots + a_n)(b_0 + \cdots + b_n) = \sum_{x \in S_n} f(x)$$

and

$$c_0 + \cdots + c_n = \sum_{x \in T_n} f(x),$$

where

$$S_n = \{(i, j) \in E : 0 \leqslant i \leqslant n, 0 \leqslant j \leqslant n\}$$

and

$$T_n = \{(i, j) \in E : 0 \leqslant i + j \leqslant n\}.$$

The reader should sketch the sets $S_n$ and $T_n$; they are 'square' and 'triangular' subsets of $E$.

As

$$T_n \subset S_n \subset T_{2n}$$

and as the $a_j$ and the $b_j$ are non-negative, we deduce that

$$\sum_{k=0}^{n} c_k \leqslant \left(\sum_{k=0}^{n} a_k\right)\left(\sum_{k=0}^{n} b_k\right) \leqslant \sum_{k=0}^{2n} c_k$$

The first inequality yields (using Theorem 2.2.1)

$$\sum_{k=0}^{n} c_k \leqslant ab$$

and so $c \leqslant ab$.

The second inequality gives

$$\left(\sum_{k=0}^{n} a_k\right)\left(\sum_{k=0}^{n} b_k\right) \leqslant c.$$

Given any positive $\epsilon$ the left-hand side is, for sufficiently large $n$, at least $(a - \epsilon)(b - \epsilon)$. Thus

$$ab \leqslant c + \epsilon(a + b)$$

and so $ab \leqslant c$. This proves that $c = ab$.

To establish the result when $a_j$ and $b_j$ are real we write

$$a_j^+ = \tfrac{1}{2}(|a_j| + a_j), \qquad a_j^- = \tfrac{1}{2}(|a_j| - a_j)$$

and similarly for $b_j$. Note that $a_j^+$ and $a_j^-$ are both non-negative and not greater than $|a_j|$. Thus, by the absolute convergence of the given series, the infinite sum of, for example, $a_j^+$ converges. The same reasoning holds for the $b_j$.

For brevity, let $\Sigma$ denote summation of $k$ over all non-negative integers. Then

$$\begin{aligned}(\Sigma a_k)(\Sigma b_k) &= (\Sigma a_k^+ - \Sigma a_k^-)(\Sigma b_k^+ - \Sigma b_k^-) \\ &= \Sigma a_k^+ \Sigma b_k^+ - \Sigma a_k^+ \Sigma b_k^- - \Sigma a_k^- \Sigma b_k^+ + \Sigma a_k^- \Sigma b_k^- \\ &= \Sigma\left[\sum_{j=0}^{k} a_j^+ b_{k-j}^+ - \sum_{j=0}^{k} a_j^+ b_{k-j}^- - \sum_{j=0}^{k} a_j^- b_{k-j}^+ + \sum_{j=0}^{k} a_j^- b_{k-j}^-\right] \\ &= \Sigma\left[\sum_{j=0}^{k} a_j b_{k-j}\right] \\ &= \Sigma c_k\end{aligned}$$

and the result follows.

Finally, the extension to the case when $a_j$ and $b_j$ are complex is achieved in a similar manner by writing $a_j = x_j + iy_j$ and $b_j = u_j + iv_j$: the details are omitted.

We give just one application now: more important applications occur in Chapter 4.

*Example 2.3.1* If $|z| < 1$, then

$$\sum_{n=1}^{\infty} nz^{n-1} = (1-z)^{-2}.$$

To see this, we simply take $a_j = b_j = z^j$.

**Exercise 2.3**

1. Express $(1-z^2)^{-1}$ as a series $\Sigma_{n=0}^{\infty} a_n z^n$
   (a) using Example 2.1.1,
   (b) using Theorem 2.3.1,
   (c) using $(1-z)^{-1} + (1+z)^{-1}$.
2. Suppose that $\Sigma_{n=0}^{\infty} a_n z^n$ converges where $|z| < 1$ and let

   $$a_0 + \cdots + a_n = s_n.$$

   Prove that

   $$\sum_{n=0}^{\infty} s_n z^n = (1-z)^{-1} \sum_{n=0}^{\infty} a_n z^n.$$

3. Use Question 2 to obtain $(1-z)^{-3}$ as a series.
4. Use Theorem 2.3.1 to verify that for all $z$,

   $$\left(\sum_{n=0}^{\infty} \frac{z^n}{n!}\right)\left(\sum_{n=0}^{\infty} \frac{(-z)^n}{n!}\right) = 1.$$

5. Use Theorem 2.3.1 to verify that for all $z$,

   $$2\left(\sum_{n=0}^{\infty} \frac{z^{2n}}{(2n)!}\right)^2 = 1 + \sum_{n=0}^{\infty} \frac{(2z)^{2n}}{(2n)!}.$$

# Chapter 3

## 3.1 CONTINUITY

In terms of approximation we shall say that $f$ is continuous at a point $w$ if, given any 'permissible error' $\epsilon$, there is a region around $w$ in which $f(z)$ can be approximated by the value $f(w)$ to within this error. It is not necessary to assume that $f$ is defined on all of $\mathbf{C}$: indeed, for reasons of economy it is best to permit $f$ to be defined only on a given subset $E$ of $\mathbf{C}$ (for example on an interval in $\mathbf{R}$).

*Definition 3.1.1* *Let $E$ be any subset of $\mathbf{C}$, let $f : E \to \mathbf{C}$ be any function and let $w$ be a point in $E$. We say that $f$ is continuous at $w$ if and only if for each positive $\epsilon$ there is a positive $\delta$ such that*

$$|f(z) - f(w)| < \epsilon$$

*whenever $z$ is in $E$ and satisfies $|z - w| < \delta$.*

We say that $f$ *is continuous on $E$* or $f : E \to \mathbf{C}$ is continuous when $f$ is continuous at each $w$ in $E$: this is simply convenient terminology.

We now introduce the notation

$$\mathbf{C}(w, \delta) = \{z \in \mathbf{C} : |z - w| < \delta\}:$$

this is the disc (without its boundary) in $\mathbf{C}$ with centre $w$ and radius $\delta$. In terms of this notation the essential feature of Definition 3.1.1 is that

$$f(z) \in \mathbf{C}(f(w), \epsilon)$$

whenever $z \in E \cap \mathbf{C}(w, \delta)$.

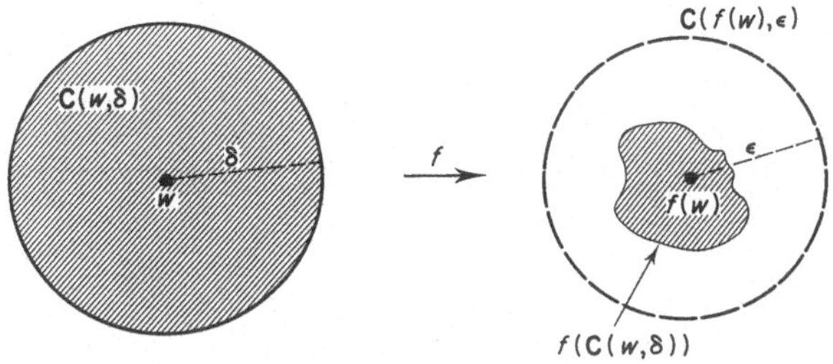


Diagram 3.1.1

Diagram 3.1.1 illustrates this in the case $E = \mathbf{C}$.

As trivial examples of functions which are continuous on **C** we mention the constant functions, the identity function defined by $f(z) = z$ (given $\epsilon$, take $\delta = \epsilon$) and finally, any function $f : E \to \mathbf{C}$ where $E$ is any finite subset of **C**.

The simplest way to construct more complicated examples of continuous functions is to combine known continuous functions by means of the following two results.

*Theorem 3.1.1* *Let $f : E \to \mathbf{C}$ and $g : E \to \mathbf{C}$ be continuous at the point $w$ in $E$. Then* (a) $f + g$, (b) $fg$ *and* (c) $|f|$ *are each continuous at $w$.*

*If, in addition, $f(w) \neq 0$ then there is a positive $\delta$ such that $1/f$ is defined on $E \cap \mathbf{C}(w, \delta)$ and continuous at $w$.*

*Theorem 3.1.2* *Let $E_1$ and $E_2$ be subsets of* **C**, *let $g : E_1 \to E_2$ be continuous at $z_0$ in $E_1$ and let $f : E_2 \to \mathbf{C}$ be continuous at $g(z_0)$. Then $f \circ g : E_1 \to \mathbf{C}$ defined by $(f \circ g)(z) = f(g(z))$ is continuous at $z_0$.*

*Example 3.1.1* The identity function is continuous on **C**, hence so is $g$, $g(z) = z^n$ ($n$ a positive integer). Each constant function is continuous, hence so is $h$, $h(z) = kz^n$ ($k$ a constant). Any finite sum of such terms is continuous: thus each polynomial

$$p(z) = a_0 + a_1 z + \cdots + a_n z^n$$

is continuous on **C**.

More generally,

$$a_{-n} z^{-n} + \cdots + a_{-1} z^{-1} + a_0 + a_1 z + \cdots + a_n z^n$$

is continuous on $\mathbf{C} - \{0\}$.

*Proof of Theorem 3.1.1* Let $\epsilon$ be any positive number: we denote by $\epsilon_1$ a positive number which will be determined (in terms of $\epsilon$) later. As $f$ is continuous at $w$ there is a positive $\delta_f$ such that

$$|f(z) - f(w)| < \epsilon_1$$

on $E \cap \mathbf{C}(w, \delta_f)$. Similarly, there is a positive $\delta_g$ such that

$$|g(z) - g(w)| < \epsilon_1$$

on $E \cap \mathbf{C}(w, \delta_g)$. If $\delta^* = \min\{\delta_f, \delta_g\}$ and $z \in E \cap \mathbf{C}(w, \delta^*)$, then

(a) $$\begin{aligned} |[f(z) + g(z)] - [f(w) + g(w)]| &\leqslant |f(z) - f(w)| + |g(z) - g(w)| \\ &< 2\epsilon_1; \end{aligned}$$

(b) $$\begin{aligned} |f(z)g(z) - f(w)g(w)| &\leqslant |g(z)|\,|f(z) - f(w)| + |f(w)|\,|g(z) - g(w)| \\ &\leqslant (|g(z) - g(w)| + |g(w)|)\epsilon_1 + |f(w)|\,\epsilon_1 \\ &\leqslant \epsilon_1(\epsilon_1 + |g(w)| + |f(w)|); \end{aligned}$$

(c) $$||f(z)|-|f(w)|| \leqslant |f(z)-f(w)| \leqslant \epsilon_1.$$

If we now select $\epsilon_1$ subject to $\epsilon_1 < 1$ and

$$\epsilon_1 \max\{2, (1 + |f(w)| + |g(w)|)\} < \epsilon$$

we find that $f+g$, $fg$ and $|f|$ are continuous at $w$.

Suppose now that, in addition, $f(w) \neq 0$. Then (by continuity with $\epsilon = |f(w)|/2$) there is a positive $r$ such that

$$|f(z)-f(w)| < |f(w)|/2$$

and hence

$$|f(z)| > |f(w)|/2$$

on $E \cap \mathbf{C}(w, r)$. We conclude that on $E \cap \mathbf{C}(w, \delta)$, $\delta = \min\{r, \delta^*\}$, $f(z) \neq 0$ and

$$|1/f(z) - 1/f(w)| = |f(z)-f(w)|/|f(z)f(w)| < 2\epsilon_1/|f(w)|^2.$$

If we now reselect $\epsilon_1$ subject to $2\epsilon_1 < \epsilon|f(w)|^2$, we find that $1/f$ is continuous at $w$.

*Proof of Theorem 3.1.2* This is illustrated in Diagram 3.1.2 where we have written $w_0$ for $g(z_0)$.

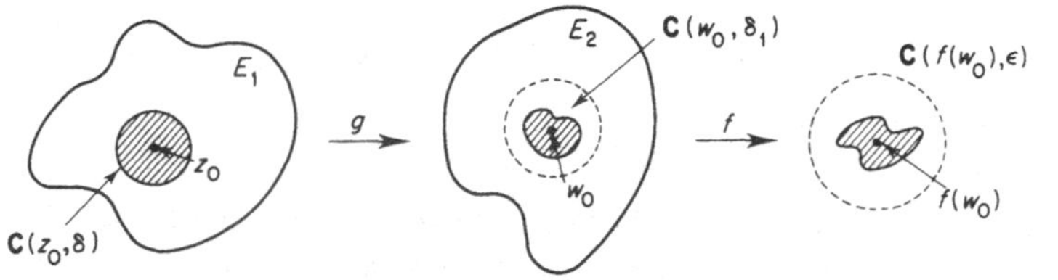


Diagram 3.1.2

Given any positive $\epsilon$ select a positive $\delta_1$ such that $|f(w)-f(w_0)| < \epsilon$ when $w \in E_2 \cap \mathbf{C}(w_0, \delta_1)$. The continuity of $g$ implies that there is a positive $\delta$ such that $|g(z) - w_0| < \delta_1$ when $z \in E_1 \cap \mathbf{C}(z_0, \delta)$.

If $z \in E_1 \cap \mathbf{C}(z_0, \delta)$, then $g(z) \in E_2 \cap \mathbf{C}(w_0, \delta_1)$ and so

$$|(f \circ g)(z) - (f \circ g)(z_0)| = |f(g(z)) - f(w_0)| < \epsilon.$$

**Exercise 3.1**

1. Prove that $f(z) = \max\{x, y\}$, $z = x + iy$, is continuous on $\mathbf{C}$.
2. Prove that every function $f : \mathbf{Z} \to \mathbf{C}$ is continuous on $\mathbf{Z}$.

3. Let $f: E \to \mathbf{C}$ be continuous on $E$. Prove that

$$R(z) = \text{Re } [f(z)], \quad I(z) = \text{Im } [f(z)]$$

are also continuous on $E$.

4. Prove that

$$f(z) = \frac{z^3 + 5}{z^2 + 1}$$

is continuous on $\mathbf{C} - \{i, -i\}$, (a) using Theorem 3.1.1, and (b) directly from Definition 3.1.1.

5. Let the functions $f_j : E \to \mathbf{R}, j = 1, 2, \ldots,$ be continuous on $E$. Prove that for each $n$

$$m_n(z) = \max\{f_1(z), \ldots, f_n(z)\}$$

is continous on $E$.

Suppose also that $f_n(z) \leqslant M$ on $E$ and define

$$s(z) = \sup\{f_n(z) : n = 1, 2, \ldots\}:$$

is this necessarily continuous on $E$?

## 3.2 REAL CONTINUOUS FUNCTIONS

We are concerned now with continous real-valued functions of a real variable and in this case there is the following consequence of continuity which is as useful as Definition 3.1.1 itself.

*Theorem 3.2.1* *Let E be a subset of* $\mathbf{R}$ *and let* $f: E \to \mathbf{R}$ *be continuous at w in E. If* $w \in E$ *and* $f(w) > 0$, *then there is a positive* $\delta$ *such that* $f(x) > 0$ *when* $x \in E \cap \mathbf{C}(w, \delta)$.

Of course, we can replace the inequalities $f(w) > 0$ and $f(x) > 0$ by $f(w) > \mathrm{k}$ and $f(x) > \mathrm{k}$ respectively (consider $f(x) - k$) or even by $f(w) < k$ and $f(x) < k$ (consider $k - f(x)$). The reader should note that we shall use this result frequently and often without explicit mention save for a reference to continuity: examples of this occur later in this section.

*Proof of Theorem 3.2.1* By assumption, $f(w) > 0$ and so, by Definition 3.1.1, there is a positive $\delta$ such that if $|z - w| < \delta$ and $z \in E$, then

$$|f(x) - f(w)| < f(w).$$

Because

$$f(w) - f(x) \leqslant |f(x) - f(w)| < f(w),$$

we may conclude that $f(x) > 0$ on $E \cap \mathbf{C}(w, \delta)$.

The next result is a direct consequence of the fact that **R** satisfies the Least Upper Bound Axiom: its importance lies in the fact that it guarantees the *existence* of a certain number. We first need the concept of an interval. A subset $I$ of **R** is an *interval* if and only if $\{t \in \mathbf{R}: x \leqslant t \leqslant y\} \subset I$ whenever $x \in I, y \in I$ and $x \leqslant y$. As examples, we have the *open intervals*

$$(a, b) = \{x \in \mathbf{R}: a < x < b\},$$

$$(a, +\infty) = \{x \in \mathbf{R}: a < x\},$$

$$(-\infty, a) = \{x \in \mathbf{R}: x < a\},$$

$$(-\infty, +\infty) = \mathbf{R}$$

and the *closed interval*

$$[a, b] = \{x \in \mathbf{R}: a \leqslant x \leqslant b\},$$

where $a$ and $b$ are real numbers with $a \leqslant b$. The intervals $(a, b)$ and $[a, b]$ have *end-points* $a$ and $b$: the end-points are in $[a, b]$ but not in $(a, b)$. If $a = b$ then $(a, b) = \emptyset$ and $[a, b] = \{a\}$.

*Theorem 3.2.2 The Intermediate Value Theorem. Let* $f$: $[a, b] \to \mathbf{R}$ *be continuous on* $[a, b]$. *If* $f(a) \leqslant y \leqslant f(b)$, *then the equation* $f(x) = y$ *has a solution in* $[a, b]$.

In geometric terms this says that as $x$ changes continuously from $a$ to $b$ so $f(x)$ changes continuously from $f(a)$ to $f(b)$ and necessarily passes through all values between $f(a)$ and $f(b)$.

*Proof* The conclusion is obviously true if $y$ is either $f(a)$ or $f(b)$, thus we may assume that

$$f(a) < y < f(b).$$

Now let

$$Y = \{x \in [a, b] : f(x) \leqslant y\}.$$

Obviously $a \in Y$ and as $Y$ is bounded, sup $(Y)$ exists. For brevity we denote sup $(Y)$ by $t$ and clearly $a \leqslant t \leqslant b$.

As $f$ is continuous and as $f(a) < y$, then $f(x) < y$ on some interval $[a, a + \delta]$ and so $t > a$. Similarly, $t < b$.

If $f(t) < y$, then $f(x) < y$ on some interval $(t - \delta, t + \delta)$ and this contradicts the fact that $t$ is an upper bound of $Y$. If $f(t) > y$, then $f(x) > y$ on some interval $(t - \delta, t + \delta)$ and $t - \delta$ is an upper bound of $Y$: this contradicts the fact that $t$ is the *least* upper bound of $Y$. We conclude that $f(t) = y$ and the proof is complete.

*Example 3.2.1* We can now easily improve upon Theorem 1.3.1. Let $n$ be an integer, $n \geqslant 2$, let $y$ be any positive number and let $f(x) = x^n, x \in \mathbf{R}$. Then $f$ is continuous on $[0, y+1]$ and

$$f(0) < y < (1+y)^n = f(y+1)$$

and so there is a positive $x$ with $x^n = y$. This is the unique positive $n$th root of $y$ (Exercise 3.2.1).

It is worthwhile to pause here and consider whether or not we could have used Example 3.2.1 earlier to establish the existence of a square root. This could have been done but only at the cost of developing the theory of real-valued continuous functions first, then the square root of positive numbers and finally, with this available, the ideas of distance between complex numbers and continuity of complex functions. We have preferred to relate the square root directly to the more primitive concept of upper bounds and to develop the ideas of continuity of real-valued and complex-valued functions simultaneously.

The last result in this section is concerned with the continuity of the inverse function $f^{-1}$: *it does not require that $f$ itself be continuous.* We say that a function $f$ defined on a subset $E$ of $\mathbf{R}$ is *strictly increasing* if for each $x$ and $y$ in $E$, $f(x) < f(y)$ whenever $x < y$. If this is so then $f$ is 1–1 on $E$ and so $f^{-1}$ exists: further, $f^{-1}$ is also strictly increasing (Exercise 3.2.2).

*Theorem 3.2.3* *Let $I = (a, b)$ and suppose that $f: I \to \mathbf{R}$ is strictly increasing. Then $f^{-1}$ is continuous on $f(I)$.*

Note that $f(I)$ need not be an interval (see Diagram 3.2.1).

*Proof* We select any $y$ in $f(I)$ and let $x_0$ be the unique point in $(a, b)$ with $f(x_0) = y$. Now let $\epsilon$ be any positive number and choose $x_1$ and $x_2$ in $(a, b)$ such that

$$x_0 - \epsilon < x_1 < x_0 < x_2 < x_0 + \epsilon.$$

As $f$ is strictly increasing,

$$f(x_1) < f(x_0) = y < f(x_2)$$

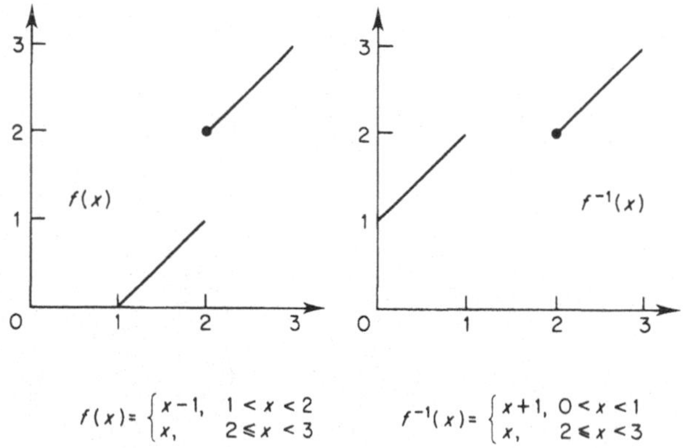


Diagram 3.2.1

and so there is some positive $\delta$ satisfying

$$f(x_1) < y - \delta < y + \delta < f(x_2).$$

If $t \in f(I) \cap (y - \delta, y + \delta)$ then, as $f^{-1}$ is strictly increasing, $x_1 < f^{-1}(t) < x_2$ and so $|f^{-1}(t) - f^{-1}(y)| < \epsilon$.

Observe that the theorem remains true if we replace $(a, b)$ by, for example, $\{x \in \mathbf{R}: x > 0\}$: the same proof is valid.

**Exercise 3.2**

1. Example 3.2.1 shows that if y is positive, then there exists a positive $n$th root, say $y^{1/n}$, of $y$. Show that there is only one positive $n$th root of $y$. Prove that $f(y) = y^{1/n}$ is continuous on $\{y \in \mathbf{R} : y > 0\}$, (a) using Theorem 3.2.3, and (b) directly from Definition 3.1.1 and (1.2.2).
2. Prove that if $f : E \to \mathbf{R}$ is strictly increasing, then so is $f^{-1} : f(E) \to \mathbf{R}$.
3. Let $f$ be defined as in Diagram 3.2.1. Prove that $f^{-1}$ is continuous at $x = 2$ but that $f$ is not continuous there.
4. Let

   $$p(z) = a_0 + a_1 z + \cdots + a_n z^n$$

   where the $a_j$ are real and $n$ is an odd positive integer. Prove that for some real $x$, $p(x) = 0$.
5. Let $I$ be an interval and let $f : I \to \mathbf{R}$ be continuous. Prove that $f(I)$ is an interval. Deduce that if $f : \mathbf{R} \to \mathbf{Q}$ is continuous, then $f$ is constant.
6. Let $f : \mathbf{R} \to \mathbf{R}$ be continuous and suppose that for each real $x$ there is a positive $\delta$ such that $f$ is constant on the interval $(x - \delta, x + \delta)$. Prove that $f$ is constant.
7. Let $f: \mathbf{R} \to \mathbf{R}$ be continuous. Prove that $\{x \in \mathbf{R} : f(x) \neq 0\}$ is a countable union of open intervals.

# Chapter 4

## 4.1 THE EXPONENTIAL FUNCTION

We now have sufficient techniques available to embark on a rigorous development of trigonometry. In order to do this we must first develop the properties of the function exp defined in Chapter 2. For emphasis, we repeat the definition.

*Definition 4.1.1* *For each complex number z define*

$$\exp(z) = \sum_{n=0}^{\infty} \frac{z^n}{n!}.$$

We have seen that this is absolutely convergent for each $z$ in $\mathbf{C}$ (Example 2.2.2) : exp is therefore a function defined on $\mathbf{C}$. The fundamental property of exp described in the next result is the point from which all of our later work stems.

*Theorem 4.1.1* *For all complex numbers z and w*, $\exp(z + w) = \exp(z)\exp(w)$.

*Proof* For each $z$ and $w$ the series

$$\sum_{n=0}^{\infty} \frac{z^n}{n!}, \quad \sum_{n=0}^{\infty} \frac{w^n}{n!}$$

are absolutely convergent. According to Theorem 2.3.1,

$$\exp(z)\exp(w) = \sum_{n=0}^{\infty} \left( \sum_{j=0}^{n} \frac{z^j}{j!} \frac{w^{n-j}}{(n-j)!} \right)$$

$$= \sum_{n=0}^{\infty} (z + w)^n / n!$$

and this is the required result.

The next two results are best considered together as they describe the behaviour of exp on $\mathbf{R}$ and on $\mathbf{C}$.

*Theorem 4.1.2* *The function* exp *is positive and strictly increasing on* **R**. *Further, for each positive k there is a unique real x with* $\exp(x) = k$.

It is clear from the definition that $\exp(0) = 1$, thus for real $x$, $\exp(x) = 1$ if and only if $x = 0$.

*Theorem 4.1.3* *For each complex number z,*

(a) exp *is continuous at* $z$;

(b) $\exp(z) \neq 0$;

(c) $\exp(-z) = 1/\exp(z)$;

(d) $\overline{\exp(z)} = \exp(\bar{z})$;

(e) *if z is real, then* $|\exp(iz)| = 1$;

(f) $|\exp(z)| = \exp(\mathrm{Re}[z]) \leqslant \exp(|z|)$.

*Proof of Theorem 4.1.2* If $x > 0$ we may conclude from Theorem 2.2.1 that

$$\exp(x) > 1 + x > 1.$$

Moreover, from Theorem 4.1.1,

$$\exp(x)\exp(-x) = \exp(0) = 1$$

and as $\exp(x)$ is positive so is $\exp(-x)$. Thus exp is positive on **R**.

Next, if $x < y$, then

$$\begin{aligned}\exp(y) &= \exp(x)\exp(y - x)\\ &\geqslant [1 + (y - x)]\exp(x)\\ &> \exp(x)\end{aligned}$$

and exp is strictly increasing on **R**.

To prove the last assertion in Theorem 4.1.2 we shall assume (temporarily) the validity of Theorem 4.1.3 (a) and (c): thus exp is continuous on **R**. For each positive $k$ we have $\exp(k) > k$, and so

$$\begin{aligned}\exp(-1/k) &= 1/\exp(1/k)\\ &< 1/(1/k)\\ &= k\\ &< \exp(k).\end{aligned}$$

The existence of a real $x$ satisfying $\exp(x) = k$ follows from the Intermediate Value Theorem applied to the function exp on $[-1/k, k]$ : this $x$ is unique as exp is strictly increasing on **R**.

*Proof of Theorem 4.1.3* We begin by proving that exp is continuous at a given complex number $w$. Let $\epsilon$ be any positive number and select a real $\delta$ satisfying

$$0 < \delta < 1, \qquad \delta\exp(1 + 2|w|) < \epsilon.$$

If $|z-w|<\delta$, then $|z|<|w|+1$, and using (1.2.2),

$$
\begin{aligned}
|\exp(z)-\exp(w)| &\leqslant \sum_{n=1}^{\infty} \frac{|z^n-w^n|}{n!} \\
&\leqslant |z-w| \sum_{n=1}^{\infty} \frac{|z^{n-1}|+|z^{n-2}w|+\cdots+|w^{n-1}|}{n!} \\
&\leqslant |z-w| \sum_{n=1}^{\infty} \frac{n \max\{|z|^{n-1},|w|^{n-1}\}}{n!} \\
&\leqslant |z-w| \sum_{n=1}^{\infty} \frac{(|z|+|w|)^{n-1}}{(n-1)!} \\
&< \delta \exp(1+2|w|) \\
&< \epsilon
\end{aligned}
$$

and this proves (a).

For each $z$,

$$\exp(z)\exp(-z)=\exp(0)=1$$

and this proves both (b) and (c).

Next, (d) is an immediate consequence of Theorem 2.1.1 (a).

If $z$ is real we may use (d) and Theorem 4.1.1; thus

$$
\begin{aligned}
|\exp(iz)|^2 &= \exp(iz)\overline{\exp(iz)} \\
&= \exp(iz)\exp(-iz) \\
&= 1,
\end{aligned}
$$

and so (e) holds.

Finally, we write $z=x+iy$ ($x$ and $y$ real) and note that by Theorem 4.1.1, Theorem 4.1.2 and (e),

$$
\begin{aligned}
|\exp(z)| &= |\exp(x)|\,|\exp(iy)| \\
&= \exp(x) \\
&\leqslant \exp(|z|).
\end{aligned}
$$

This completes the proof of Theorem 4.1.3.

**Exercise 4.1**

1. Let $e=\exp(1)$. Prove that for all positive integers $N$

$$\sum_{n=0}^{N} \frac{1}{n!} \leqslant e \leqslant \sum_{n=0}^{N} \frac{1}{n!} + \frac{1}{N(N!)} .$$

Using a calculator, show that $e=2.718\cdots$.

2. Use Theorem 2.3.1 to prove Theorem 4.1.3(f) in the form $|\exp(z)|^2 = \exp(2x)$.
3. Given a positive $y$, there is a unique real $x$ with $y = \exp(x)$ (Theorem 4.1.2). Prove that $\exp(x/n)$ is the unique positive $n$th root of $y$.
4. Let $x$ and $y$ be as in Question 3. For each real $t$ define

$$y^t = \exp(tx).$$

Prove that $y^t > 0, y^0 = 1$,

$$y^a y^b = y^{a+b}, (y^a)^b = y^{ab}$$

and

$$y^{-a} = 1/y^a = (1/y)^a.$$

5. Let $n$ be a positive integer. Prove that there exists a real number $x_0$ such that if $x > x_0$, then

$$\exp(x) > x^n.$$

## 4.2 THE TRIGONOMETRIC FUNCTIONS

In an elementary approach to trigonometry one treats the functions sin and cos as functions of the measurement of angles (or even of angles). In fact, they are more properly considered as functions on **C**, and most of the well-known trigonometric results are valid in this more general setting. To do this we require an analytic definition of these functions and we select the definition which leads to the greatest economy of effort without sacrificing rigour. In view of our extensive knowledge of the function exp the definition that we use is based on this function.

*Definition 4.2.1 For each complex z define*

$$\cos(z) = [\exp(iz) + \exp(-iz)]/2$$

*and*

$$\sin(z) = [\exp(iz) - \exp(-iz)]/(2i)$$

We can now use our knowledge of exp to derive (with very little effort) the basic properties of sin and cos. The next result and its proof is a typical example of this.

*Theorem 4.2.1 For all complex numbers z and w,*

(a) $\exp(iz) = \cos(z) + i\sin(z)$;
(b) $\cos(-z) = \cos(z)$, $\sin(-z) = -\sin(z)$;
(c) $1 = \cos^2(z) + \sin^2(z)$;
(d) $\sin(z + w) = \sin(z)\cos(w) + \cos(z)\sin(w)$;
(e) $\cos(z + w) = \cos(z)\cos(w) - \sin(z)\sin(w)$;

(f) $\sin(z) = \sum_{n=0}^{\infty} \dfrac{(-1)^n z^{2n+1}}{(2n+1)!}$;

(g) $\cos(z) = \sum_{n=0}^{\infty} \frac{(-1)^n z^{2n}}{(2n)!}$ ;

(h) sin *and* cos *are both continuous at* $w$.

*Proof* The proofs of these results are very easy. First, (a) and (b) follow directly from Definition 4.2.1.

To prove (c), (d) and (e) the reader may simply express the right-hand side of these equations in terms of the function exp and simplify, using Theorem 4.1.1.

Next, (f) and (g) follow immediately from the definitions of exp, sin and cos in conjunction with Theorem 2.1.1 (b).

Finally, (h) is an easy consequence of Theorems 3.1.1, 3.1.2 and 4.1.3 (a), and this completes the proof.

At this point the reader should be warned against assuming that *all* of the familiar properties of sin and cos as functions on **R** remain true when these are viewed as functions on **C**. If $x$ is real, then so are $\sin(x)$ and $\cos(x)$ (Exercise 4.2.1), and Theorem 4.2.1 (c) now implies that

$$|\sin(x)| \leqslant 1, \qquad |\cos(x)| \leqslant 1.$$

In general, however, sin and cos are complex-valued and this application of Theorem 4.2.1 (c) cannot be justified. In fact, no such result holds on **C**: for example, if $y$ is positive, then

$$\begin{aligned}\cos(iy) &= [\exp(y) + \exp(-y)]/2\\ &\geqslant (1+y)/2\end{aligned}$$

and so there is no constant $m$ for which $|\cos(z)| \leqslant m$. This, then, is one property which holds on **R** but fails on **C**.

We remark that we shall consistently use Theorem 4.2.1 (c), (d) and (e) and the consequent elementary trigonometric identities (for example, for $\sin(2z)$) without giving an explicit reference to these results.

**Exercise 4.2**

1. Prove that if $x$ is real, then so are $\cos(x)$ and $\sin(x)$.
2. Prove that for all real $x$ and all positive integers $n$,

   $$|\sin(nx)| \leqslant n\,|\sin(x)|.$$

   Show that this is false for some complex $x$.
3. Prove De Moivre's Theorem: for all positive integers $n$,

   $$(\cos(z) + i\sin(z))^n = \cos(nz) + i\sin(nz).$$

   Deduce that for some constants $a_j$,

   $$\cos(nz) = \sum_{k=0}^{n} a_k[\cos(z)]^k.$$

Show that

$$\sum_{k=0}^{n} \cos(kz) = \tfrac{1}{2} + \frac{\cos(nz) - \cos([n+1]z)}{4\sin^2(\tfrac{1}{2}z)}.$$

4. Prove that $f$ defined by $f(0) = 1$ and $f(z) = z^{-1}\sin(z)$ if $z \neq 0$ is continuous on **C**.

## 4.3 PERIODICITY

A function $f$ has a *period* $p$ if and only if each $z$ in **C**, $f(z+p) = f(z)$. Obviously, if $p$ is a period of $f$ then so is $np$ for every integer $n$. The function $f$ is said to be *periodic* if and only if it has a non-zero period.

There is no indication at all from the series for exp, sin and cos that any of these functions are periodic and yet it is surprisingly easy to establish that they are.

Let us suppose for the moment that there is a non-zero $t$ with $\sin(t) = 0$. Then

$$\cos(2t) = 1 - 2\sin^2(t) = 1 \qquad \text{and} \qquad \sin(2t) = 0.$$

This implies that for all $z$, $\sin(z + 2t) = \sin(z)$ and so $2t$ is a non-zero period of sin. The existence of a non-zero period of sin follows, then, from the existence of a zero of sin. A similar argument shows that $2t$ is also a period of cos and hence, by Theorem 4.2.1 (a), of $\exp(iz)$.

In order that we may completely describe the set of periods of sin, cos and exp we shall first describe the set of zeros of sin and cos. As the reader well knows, this involves the number $\pi$ and in order to maintain a self-contained development which is free from geometric 'proof' we shall introduce the number $\pi$ in this way. There is no harm in doing this (indeed, $\pi$ has to be introduced in some way) provided that each subsequent use of $\pi$ is based on this definition. To reassure the reader, we shall obtain a reasonable approximation to the numerical value of $\pi$ (an exercise which is valuable in its own right).

For those readers who are familiar with the definition of a group we remark that the set $S_0$ of real zeros of sin is an additive subgroup of **R** (for example, if $z$ and $w$ are zeros of sin then so, by Theorem 4.2.1 (d), is $z + w$). It is not hard to see that any additive subgroup of **R**, say $G$, is either 'dense' in **R** (every interval of positive length contains points of $G$) or is cyclic and generated, say, by $\lambda$ so

$$G = \{n\lambda : n \in \mathbf{Z}\}.$$

If $S_0$ were dense in **R** the continuity of sin would imply that $\sin(x) = 0$ for all real $x$ and we shall prove that this is not so. Thus for some $\lambda$ (possibly equal to zero)

$$S_0 = \{n\lambda : n \in \mathbf{Z}\}.$$

We shall also prove that for some non-zero $x$, $\sin(x) = 0$; thus $\lambda \neq 0$. We can then define $\pi$ to be $|\lambda|$, the positive generator of $S_0$.

*Theorem 4.3.1* *There is a positive number $\pi$ such that*

(a) $\sin(z) = 0$ *if and only if* $z \in \{n\pi : n \in \mathbf{Z}\}$;
(b) $\cos(z) = 0$ *if and only if* $z \in \{n\pi + \pi/2 : n \in \mathbf{Z}\}$.

Note that this result includes the statement that *every* zero of sin is real.

*Proof* Let $S$ denote the set of complex numbers $z$ satisfying $\sin(z) = 0$. Clearly $0 \in S$ and so $S$ is not empty. Next, if $z \in S$, then

$$\exp(iz) - \exp(-iz) = 0$$

and so $\exp(2iz) = 1$. Now write $z = x + iy$ and use Theorem 4.1.2 and Theorem 4.1.3 (f) : thus

$$\exp(-2y) = |\exp(2iz)| = 1$$

and so $y = 0$. This proves that every zero of sin is real.

To proceed further we need more information about sin (the reader may find it helpful to sketch a graph showing, at each stage of the following discussion, the increased information available about sin). First, if $0 < x < 2$, then from Theorem 4.2.1 (f),

$$\begin{aligned} |\sin(x) - x| &\leqslant \frac{x^3}{3!} + \frac{x^5}{5!} + \cdots \\ &\leqslant \frac{x^3}{3!}\left(1 + \left[\frac{x^2}{20}\right] + \left[\frac{x^2}{20}\right]^2 + \cdots\right) \\ &\leqslant \frac{x^3}{3!}\left(1 + \frac{1}{5} + \frac{1}{5^2} + \cdots\right) \\ &< x^3/4. \end{aligned}$$

This proves that if $0 < x < 2$, then

$$\sin(x) > x(4 - x^2)/4 > 0. \tag{4.3.1}$$

It is now apparent that $\sin(1) > \frac{3}{4}$ and, from Theorem 4.2.1 (c), that $\cos(1) < \frac{3}{4}$. The Intermediate Value Theorem is applicable to (sin − cos) on $[0, 1]$ and we deduce that for some $t$ in $(0, 1)$, $\cos(t) = \sin(t)$. Thus $\cos(2t) = 0$, $\sin(4t) = 0$ and therefore sin has at least one zero in $(0, 4)$.

We now *define* $\pi$ by

$$\pi = \inf\{x > 0 : \sin(x) = 0\}.$$

The preceding remarks show that $\pi < 4$ and (4.3.1) implies that $\pi \geqslant 2$. Next, if $\sin(\pi) \neq 0$, then the continuity of sin shows that on some interval $(\pi - \delta, \pi + \delta)$, sin is never zero. This contradicts the definition of $\pi$; thus

$$0 = \sin(\pi) = 2\sin(\pi/2)\cos(\pi/2).$$

As $0 < \pi/2 < 2$, we find from (4.3.1) that $\sin(\pi/2) > 0$: thus

$$\cos(\pi/2) = 0, \qquad \sin(\pi/2) = 1. \tag{4.3.2}$$

In fact, (4.3.1) implies that sin is positive on $(0, \pi/2)$ and as

$$\sin(\pi/2 + x) = \sin(\pi/2 - x)$$

we find that sin is positive on $(0, \pi)$. Now suppose that $0 < x < y < \pi/2$. Then the identity

$$\sin(y + x)\sin(y - x) = \sin^2(y) - \sin^2(x)$$

implies that $\sin^2(y) > \sin^2(x)$ or, as each term is positive, $\sin(y) > \sin(x)$.

We have now shown that sin is strictly increasing from 0 to 1 on $[0, \pi/2]$, and this, together with (4.3.2) and the addition formulae, provides the general character of the graphs of sin and cos.

Of course, this contains a proof of Theorem 4.3.1 and much more besides. If $p$ is a period of sin then for all $z$,

$$\sin(z) = \sin(z + p) = \sin(z)\cos(p) + \cos(z)\sin(p)$$

Put $z = 0$, so $p = n\pi$, say. Put $z = \pi/2$, so $n$ is even. In this way we can easily prove the next result (the details are omitted).

*Theorem 4.3.2* *The set of periods of both* sin *and* cos *is* $\{2n\pi : n \in \mathbf{Z}\}$.

We have promised to obtain a numerical approximation to $\pi$ : we shall prove that

$$3.1 \leqslant \pi \leqslant 3.2.$$

If $0 \leqslant x \leqslant 2$, then

$$\begin{aligned}\cos(x) - \left[1 - \frac{x^2}{2!} + \frac{x^4}{4!} - \frac{x^6}{6!}\right] &\leqslant \frac{x^8}{8!} + \frac{x^{10}}{10!} + \dots \\ &\leqslant \frac{x^6}{6!}\left(\sum_{n=1}^{\infty} \frac{1}{3^{2n}}\right) \\ &< \frac{x^6}{6!}\end{aligned}$$

and so

$$\cos(x) < 1 - \frac{x^2}{2} + \frac{x^4}{24}.$$

This shows that $\cos(1.6) < 0$ and so $\pi/2 < 1.6$.

A similar argument shows that

$$\left[1 - \frac{x^2}{2!} + \frac{x^4}{4!} - \frac{x^6}{6!} + \frac{x^8}{8!}\right] - \cos(x) < \frac{x^8}{8!}$$

and so

$$\begin{aligned}\cos(x) &\geqslant 1 - \frac{x^2}{2!} + \frac{x^4}{4!} - \frac{x^6}{6!} \\ &\geqslant 1 - \frac{x^2}{2} + \frac{x^4}{28}.\end{aligned}$$

This polynomial is positive when $|x| \leqslant 1.55$ (it has four real zeros which can easily be found) and so $\pi/2 \geqslant 1.55$.

We complete this section by describing the periodicity of exp.

*Theorem 4.3.3* *The following are equivalent:*

(a) *p is a period of* exp;
(b) $\exp(p) = 1$;
(c) $p \in \{2n\pi i : n \in \mathbf{Z}\}$.

*Proof*

Theorem 4.1.1 (with $w = p$) shows that (a) and (b) are equivalent. If (b) holds, then $\sin(ip) = 0$ and $\cos(ip) = 1$ and (c) follows: if (c) holds, then (b) follows from Theorem 4.2.1 (a) (with $z = p/i$).

**Exercise 4.3**

1. Prove (analytically) that

$$\sin(\pi/4) = \cos(\pi/4) = 1/\sqrt{2};$$
$$\sin(\pi/6) = \tfrac{1}{2}, \qquad \cos(\pi/6) = \sqrt{3}/2.$$

2. Prove that the set of $z$ in $\mathbf{C}$ for which $\sin(z)$ is real is the union of one horizontal line and infinitely many vertical lines.
3. Define (for appropriate $z$) $\tan(z) = \sin(z)/\cos(z)$. Prove

(a) tan has period $\pi$;
(b) $\tan(-z) = -\tan(z)$;
(c) $\tan(0) = 0$;
(d) $\tan(x)$ is continuous and strictly increasing for $0 \leqslant x < \pi/2$;
(e) $\tan(x) \geqslant [3(\pi/2 - x)]^{-1}$ if $\pi/6 \leqslant x < \pi/2$ (write $y = \pi/2 - x$).

4. Prove that if $0 \leqslant x \leqslant y \leqslant \pi/2$, then

$$\tfrac{1}{2}[\sin(x) + \sin(y)] \leqslant \sin\left(\frac{x+y}{2}\right)$$

and interpret this in terms of the graph of sin.

Let

$$E_n = \{k\pi/2^n : k = 0, 1, \ldots, 2^{n-1}\}, \qquad n = 1, 2, \ldots .$$

Show that $E_{n+1} \subset E_n$ and that between two consecutive points of $E_n$ lies exactly one point of $E_{n+1}$.

Prove (by induction) that $\sin(x) \geqslant 2x/\pi$ on $E_n$. Use the continuity of sin to conclude that this inequality is valid on $[0, \pi/2]$.

Modify this argument to show that sin is concave on $[0, \pi/2]$, that is the graph of sin lies above every segment whose endpoints are on the graph.

5. Show that $\sin(x) = x, x \in R$, has exactly one solution.

## 4.4 THE HYPERBOLIC FUNCTIONS

The hyperbolic functions cosh and sinh are closely related to the functions cos and sin and are defined by

$$\cosh(z) = \cos(iz) = \tfrac{1}{2}[\exp(z) + \exp(-z)]$$

and

$$\sinh(z) = -i\sin(iz) = \tfrac{1}{2}[\exp(z) - \exp(-z)].$$

With these definitions the basic properties of these functions can be derived from the corresponding properties of cos and sin or exp. The reader should have no difficulty in proving, for example,

(a) $\cosh(z) = \sum_{n=0}^{\infty} \dfrac{z^{2n}}{(2n)!}$;

(b) $\sinh(z) = \sum_{n=0}^{\infty} \dfrac{z^{2n+1}}{(2n+1)!}$;

(c) $\cosh^2(z) - \sinh^2(z) = 1$;
(d) $\cosh(-z) = \cosh(z)$;
(e) $\sinh(-z) = -\sinh(z)$.

Observe that (a) and (b) show that both cosh and sinh are positive and strictly increasing when restricted to the set of positive real numbers.

The real reason for introducing cosh and sinh is that they arise naturally in the following identities: if $z = x + iy$, then

$$\begin{aligned} |\sin(z)|^2 &= \sin^2(x) + \sinh^2(y), \\ |\cos(z)|^2 &= \cos^2(x) + \sinh^2(y). \end{aligned} \tag{4.4.1}$$

Observe that these identities confirm that the only zeros of cos and sin are real for $\sinh(y) = 0$ if and only if $y = 0$. The identities are easily proved. For example,

$$\begin{aligned} |\sin(z)|^2 &= |\sin(x)\cos(iy) + \cos(x)\sin(iy)|^2 \\ &= |\sin(x)\cosh(y) + i\cos(x)\sinh(y)|^2 \\ &= \sin^2(x)\cosh^2(y) + \cos^2(x)\sinh^2(y) \\ &= \sin^2(x)[1 + \sinh^2(y)] + [1 - \sin^2(x)]\sinh^2(y) \end{aligned}$$

and this gives (4.4.1).

In Chapter 9 we shall need explicit estimates of the functions cot (defined as cos/sin) and cosec (defined as 1/sin and it is convenient to obtain these now.

*Theorem 4.4.1* *Let z satisfy* $|z - n| \geqslant \tfrac{1}{4}$ *for every integer n. Then*

(a) $|\operatorname{cosec}(\pi z)| \leqslant 2$ *and*
(b) $|\cot(\pi z)| \leqslant 3$.

*Proof* Because

$$\cot^2(\pi z) = \operatorname{cosec}^2(\pi z) - 1$$

it is only necessary to prove (a): equivalently we must prove that for the given $z$, $|\sin(\pi z)| \geqslant \frac{1}{2}$.

If $|\sin(\pi z)| \leqslant \frac{1}{2}$, we write $z = x + iy$ and deduce that

$$\sin^2(\pi x) \leqslant \tfrac{1}{4}, \qquad \sinh^2(\pi y) \leqslant \tfrac{1}{4}.$$

It is easy to see that the first of these inequalities implies that for some integer $n$, $|\pi x - \pi n| \leqslant \pi/6$. Further, as $\pi > 3$ and

$$(\pi y)^2 = (|\pi y|)^2 \leqslant \sinh^2(|\pi y|) = \sinh^2(\pi y)$$

we find that $y^2 < \frac{1}{36}$ Thus for this $n$,

$$|z - n|^2 = (x - n)^2 + y^2 < (\tfrac{1}{4})^2$$

contrary to our assumption.

**Exercise 4.4**

1. Prove that

$$|\sinh(z)|^2 = \sinh^2(x) + \sin^2(y);$$
$$|\cosh(z)|^2 = \cosh^2(x) - \sin^2(y);$$
$$\sinh(y) \leqslant |\sin(z)| \leqslant \cosh(y).$$

2. Find all zeros and all periods of cosh.
3. Show $\sin(z) = \sin(w)$ if and only if for some integer $m$, $z = (-1)^m w + m\pi$.
4. Find all solutions of $\cos(z) = 3$.
5. Find the image of the following sets by the function cos:

   (a) the imaginary axis $\{iy : y \in \mathbf{R}\}$;
   (b) $\{x : 0 \leqslant x \leqslant \pi\}$;
   (c) $\{\pi + iy : y \in \mathbf{R}\}$.
   [The image of $E$ by the function $f$ is $f(E)$.]

# Chapter 5

## 5.1 THE ARGUMENT OF A COMPLEX NUMBER

We are now in a position to give a precise definition of the argument of a non-zero complex number. Our knowledge of elementary geometry (although not rigorously developed) clearly leads us to expect that the *argument* (or angle in polar co-ordinates) is only defined to within a multiple of $2\pi$ : we shall see that this is simply another way of expressing the periodicity of the exponential function.

The fact that the argument will not be a single number but a collection of numbers (differing from each other by integral multiples of $2\pi$) presents us with a minor problem of terminology. To call the argument of $z$ a function would be in direct contradiction with the basic notion that a function is single valued. To overcome this problem we simply define the argument of $z$ as the *set* of values that we would normally assign to the polar angle in the usual way. This does not lead to any contradiction and is in keeping with our desire for a simple yet precise development.

*Definition 5.1.1* *For each non-zero complex number z we define* Arg $(z)$ *to be the set*

$$\{\theta \in \mathbf{R} : z = |z| \exp(i\theta)\}.$$

In recognition of the existing terminology we make the following definition.

*Definition 5.1.2* *We shall say that $\theta$ is a choice or value of* Arg $(z)$ *if and only if* $z = |z| \exp(i\theta)$, *that is if and only if $\theta$ is an element of* Arg $(z)$.

It is not obvious that the set Arg $(z)$ has any elements at all. Our first task, then, is to show that Arg $(z)$ is not empty and to obtain the general form of this set.

*Theorem 5.1.1* *Let z be any non-zero complex number. Then there is a real $\theta$ such that*

$$\text{Arg}(z) = \{\theta + 2n\pi : n \in \mathbf{Z}\}.$$

*Proof.* As $z$ is non-zero, the complex number $z/|z|$ is defined and we may write $z = |z|(u + iv)$, where $u$ and $v$ are real and $u^2 + v^2 = 1$.

As cos is strictly decreasing from 1 to $-1$ on the interval $[0, \pi]$, we can find a unique real $\theta$ in this interval with $u = \cos\theta$. Then

$$v^2 = 1 - u^2 = \sin^2\theta.$$

If $v = \sin\theta$, then

$$z = |z|(\cos\theta + i\sin\theta):$$

if $v = -\sin\theta$ then

$$z = |z|(\cos(-\theta) + i\sin(-\theta)).$$

Thus either $\theta$ or $-\theta$ is in Arg $(z)$ and so Arg $(z) \neq \emptyset$.

Suppose now that $\theta_1$ is in Arg $(z)$. Then $\varphi$ is in Arg $(z)$ if and only if

$$\exp(i\varphi) = \exp(i\theta_1)$$

or, equivalently,

$$\exp(i[\theta_1 - \varphi]) = 1$$

and the required result follows from Theorem 4.3.3.

The familiar rules for Arg hold when suitably interpreted and we next verify some some of these.

*Theorem 5.1.2* *Let z and w be non-zero complex numbers and let $\theta$ and $\varphi$ be a choice of* Arg $(z)$ *and* Arg $(w)$ *respectively. Then*

(a) $-\theta$ *is a choice of* Arg $(z^{-1})$;
(b) $\theta + \varphi$ *is a choice of* Arg $(zw)$;
(c) $\theta - \varphi$ *is a choice of* Arg $(z/w)$.

*Proof* We simply observe that

$$z^{-1} = [\,|z|\exp(i\theta)]^{-1}$$
$$= |z^{-1}|\exp(-i\theta)$$

and

$$zw = |z|\exp(i\theta)\,|w|\exp(i\varphi)$$
$$= |zw|\exp(i[\theta + \varphi])$$

and this proves (*a*) and (b). Finally, (c) follows by combining (a) and (b).

We end this section with three examples.

*Example 5.1.1* We have seen that $|\exp(z)| = \exp(x)$ : let us find Arg (exp $(z)$). If $z = = x + iy$, then

$$\exp(z) = \exp(x)\exp(iy)$$
$$= |\exp(z)|\exp(iy)$$

and so $y \in$ Arg (exp $(z)$). Thus

$$\text{Arg}(\exp(z)) = \{y + 2n\pi : n \in \mathbf{Z}\}$$

*Example 5.1.2* Let us relate our development of trigonometry to the more elementary approach. We suppose, for example, that $z = x + iy$, where $x$ and $y$ are positive. Then

$$x + iy = |z| (\cos \theta + i \sin \theta) \qquad (\theta \in \text{Arg}(z))$$

and so

$$\cos \theta = x/|z|, \qquad \sin \theta = y/|z|.$$

This is the customary definition in terms of the sides of a triangle, $\theta$ being the 'angle' (which has not yet been defined) between the positive real axis and the segment joining 0 and $z$.

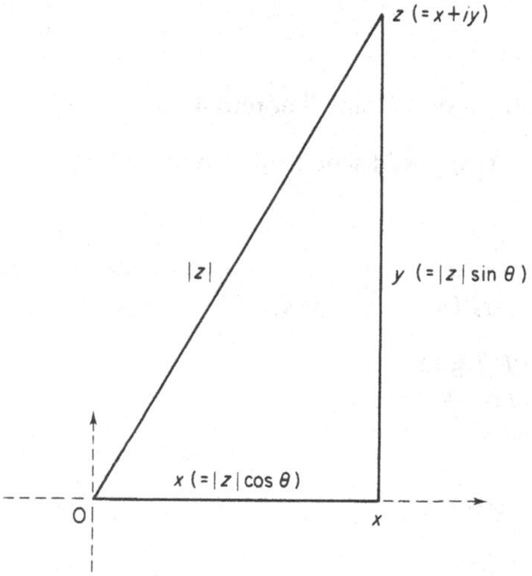


Diagram 5.1.1

Observe that

$$\tan(\theta) = y/x;$$

however this *cannot* be used alone to determine $\theta$ from $x$ and $y$ $\tan(\theta) = \tan(\theta + \pi)$.

We have still to define an angle. To do so we need the notion of a *half-line*, and this is a set of the form

$$L(z_0, \alpha) = \{z_0 + t\alpha : t \geqslant 0\}$$

where $\alpha \neq 0$.

We need to be able to speak of the *initial point* $z_0$ and the *direction* Arg $(\alpha)$ of the half-line and this would not be allowed if

$$L(z_1, \alpha_1) = L(z_2, \alpha_2) \tag{5.1.1}$$

were possible with either $z_1 \neq z_2$ or $\text{Arg}(\alpha_1) \neq \text{Arg}(\alpha_2)$. We must prove, then, that (5.1.1) holds if and only if $z_1 = z_2$ and $\text{Arg}(\alpha_1) = \text{Arg}(\alpha_2)$.

The latter two equations clearly imply that (5.1.1) holds. Now suppose that (5.1.1) holds. The half-line $L(z_1, \alpha_1)$ meets the circle

$$Q = \{z : |z - z_1| = r\}, r > 0,$$

at points $z_1 + t\alpha_1$, where $t \geqslant 0$ and $|t\alpha_1| = r$. Thus $L(z_1, \alpha_1)$ meets $Q$ at one point only, namely when $t = r/|\alpha_1|$. In contrast to this, the circle $Q^*$ with centre $z_1 + s\alpha_1$, $s > 0$, and radius $\frac{1}{2} s |\alpha_1|$ meets the half-line at two points, namely $z_1 + s\alpha_1 \pm \frac{1}{2}s\alpha_1$. Thus $z_1$ is the unique point $w$ on $L(z_1, \alpha_1)$ with the property that $L(z_1, \alpha_1)$ meets every circle centre $w$ and positive radius $r$ in exactly one point.

As (5.1.1) holds, this assertion remains true if we replace $L(z_1, \alpha_1)$ by $L(z_2, \alpha_2)$. It also remains true if we replace both $z_1$ and $L(z_1, \alpha_1)$ by $z_2$ and $L(z_2, \alpha_2)$. Thus both $z_1$ and $z_2$ satisfy this criterion with respect to $L(z_2, \alpha_2)$ and so $z_1 = z_2$.

Finally, the point $z_1 + \alpha_1$ is in $L(z_1, \alpha_1)$ and hence in $L(z_2, \alpha_2)$, and so there is some positive $t$ with

$$z_1 + \alpha_1 = z_1 + t\alpha_2.$$

Thus $\alpha_1 = t\alpha_2$ and so Arg $(\alpha_1)$ = Arg $(\alpha_2)$. We can now talk freely of the initial point and the direction of a given half-line.

An *angle at* $w$ *is an ordered pair of half-lines with the same initial point* $w$, say

$$(L(w, \alpha_1), L(w, \alpha_2)):$$

the *measurement* of this angle is, by definition, Arg $(\alpha_2/\alpha_1)$.

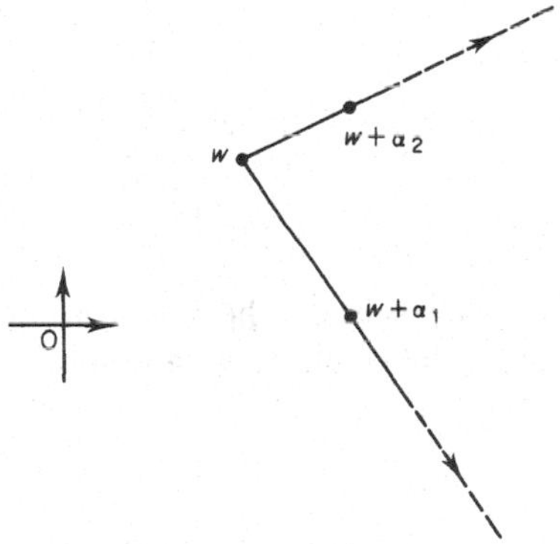


Diagram 5.1.2

For example, the positive $x$-axis and the positive $y$-axis are half-lines giving the two angles

$$(L(0, 1), L(0, i)), \qquad (L(0, i), L(0, 1))$$

at the origin : the measurements of these angles are

$$\{\pi/2 + 2n\pi : n \in \mathbf{Z}\}, \qquad \{3\pi/2 + 2n\pi : n \in \mathbf{Z}\}$$

respectively.

Quite generally, if $\theta$ is one choice of the measurement of the angle

$$(L(w, \alpha_2), L(w, \alpha_1)),$$

then $2\pi - \theta$ is one choice of the measurement of the angle

$$(L(w, \alpha_1), L(w, \alpha_2)).$$

Returning to the discussion at the beginning of this example we can consider any non-zero $z$ and form the angle

$$(L(0, 1), L(0, z)),$$

These half-lines are the positive real axis and the half-line from 0 through $z$ : the measurement of this angle is Arg $(z)$.

*Example 5.1.3* We shall find all $n$th roots of a given complex number $w$ : equivalently, we shall seek all $z$ satisfying $z^n = w$. If $w = 0$, the only solution is 0 itself.

Next we consider the case $w = 1$. If $\theta \in \text{Arg}(z)$ and $z^n = 1$, then

$$1 = z^n = |z|^n \exp(in\theta).$$

Using Theorem 4.1.3 (e) and Theorem 4.3.3, we find that $|z| = 1$ and

$$n\theta \in \{2k\pi i : k \in \mathbf{Z}\}.$$

This shows that

$$z \in \{\exp(2k\pi i/n) : k \in \mathbf{Z}\},$$

and so the only possible values of $z$ are $1, \omega, \omega^2, \ldots, \omega^{n-1}$ (as $k$ varies these values are repeated) where

$$\omega = \exp(2\pi i/n).$$

It is clear that $\omega^n = 1$, thus these values are the only $n$th roots of 1.

Now let $w$ be any non-zero number and select a value $\varphi$ of Arg $(w)$ and a positive $r$ with $r^n = |w|$. If

$$s = r \exp(i\varphi/n),$$

then $s^n = w$ and so $z^n = w$ if and only if $(z/s)^n = 1$. The complete set of $n$th roots of $w$ is therefore

$$s, s\omega, s\omega^2, \ldots, s\omega^{n-1}.$$

We can now indicate briefly how our definition of $\pi$ relates to the length of the circumference of a circle. The points $1, \omega, \ldots, \omega^{n-1}$ lie on the circle $\{z : |z| = 1\}$ and we can form a closed polygon $P_n$ by joining the points $\omega^k$ to $\omega^{k+1}$ ($k = 0, 1, \ldots, n-1 : \omega^n = 1$).

This is, of course, a regular $n$-sided polygon whose vertices lie on the given circle.

The length of $P_n$ is

$$L_n = \sum_{k=0}^{n-1} | \omega^k - \omega^{k+1} |$$

$$= \sum_{k=0}^{n-1} | \omega |^k \, | 1 - \omega |$$

$$= n \, | 1 - \omega |$$

as $| \omega | = 1$. However,

$$| 1 - \omega |^2 = | [1 - \cos (2\pi/n)] + i \sin (2\pi/n) |^2$$

$$= 2 - 2 \cos (2\pi/n)$$

$$= 4 \sin^2 (\pi/n).$$

Thus

$$L_n = 2n \sin (\pi/n).$$

We can show that for large $n$, $L_n$ is approximately $2\pi$. If $0 < x < 2$ then, as we have seen before,

$$| \sin (x) - x | < x^3/4.$$

If $n \geqslant 2$ we can put $x = \pi/n$: then

$$| L_n - 2\pi | < \pi^3/2n^2.$$

We have not yet defined the length of a curve, nevertheless this certainly suggests that the circumference of the circle $| z | = 1$ has length $2\pi$.

**Exercise 5.1**

1. Find Arg $(1 - i)$.
2. Show that if $z \neq 0$ and $t > 0$, then

   $$\text{Arg}\,(tz) = \text{Arg}\,(z),$$

   $$\text{Arg}\,(-tz) = \{\theta + \pi : \theta \in \text{Arg}\,(z)\}.$$

3. Suppose that $zw \neq 0$. Prove that Re $[z\overline{w}] = 0$ if and only if

   $$\text{Arg}\,(w) = \{\theta + \pi/2 : \theta \in \text{Arg}\,(z)\} \cup \{\theta - \pi/2 : \theta \in \text{Arg}\,(z)\}.$$

   This shows that Re $[z\overline{w}] = 0$ is a condition for $z$ and $w$ to be orthogonal.
4. Suppose that $zw \neq 0$ and let $\varphi$ and $\psi$ be choices of Arg $(z)$ and Arg $(w)$ respectively. Prove that

   $$\text{Re}\,[z\overline{w}] = | z | \, | w | \cos (\varphi - \psi).$$

   Derive the cosine rule:

   $$| z - w |^2 = | z |^2 + | w |^2 - 2 \, | z | \, | w | \cos (\varphi - \psi).$$

5. Prove that $|z| = 1$ if and only if

$$\text{Re}\,[(z-1)(\bar{z}+1)] = 0$$

and interpret geometrically. [Draw the circle $x^2 + y^2 = 1$ and use Question 3.]

6. Let $k$ be an integer satisfying $0 \leqslant k \leqslant n-1$ and let

$$\omega = \exp(2\pi ik/n).$$

The *order* of $\omega$ is the least positive integer $p$ such that $\omega^p = 1$. Find the order of $\omega$ in terms of $k$ and $n$. [*Hint* : let $d$ be the greatest common divisor of $k$ and $n$.]

Prove that for each positive integer $m$

$$1 + \omega^m + \cdots + \omega^{m(n-1)} = \begin{cases} n & \text{if} \quad n \mid mk; \\ 0 & \text{if} \quad n \nmid mk. \end{cases}$$

[$a \mid b$ means that a divides $b$ : $a \nmid b$ means that $a$ does not divide $b$.]

## 5.2 LOGARITHMS

Recall that $\mathbf{R}^+$ is the set of positive numbers. According to Theorem 4.1.2, exp is a strictly increasing function which maps $\mathbf{R}$ onto $\mathbf{R}^+$: this means that the inverse of exp exists on $\mathbf{R}^+$.

*Definition 5.2.1* *The inverse of* $\exp : \mathbf{R} \to \mathbf{R}^+$ *is denoted by* $\log_e$.

The function $\log_e$ is the *natural logarithm* and if $x \in \mathbf{R}$ and $y \in \mathbf{R}^+$, then

$$\log_e(\exp(x)) = x, \qquad \exp(\log_e(y)) = y.$$

Note that $\exp(\log_e(y))$ is not defined if $y \leqslant 0$. As $\log_e$ is the inverse of the strictly increasing function exp, $\log_e$ is continuous and strictly increasing on $\mathbf{R}^+$ (Theorem 3.2.3). Now we establish the familiar properties of $\log_e$.

*Theorem 5.2.1* *If x and y are positive, then*

$$\log_e(xy) = \log_e(x) + \log_e(y)$$

*and*

$$\log_e(1/x) = -\log_e(x).$$

*Proof* Write

$$\alpha = \log_e(x), \qquad \beta = \log_e(y).$$

Then using Theorem 4.1.1,

$$\begin{aligned} \log_e(xy) &= \log_e(\exp(\alpha)\exp(\beta)) \\ &= \log_e(\exp(\alpha+\beta)) \\ &= \alpha + \beta. \end{aligned}$$

The second result follows from this by putting $y = 1/x$ and noting that $\log_e(1) = 0$ because $\exp(0) = 1$.

In view of the definition of $\log_e$ it is natural to consider defining the function log on **C** as the inverse function of $\exp : \mathbf{C} \to \mathbf{C}$. As exp is not 1–1 no such function exists and, as in the case of Arg, we overcome this difficulty by defining the logarithm as a set.

*Definition 5.2.2* *For each non-zero complex number z,*

$$\text{Log}(z) = \{w \in \mathbf{C} : \exp(w) = z\}.$$

Again we say that $w$ in $\text{Log}(z)$ is a *value* or *choice* of $\text{Log}(z)$: thus $w$ is a choice of $\text{Log}(z)$ if and only if $\exp(w) = z$. In comparison with $\log_e$ we now have

$$\begin{aligned}\exp(\text{Log}(z)) &= \{\exp(w) : w \in \text{Log}(z)\}\\ &= \{\exp(w) : \exp(w) = z\}\\ &= \{z\}\end{aligned}$$

(this is a *set*) and

$$\begin{aligned}\text{Log}(\exp(z)) &= \{w \in C : \exp(w) = \exp(z)\}\\ &= \{z + 2n\pi i : n \in \mathbf{Z}\}.\end{aligned}$$

Observe that the periodicity of exp 'annihilates' the many-valued character of Log when computing $\exp(\text{Log}(z))$ : the many-valued character survives, however, in $\text{Log}(\exp(z))$, where Log is applied after exp.

The reader should note the use of the capital letters A and L in $\text{Arg}(z)$ and $\text{Log}(z)$ to denote sets rather than choices of the argument and logarithm.

There is another characterization of $\text{Log}(z)$ which is suggested by the 'polar form'

$$z = |z| \exp(i\theta), \qquad \theta \in \text{Arg}(z),$$

of the complex number $z$.

*Theorem 5.2.2* *For each non-zero complex number z*

$$\text{Log}(z) = \{\log_e(|z|) + i\theta : \theta \in \text{Arg}(z)\}.$$

*Proof* Let $w = u + iv$, where $u$ and $v$ are real. Then $w \in \text{Log}(z)$ if and only if

$$\begin{aligned}z &= \exp(w)\\ &= \exp(u)\exp(iv).\end{aligned}$$

This is equivalent to

$$|z| = \exp(u), \qquad v \in \text{Arg}(z)$$

or

$$u = \log_e(|z|), \qquad v \in \text{Arg}(z)$$

and the proof is complete.

If we combine Theorems 5.1.2, 5.2.1 and 5.2.2 we immediately obtain the following result.

*Theorem 5.2.3* *Let $z_1$ and $z_2$ be non-zero complex numbers and let $w_1$ and $w_2$ be choices of* Log $(z_1)$ *and* Log $(z_2)$ *respectively. Then*

(a) $w_1 + w_2$ *is a choice of* Log $(z_1 z_2)$;
(b) $-w_1$ *is a choice of* Log $(1/z_1)$.

We end this section with three examples.

*Example 5.2.1* Let us examine Log $(x)$ where $x$ is a positive number. First, one choice of Arg $(x)$ is 0 (because $x = |x| \exp(i0)$). Therefore

$$\begin{aligned}\text{Log}(x) &= \{\log_e(|x|) + 2n\pi i : n \in \mathbf{Z}\}\\ &= \{\log_e(x) + 2n\pi i : n \in \mathbf{Z}\}.\end{aligned}$$

This example prompts the following remark. For each non-zero $z$, Log $(z)$ is an infinite set. In the case when $z$ is real and positive, one choice of Arg $(z)$ is 0 and *one* choice of Log $(z)$ is $\log_e(z)$. Even in this case, however, there are infinitely many elements in the set Log $(z)$.

*Example 5.2.2* We shall compute Log $(-1)$. We have $-1 = |-1| \exp(i\pi)$ and so

$$\begin{aligned}\text{Log}(-1) &= \{\log_e(|-1|) + i\theta : \theta \in \text{Arg}(-1)\}\\ &= \{(2n+1)\pi i : n \in \mathbf{Z}\}.\end{aligned}$$

*Example 5.2.3* We now compute Log $(3i)$. As

$$3i = |3i| \exp(i\pi/2)$$

we have

$$\text{Log}(3i) = \{\log_e(3) + (4n+1)\pi i/2 : n \in \mathbf{Z}\}.$$

**Exercise 5.2**

1. Prove that if $x > 0$, then $\log_e(x) < x$.
2. Prove that

$$\sinh^{-1}(z) = \text{Log}(z + \sqrt{(1+z^2)}).$$

*Note*: this is an identity of *sets* and is to be interpreted as follows:

$$\sinh^{-1}(z) = \{w : \sinh(w) = z\},$$
$$\text{Log}(z + \sqrt{(1+z^2)}) = \text{Log}(z + \zeta) \cup \text{Log}(z - \zeta)$$

where $\zeta$ is any square root of $1 + z^2$.
3. Formulate and prove a result similar to that in Question 2 for sin.

4. Show that for every positive $\delta$ and for every non-zero $w$ there is a solution of $\exp(1/z) = w$ in $C(0, \delta)$.
5. For each complex number $z$ and each positive integer $n$ define

$$n^z = \exp(z \log_e (n)).$$

Prove that

$$\sum_{n=1}^{\infty} \frac{1}{n^z}$$

is absolutely convergent if Re $[z] > 1$. [Assume (or prove) that the series

$$\sum_{n=1}^{\infty} \frac{1}{n^{1+\epsilon}},$$

$\epsilon > 0$, is convergent.]

## 5.3 EXPONENTS

For any positive $x$ and any positive integer $n$ we use $x^{1/n}$ to denote the unique positive $n$th root of $x$.
Then

$$x^{1/n} = \exp\left(\frac{1}{n} \log_e (x)\right)$$

as this is a positive number whose $n$th power is $x$. Observe that for any positive integer $m$,

$$\begin{aligned}(x^{1/n})^m &= \left[\exp\left(\frac{1}{n} \log_e (x)\right)\right]^m \\ &= \exp\left(\frac{m}{n} \log_e (x)\right) \\ &= \exp\left(\frac{1}{n} \log_e (x^m)\right) \\ &= (x^m)^{1/n}.\end{aligned}$$

It is natural to write this as $x^{m/n}$ and to extend this to complex numbers as follows.

*Definition 5.3.1 If $z$ and $w$ are complex numbers with $z \neq 0$, then $z^{[w]}$ is the set of numbers of the form*

$$\exp(w \log(z)),$$

*where* log $(z)$ *is any choice of* Log $(z)$. *Equivalently,*

$$z^{[w]} = \{\exp(w\,[\log_e(|z|) + i\theta]) : \theta \in \text{Arg}(z)\}.$$

This is the formal definition of '$z$ raised to the power $w$' : the fact that it is a set reflects the many-valued nature of the operation. To clarify this definition we give two examples.

*Example 5.3.1* We shall find $i^{[i]}$. We have

$$\log_e(|i|) = 0, \qquad \pi/2 \in \text{Arg}(i)$$

and so

$$\begin{aligned} i^{[i]} &= \{\exp(i[i(2n\pi + \pi/2)]) : n \in \mathbf{Z}\} \\ &= \{\exp(-[2n\pi + \pi/2]) : n \in \mathbf{Z}\}. \end{aligned}$$

Observe that each element of this set is real.

*Example 5.3.2* By definition,

$$\begin{aligned} z^{[0]} &= \{\exp(0 \cdot \log(z)) : \log(z) \in \text{Log}(z)\} \\ &= \{1\}. \end{aligned}$$

Before stating the next result we remind the reader that for a non-zero $z$ and a positive integer $n$, $z^n$ is defined inductively by $z^0 = 1$ and $z^{n+1} = z \cdot z^n$ (so $z^n$ is a product of $z$ with itself $n$ times) : if $n$ is a negative integer, then $z^n = 1/z^{|n|}$.

*Theorem 5.3.1* *Let $z$ and $w$ be complex numbers with $z \neq 0$. Then*

(a) *if $w$ is an integer, say $w = n$, then*

$$z^{[n]} = \{z^n\}$$

(b) *if $w$ is a rational, say $w = p/q$, where $p$ and $q$ are integers, then*

$$z^{[p/q]} = \{\zeta_1^p, \ldots, \zeta_q^p\}$$

*where $\zeta_1, \ldots, \zeta_q$ are the $q$ $q$th roots of $z$;*

(c) *in all other cases, $z^{[w]}$ is a countably infinite set.*

This means that for a positive integer $n$, $z^{[n]}$ is the set whose only element is $z^n$ and $z^{[1/n]}$ is the complete set of $n$th roots of $z$.

*Proof* We select a value $\theta$ of Arg $(z)$ which will remain unaltered throughout the proof.

If $w = n$, an integer, then

$$\text{Arg}(z) = \{\theta + 2k\pi : k \in \mathbf{Z}\}$$

and so

$$z^{[n]} = \{\exp(n[\log_e(|z|) + i\theta + 2k\pi i]) : k \in \mathbf{Z}\}.$$

The result now follows as for each integer $k$,

$$\begin{aligned}\exp(n[\log_e(|z|) + i\theta + 2k\pi i]) &= \exp(n\log_e(|z|))\exp(in\theta)\exp(2kn\pi i)\\ &= |z|^n\exp(in\theta)\\ &= z^n.\end{aligned}$$

Now let $w = p/q$ as in (b). Then for each integer $k$,

$$\begin{aligned}\exp\left(\frac{p}{q}[\log_e(|z|) + i\theta + 2k\pi i]\right)\\ &= (|z|^{1/q})^p\exp(pi\theta/q)\exp(2kp\pi i/q)\\ &= [|z|^{1/q}\exp(i\theta/q)]^p\exp(2kp\pi i/q)\\ &= [\zeta\exp(2k\pi i/q)]^p,\end{aligned}$$

where

$$\zeta = |z|^{1/q}\exp(i\theta/q).$$

The result follows as

$$\{\zeta\exp(2k\pi i/q) : k \in \mathbf{Z}\}$$

is the complete set of $q$th roots of $z$.

To prove (c) we suppose that $z^{[w]}$ is a finite set. This means that

$$\{\exp(w[\log_e(|z|) + i\theta + 2k\pi i]) : k \in \mathbf{Z}\}$$

and hence

$$\{\exp(2wk\pi i) : k \in \mathbf{Z}\}$$

is a finite set. Thus there are two distinct integers $k$ and $k'$ such that

$$\exp(2wk\pi i) = \exp(2wk'\pi i)$$

and this implies that for some integer $m$,

$$2w(k - k')\pi i = 2m\pi i.$$

We conclude that $w$ is rational.

If $w$ is not rational then $z^{[w]}$ is infinite: it is countable because $\mathrm{Arg}(z)$ is countable.

*Example 5.3.3* Let $e = \exp(1)$ (which is real). Then

$$e^{[z]} = \{\exp(z[1 + 2n\pi i]) : n \in \mathbf{Z}\}$$

because $\log_e(e) = 1$ and $0 \in \mathrm{Arg}(e)$. Thus

$$\exp(z) \in e^{[z]}.$$

We have taken great care to distinguish between the 'many-valued expression' $e^{[z]}$ ($e$ 'raised to the power' $z$) and the (single-valued) function exp $(z)$, and the previous example shows that exp $(z)$ is always one value in the set $e^{[z]}$. It is customary (mainly for typographical reasons) to use $e^z$ in place of exp $(z)$ and we shall use both notations interchangeably. We stress, however, that $e^z$ is *not* $e$ raised the power $z$: it is simply an alternative notation for exp $(z)$.

We shall have no further use for the general expression for $z^{[w]}$: without doubt the most important case is when $w$ is rational and this will be used later in the text.

**Exercise 5.3**

1. Show that $1 \in 1^{[z]}$ and find $1^{[z]}$.
2. Show that $-1 \in e^{[i\pi]}$ and find $e^{[i\pi]}$.
3. Show that $e^{\pi/2} \in i^{[-i]}$ and find $i^{[-i]}$.
4. Prove that
   (a) $w \in \text{Log}\,(z)$ if and only if $\bar{w} \in \text{Log}\,(\bar{z})$;
   (b) $w \in a^{[b]}$ if and only if $\bar{w} \in \bar{a}^{[\bar{b}]}$.
5. Suppose that $a \neq 0$ and that $b$ is not real and let

$$S = \{|z| : z \in a^{[b]}\}.$$

   Prove that

$$\inf(S) = 0, \qquad \sup(S) = +\infty$$

   (this means that $S$ is not bounded above).
   What can be said of $S$ if $b$ is real?
6. Define

$$z^{[a]}z^{[b]} = \{w_1 w_2 : w_1 \in z^{[a]}, w_2 \in z^{[b]}\}.$$

   Show that this need not be the same as $z^{[a+b]}$ (take $z = 1, a = -b = \frac{1}{2}$).
7. Express the following sets explicitly:

$$\exp\,(\text{Log}\,(z)), \quad e^{[\text{Log}\,(z)]}, \quad \text{Log}\,(e^{[z]}), \quad \text{Log}\,(\exp\,(z)).$$

8. Let $p$ and $q$ be positive integers. How many elements are there in the set $z^{[p/q]}$? [There may be less than $q$.] Does the number depend on $z$?

## 5.4 CONTINUITY OF THE LOGARITHM

In this section we shall consider the possibility of selecting a single choice, say $\theta(z)$, of Arg $(z)$ in such a way that $\theta(z)$ is a continuous function of $z$. Although it may seem clear that such a continuous choice can be made, the reader should be warned that it is necessary to exercise great care in any discussion concerning Arg. For example, it is fundamental that *no such continuous function exists on the set* **C*** *of non-zero complex numbers.* The absence of such a continuous function is not an empty remark: on the contrary, it is a pregnant observation which we shall exploit consistently in the rest of this book. Its immediate impact, however, is to raise the problem of deciding when such a continuous choice can be made.

*Example 5.4.1* We shall prove that it is impossible to select a continuous choice $\theta(z)$ of Arg $(z)$ on $\mathbf{C}^*$ by assuming that such a choice can be made and reaching a contradiction.

The function $f$ defined on $[0, 2\pi]$ by

$$2\pi f(x) = \theta(e^{ix}) + \theta(e^{-ix})$$

is the real valued and continuous on $[0, 2\pi]$ as both $\theta$ and exp are continuous. Further, $2\pi f(x)$ is a choice of

$$\text{Arg}\,(e^{ix}e^{-ix}),$$

that is, a choice of Arg (1) and this means that $f$ is actually a continuous integer-valued function on $[0, 2\pi]$. The Intermediate Value Theorem now implies that $f$ is necessarily constant and so $f(0) = f(\pi)$. This gives $\theta(1) = \theta(-1)$, which is false as no choice of Arg (1) is a choice of Arg $(-1)$. We conclude that no such choice of $\theta$ exists.

From a geometric viewpoint it is impossible to select a continuous choice of $\theta$ on $\mathbf{C}^*$ because we have allowed $z$ to move around the origin and to return to its original value (by which time $\theta$ must have changed by at least $2\pi$). To reassure the reader that this is the reason and to prepare for later work we shall now show that we can make a continuous choice of $\theta$ if we prevent $z$ from moving around the origin. We achieve this by deleting from $\mathbf{C}$ a half-line $L_\alpha$ from the origin in the direction $\alpha$. The remaining material is the formal expression of these ideas.

Let $\alpha$ be any real number and let

$$L_\alpha = \{te^{i\alpha} : t \geqslant 0\}.$$

If $z \notin L_\alpha$, then $\alpha \notin \text{Arg}\,(z)$ and so there is a unique choice of Arg $(z)$ in $(\alpha, \alpha + 2\pi)$. In fact, $z \in \mathbf{C} - L_\alpha$ if and only if $\alpha \notin \text{Arg}\,(z)$.

*Definition 5.4.1* *For each real* $\alpha$ *and each* $z$ *in* $\mathbf{C} - L_\alpha$, $\text{Arg}_\alpha\,(z)$ *denotes the unique choice of* Arg $(z)$ *in* $(\alpha, \alpha + 2\pi)$ *and*

$$\text{Log}_\alpha\,(z) = \log_e\,(\,|z|\,) + i\,\text{Arg}_\alpha(z).$$

*Theorem 5.4.1* $\text{Arg}_\alpha$ *and* $\text{Log}_\alpha$ *are continuous on* $\mathbf{C} - L_\alpha$.

*Proof* We first consider $\text{Arg}_0$ and, for brevity, write $\theta(z)$ for $\text{Arg}_0\,(z)$. If $z$ is in the half-plane $H_0 = \{x + iy : y > 0\}$, then

$$\text{Im}\,[z] = |z|\sin(\theta(z))$$

and so $\sin(\theta(z)) > 0$. As $\theta(z) \in (0, 2\pi)$, we now see that $\theta(z) \in (0, \pi)$.

The function cos: $(0, \pi) \to (-1, 1)$ is strictly decreasing and so has a continuous inverse $\cos^{-1} : (-1, 1) \to (0, \pi)$ (Theorem 3.2.3 is, of course, equally applicable to strictly decreasing functions). Thus

$$\begin{aligned}\theta(z) &= \cos^{-1}(\cos(\theta(z)))\\ &= \cos^{-1}(\text{Re}\,[z]/|z|)\end{aligned}$$

and this is continuous on $H_0$.

A similar argument shows that $\mathrm{Arg}_0$ is continuous on the half-planes

$$H_1 = \{x + iy : x < 0\}, \qquad H_2 = \{x + iy : y < 0\}.$$

If $z \in \mathbf{C} - L_0$ (which is $H_0 \cup H_1 \cup H_2$), then there is a positive $\delta$ such that $\mathbf{C}(z, \delta)$ lies in one $H_j$ and so $\mathrm{Arg}_0$ is continuous on $\mathbf{C} - L_0$.

The general case now follows easily. Given any real $\alpha$ let $g(z) = z \exp(-i\alpha)$. Then $g$ is a continuous map of $\mathbf{C} - L_\alpha$ onto $\mathbf{C} - L_0$ and so $\varphi$ defined by

$$\varphi(z) = \alpha + \mathrm{Arg}_0\,(g(z))$$

is continuous on $\mathbf{C} - L_\alpha$. However, $\alpha < \varphi(z) < \alpha + 2\pi$, and as

$$g(z) = |\,g(z)\,| \exp\,(i\ \mathrm{Arg}_0\,(g(z)))$$

we also have

$$|\,z\,| \exp\,(i\varphi(z)) = z.$$

Thus $\varphi(z) = \mathrm{Arg}_\alpha\,(z)$ and $\mathrm{Arg}_\alpha$ is continuous on $\mathbf{C} - L_\alpha$.

The following corollary is used as frequently as Theorem 5.4.1 itself (see Diagram 5.4.1).

*Corollary* *For each non-zero w there is a continuous choice of* Arg *on* $\mathbf{C}(w, |\,w\,|)$.

This follows from Theorem 5.4.1 because $\mathbf{C}(w, |\,w\,|)$ is the disc with centre $w$ whose circumference passes through the origin and so if $\alpha \in \mathrm{Arg}\,(w)$, then (Excercise 5.4.1)

$$\mathbf{C}(w, |\,w\,|) \subset \mathbf{C} - L_{\alpha+\pi}. \qquad (5.4.1)$$

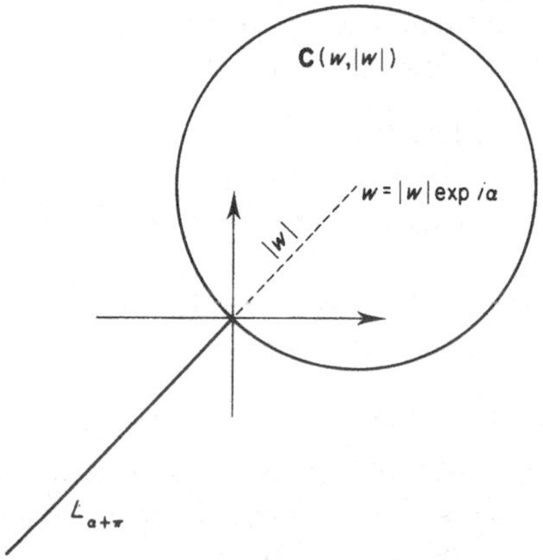


Diagram 5.4.1

The general question of when a continuous choice of Arg $(z)$ can be made on a set $E$ is considered in greater detail in Chapter 11.

**Exercise 5.4**

1. Prove (rigorously) (5.4.1).
2. Show that there exists a continuous choice of

$$\operatorname{Arg}\left(\frac{z+1}{z-1}\right)$$

on the set of $z$ satisfying $|z| > 1$.
3. Let $n$ be a positive integer. Show how to find a continuous function $h$ on $\mathbf{C} - L_\pi$ such that $[h(z)]^n = z$ (so $h$ is a continuous $n$th root of $z$).
4. Let $\alpha$ be real and define

$$S_\alpha = \{z \in \mathbf{C} : \alpha < \operatorname{Im}[z] < \alpha + 2\pi\}$$

(this is an infinite horizontal strip in the plane). Prove that $\exp : S_\alpha \to \mathbf{C}$ is a 1–1 continuous map of $S_\alpha$ onto $\mathbf{C} - L_\alpha$.

Prove also that $\operatorname{Log}_\alpha$ is a 1–1 continuous map of $\mathbf{C} - L_\alpha$ onto $S_\alpha$ and is the inverse of the above function.

## 2019 UPDATE FOR THEOREM 5.4.1 ON PAGE 59

If $z$ is a non-zero complex number then $\theta$ is a choice of $\mathrm{Arg}(z)$ if and only if $\theta$ is a real number that satisfies $z = |z|(\cos\theta + i\sin\theta)$. The negative real axis is the half-line $L_\pi$ (see p.59), and if $z \notin L_\pi$ then there is a unique choice of $\mathrm{Arg}(z)$ in the interval $(-\pi, \pi)$. We denote this choice by $\Theta(z)$, and we shall now give an explicit formula for $\Theta(z)$. Let $z = x + iy$. As $z$ is not on the negative real axis, then $|z| + x \neq 0$ and, for any choice $\theta$ of $\arg z$, we have

$$\frac{y}{|z|+x} = \frac{|z|\sin\theta}{|z|+|z|\cos\theta} = \frac{\sin\theta}{1+\cos\theta} = \tan\tfrac{1}{2}\theta.$$

In particular,

$$\frac{y}{|z|+x} = \tan\tfrac{1}{2}\Theta(z), \quad \tfrac{1}{2}\Theta(z) \in (-\tfrac{1}{2}\pi, \tfrac{1}{2}\pi).$$

If we now use $\tan^{-1}$ for the principal branch of the inverse tan function (which is defined by $-\pi/2 < \tan^{-1} < \pi/2$), then we have the following explicit formula for $\Theta(z)$:

$$\Theta(z) = 2\tan^{-1}\left(\frac{y}{|z|+x}\right).$$

For example, $\Theta(1) = 2\tan^{-1}(0) = 0$, and $\Theta(i) = 2\tan^{-1}(1) = 2(\pi/4) = \pi/2$. This formula for $\Theta(z)$ is important because (as $\tan^{-1}$ is a continuous function) it shows immediately that $\Theta(z)$ is a single-valued, *continuous choice* of the argument of $z$ on the complex plane cut along the negative real axis. Given this, we leave the reader to construct, for each real $\alpha$, a single-valued, continuous choice of the argument of $z$ on the complement of the half-line $L_\alpha$.

# PART II

# *Basic Complex Analysis*

# Chapter 6

## 6.1 OPEN AND CLOSED SETS

We recall that we have defined the *open disc* $\mathbf{C}(w, r)$ by

$$\mathbf{C}(w, r) = \{z \in \mathbf{C} : |z - w| < r\}.$$

If $z \in \mathbf{C}(w, r)$ then (draw a diagram)

$$\mathbf{C}(z, t) \subset \mathbf{C}(w, r) \tag{6.1.1}$$

where $t = r - |z - w|$ : this is because if $\zeta \in \mathbf{C}(z, t)$ then

$$\begin{aligned} |\zeta - w| &\leqslant |\zeta - z| + |z - w| \\ &< t + |z - w| \end{aligned}$$

and so $\zeta \in \mathbf{C}(w, r)$.

It is more fruitful to study the class of sets which are expressible as a union of open discs rather than the class of open discs itself.

*Definition 6.1.1* *A subset $E$ of $\mathbf{C}$ is open if and only if*

$$E = \bigcup_{\alpha} \mathbf{C}(a_\alpha, r_\alpha)$$

*for some family of open discs $\mathbf{C}(a_\alpha, r_\alpha)$, $\alpha \in A$.*

As simple examples of open sets we have

$$\mathbf{C} = \bigcup_{n=1}^{\infty} \mathbf{C}(0, n)$$

and

$$\{x + iy : x > 0\} = \bigcup_{t>0} \mathbf{C}(t, t).$$

Of course, each $\mathbf{C}(a, r)$ is open for we need only take one $\mathbf{C}(a_\alpha, r_\alpha)$, namely $\mathbf{C}(a, r)$ itself. As $\mathbf{C}(a, r) = \emptyset$ if $r < 0$ we see that the empty set $\emptyset$ is open.

Even though an open set is a union of discs, its boundary can still have ‘corners’ (and indeed be very much worse than this). There is no point in trying to make this statement precise, an example will illustrate the point. The square

$$S = \{x + iy : 0 < x < 1, 0 < y < 1\}$$

is an open set. If $z = x + iy$ is in $S$ let

$$r_z = \min\{1 - x, 1 - y, x, y\}:$$

then

$$S = \bigcup_{z \in S} \mathbf{C}(z, r_z).$$

An equivalent way of defining open sets is to say that $E$ is open if and only if for each $z$ in $E$, there is a positive $r_z$ with

$$\{z\} \subset \mathbf{C}(z, r_z) \subset E. \tag{6.1.2}$$

If this is true then certainly

$$E = \bigcup_{z \in E} \{z\} \subset \bigcup_{z \in E} \mathbf{C}(z, r_z) \subset E,$$

and so $E$ is open. Conversely, if $E$ is open and if $z \in E$, then $z \in \mathbf{C}(a_\alpha, r_\alpha)$ for some $\alpha$ and so by (6.1.1) such a positive $r_z$ exists.

It is important to understand how the class of open sets behaves with respect to the operations of union and intersection. It is immediate from Definition 6.1.1 that *if* $\{E_\beta : \beta \in B\}$ *is any family of open sets, then* $\cup_\beta E_\beta$ *is open.*

If $E_1$ and $E_2$ are open then there are positive numbers $r_1$ and $r_2$ such that

$$\{z\} \subset \mathbf{C}(z, r_j) \subset E_j \qquad (j = 1, 2).$$

Let $r$ be the minimum of $r_1$ and $r_2$, then

$$\{z\} \subset \mathbf{C}(z, r) \subset E_1 \cap E_2$$

and this shows that $E_1 \cap E_2$ is open. More generally, *if* $E_1, \ldots, E_n$ *are open then so is* $E_1 \cap \cdots \cap E_n$.

The complements of the open sets are as important as the open sets themselves: *a set E is said to be closed if and only if* $\mathbf{C} - E$ *is open.* We are not saying that 'closed' means 'not open' and it does not. The set of positive real numbers, for example, is neither open nor closed: $\mathbf{C}$ is both open and closed.

The rules governing the behaviour of closed sets with respect to unions and intersections are direct consequences of (1.1.2), (1.1.3) and the corresponding results for the open sets. *If* $\{E_\beta : \beta \in B\}$ *is a family of closed sets, then* $\cap_\beta E_\beta$ *is also closed.* If $E_1, \ldots, E_n$ *are closed then so is* $E_1 \cup \cdots \cup E_n$. The reader should verify these results (Exercise 6.1.1).

It is easy to see that each *closed disc* $\bar{\mathbf{C}}(w, r)$ defined by

$$\bar{\mathbf{C}}(w, r) = \{z \in \mathbf{C} : |z - w| \leqslant r\}$$

is closed (Exercise 6.1.2). The special case $r = 0$ shows that each set $\{w\}$ is closed and we deduce that each finite set is closed. If $E$ is open and $K$ is closed, then

$$E - K = E \cap (\mathbf{C} - K)$$

and this is open: similarly, $K - E$ is closed. In particular, if $E$ is open, so is $E - \{z_1, \ldots, z_n\}$.

Given any set $E$, let $\bar{E}$ denote the intersection of all closed sets which contain $E$, we call $\bar{E}$ the *closure* of $E$. Thus

$$\bar{E} = \bigcap_{K \in \mathscr{E}} K$$

where $K \in \mathscr{E}$ if and only if $K$ is closed and $K \supset E$. Observe that $\mathbf{C}$ is in $\mathscr{E}$ so $\mathscr{E} \neq \emptyset$. Obviously $\bar{E}$ is closed and contains $E$: moreover, every $K$ in $\mathscr{E}$ contains $\bar{E}$. Thus, in more descriptive terms, $\bar{E}$ is the smallest closed set that contains $E$.

We can give an alternative (but equivalent) definition of $\bar{E}$. If there exists a positive $r$ such that $\mathbf{C}(\zeta, r) \cap E = \emptyset$, then $E$ is a subset of the closed set $\mathbf{C} - \mathbf{C}(\zeta, r)$. Thus $\bar{E}$ is contained in this closed set and so $\zeta \notin \bar{E}$.

Conversely, if $\zeta \notin \bar{E}$, then for some closed set $K$, $E \subset K$ and $\zeta \notin K$. Thus $\zeta$ is in the open set $\mathbf{C} - K$ and so for some positive $r$, $\mathbf{C}(\zeta, r) \cap K = \emptyset$. We deduce that $\zeta \in \bar{E}$ if and only if for every positive $r$, $E \cap \mathbf{C}(\zeta, r) \neq \emptyset$.

This shows, for example, that the closure of $\mathbf{C}(\zeta, r)$ is $\bar{\mathbf{C}}(\zeta, r)$, that is

$$\overline{\mathbf{C}(\zeta, r)} = \bar{\mathbf{C}}(\zeta, r).$$

We have seen that $E \subset \bar{E}$ and that $\bar{E}$ is closed. If $E$ itself is closed, then $E \in \mathscr{E}$ and so $\bar{E} \subset E$. Thus if $E$ is closed, then $E = \bar{E}$. Of course, if $E = \bar{E}$, then $E$ is closed as $\bar{E}$ is: hence $E$ is closed if and only if $E = \bar{E}$.

Now let $E$ be any subset of $\mathbf{C}$. The *boundary*, $\partial E$, of $E$ is defined to be the set of points $\zeta$ with the property that for every positive $r$, $\mathbf{C}(\zeta, r)$ meets both $E$ and $C - E$. As $\zeta \in \bar{E}$ if and only if for every positive $r$, $\mathbf{C}(\zeta, r) \cap E \neq \emptyset$ we see that

$$\partial E = \bar{E} \cap \overline{(\mathbf{C} - E)}.$$

The boundary of $\mathbf{C}(w, r)$, for example, is $\{z : |z - w| = r\}$.

We shall need the concept of a *punctured disc* $\mathbf{C}^*(w, r)$: this is defined by

$$\mathbf{C}^*(w, r) = \{z \in \mathbf{C}: 0 < |z - w| < r\}$$

and it is the open disc $\mathbf{C}(w, r)$ with the centre $w$ removed.

A complex number $w$ is a *limit point* of $E$ if and only if for each positive $r$, $\mathbf{C}^*(w, r) \cap E \neq \emptyset$. If $w$ is a limit point of $E$ then, for example, there are points $z_1, z_2, \ldots$ in $E$ with

$$0 < |z_n - w| < \frac{1}{n}.$$

A limit point of $E$ may but need not be in $E$.

We say that $w$ is an *isolated point* of $E$ if and only if $w \in E$ and there is a positive $r$ such that $\mathbf{C}^*(w, r) \cap E = \emptyset$ or, equivalently,

$$\mathbf{C}(w, r) \cap E = \{w\}.$$

By definition, an isolated point of $E$ is in $E$.

The closure $\bar{E}$ of a set $E$ is the union of the set of isolated points of $E$ and the set of limit points of $E$. The set $E$ is closed if it contains all of its limit points (Exercise 6.1.3).

We end this section with a result which relates open sets and continuous functions.

*Theorem 6.1.1* *Let* E *be an open set* $f: E \to \mathbf{C}$ *be any function. Then* $f$ *is continuous on* $E$ *if and only if for every open subset* $A$ *of* $\mathbf{C}$, $f^{-1}(A)$ *is open.*

*Proof* Suppose that $f$ is continuous on $E$ and let $A$ be any open set. If $f^{-1}(A)$ is empty, then it is open. Otherwise, select any $w$ in $f^{-1}(A)$; so $w \in E$ and $f(w) \in A$. As $A$ is open there is a positive $\epsilon$ with $\mathbf{C}(f(w), \epsilon) \subset A$. As $f$ is continuous at $w$ and as $E$ is open there is a positive $\delta$ such that $\mathbf{C}(w, \delta) \subset E$ and $f$ maps $\mathbf{C}(w, \delta)$ into $\mathbf{C}(f(w), \epsilon)$. Thus $\mathbf{C}(w, \delta) \subset f^{-1}(A)$ and this shows that $f^{-1}(A)$ is open.

Now suppose that $f^{-1}(A)$ is open whenever $A$ is open and select $w$ in $E$. Given any positive $\epsilon$, let $A = \mathbf{C}(f(w), \epsilon)$. Then $E \cap f^{-1}(A)$ is an open set containing $w$ and so there is a positive $\delta$ such that

$$\mathbf{C}(w, \delta) \subset E \cap f^{-1}(A).$$

This means that $f$ is defined on $\mathbf{C}(w, \delta)$ and maps it into $\mathbf{C}(f(w), \epsilon)$; thus $f$ is continuous at $w$.

**Exercise 6.1**

1. Verify that the intersection of any family of closed sets is a closed set and that the union of a finite number of closed sets is a closed set.
2. Prove that each closed disc is a closed set. [Draw a diagram.]
3. Prove that a set $E$ is closed if and only if it contains all of its limit points. [*Hint*: use the alternative definition of $\bar{E}$.]
4. Prove that the punctured disc $\mathbf{C}^*(w, r)$ is open.
   Prove also that

$$\partial\mathbf{C}^*(w, r) = \{z : |z - w| = r\} \cup \{w\}.$$

5. For any set $E$ define the interior, $E^0$, of $E$ to be the union of all open subsets of $E$. Prove

(a) $E$ is open if and only if $E = E^0$;
(b) $E^0$ is an open subset of $E$;
(c) $E^0$ is the largest open subset of $E$ (make this precise);
(d) $\bar{E} = \mathbf{C} - (\mathbf{C} - E)^0$.

6. Let $E$ be the set of real numbers which are of the form

$$\frac{\epsilon_1}{1} + \frac{\epsilon_2}{2} + \cdots + \frac{\epsilon_n}{n} \qquad (n \in \{1, 2, \ldots\})$$

   where each $\epsilon_j$ is 1 or $-1$. Find $\bar{E}$, $E^0$ and $\partial E$.
   [You will need to use Example 2.2.3.]

7. Let $E$ be any set and let $f$ and $g$ be continuous on $\bar{E}$. Show that if $f \leqslant g$ on $E$ then $f \leqslant g$ on $\bar{E}$.
8. Let $A$ and $B$ be any sets such that $A \cap B \neq \emptyset$. Let $f$ be continuous on $A$, $g$ continuous on $B$ and suppose also that $f = g$ on $A \cap B$. Define $F$ on $A \cup B$ by

$$F(z) = \begin{cases} f(z) & \text{if} \quad z \in A \\ g(z) & \text{if} \quad z \in B. \end{cases}$$

Prove
(a) if $A$ and $B$ are open, then $F$ is continuous on $A \cup B$;
(b) if $A$ and $B$ are closed, then $F$ is continuous on $A \cup B$;
(c) $F$ need not be continuous on $A \cup B$.

## 6.2 CONNECTED SETS

Most readers will have some intuitive concept of connected sets; for example, we can presumably agree that according to any reasonable definition, $\mathbf{R}$ is connected whereas $\mathbf{R} - \{0\}$ is not. The concept of connectedness is essential for analysis and it is necessary to make a preliminary study of it. The definition we choose involves continuous integer-valued functions and is particularly suited to our needs: an equivalent definition based on open sets rather than on continuity is given in Exercise 6.3.7.

*Definition 6.2.1* *A subset E of* $\mathbf{C}$ *is connected if and only if each continuous integer-valued function* $f : E \to \mathbf{Z}$ *is constant on E. A set is disconnected if and only if it is not connected.*

Obviously, any set $\{a\}$ with exactly one element is connected: any finite set with more than one element is disconnected. The motivation for this definition is simply that if $f$ is both continuous and integer valued then a sufficiently small change in $x$ can only lead to a small and hence to no change in $f(x)$. The question of whether or not a continuous integer-valued function on $E$ is necessarily constant is therefore related in some sense to the ability to move throughout $E$ by a succession of small steps.

*Example 6.2.1* Let us show that each interval $I$ in $\mathbf{R}$ is connected. Let $f : I \to \mathbf{Z}$ be continuous and let $a$ and $b$, $a < b$, be points in $I$ (we may assume that $I$ has at least two points) so $[a, b] \subset I$. The Intermediate Value Theorem applied to $f$: $[a, b] \to \mathbf{Z}$ shows that $f(a) = f(b)$, as otherwise $f$ assumes all values between $f(a)$ and $f(b)$ and some of these are non-integral values. As $a$ and $b$ are any points in $I$, $f$ is constant on $I$ and $I$ is connected. The same argument shows that $\mathbf{R}$ is connected. The reader should now show that $\mathbf{R} - \{0\}$ is disconnected (Exercise 6.2.1).

The following two results are simple yet have many applications.

*Theorem 6.2.1* *Let $E$ be connected and let $f$ be continuous on $E$. Then $f(E)$ is connected.*

*Proof* Suppose that $g : f(E) \to \mathbf{Z}$ is continuous; then as $f$ is continuous, so is $g \circ f : E \to \mathbf{Z}$. As $E$ is connected, $g \circ f$ is constant on $E$. If $w_1$ and $w_2$ are points in $f(E)$ with, say, $w_j = f(z_j), j = 1, 2, z_j \in E$, then

$$g(w_1) = g \circ f(z_1) = g \circ f(z_2) = g(w_2).$$

Thus $g$ is constant on $f(E)$ and so $f(E)$ is connected.

*Theorem 6.2.2* *Let $\{E_\alpha : \alpha \in A\}$ be a non-empty family of connected sets. If $\bigcap_{\alpha \in A} E_\alpha \neq \emptyset$ then $\bigcup_{\alpha \in A} E_\alpha$ is connected.*

*Proof* Suppose that $f : \bigcup_{\alpha \in A} E_\alpha \to \mathbf{Z}$ is continuous. Then for each $\alpha$, $f : E_\alpha \to \mathbf{Z}$ is continuous and so $f$ is constant (say equal to $k_\alpha$) on each $E_\alpha$. If $w \in \bigcap_\alpha E_\alpha$, then $k_\alpha = f(w)$ for each $\alpha$. Thus $f$ is constant on $\bigcup_{\alpha \in A} E_\alpha$ and $\bigcup_{\alpha \in A} E_\alpha$ is connected.

We can now construct other examples of connected sets.

*Example 6.2.2* If $z_1$ and $z_2$ are in $\mathbf{C}$ we define the segment $[z_1, z_2]$ with end-points $z_1$ and $z_2$ by

$$[z_1, z_2] = \{z_1 + t(z_2 - z_1) : 0 \leqslant t \leqslant 1\}.$$

The function defined by $f(t) = z_1 + t(z_2 - z_1)$ is a continuous map of the connected set $[0, 1]$ onto the segment $[z_1, z_2]$: thus by Theorem 6.2.1, $[z_1, z_2]$ is connected.

A set $E$ is *convex* if and only if $[z_1, z_2] \subset E$ whenever $z_1$ and $z_2$ are in $E$. If $E$ is convex and if $z_1 \in E$, then

$$E = \bigcup_{z \in E} [z_1, z]$$

and by Theorem 6.2.2, $E$ is connected. As each disc, each half-plane and $\mathbf{C}$ itself are convex, these sets are also connected.

This idea can also be used to analyse the structure of an arbitrary set $E$. For each $z$ in $E$ denote by $E(z)$ the union of all connected subsets of $E$ which contain $z$ (such subsets do exist, for example $\{z\}$). A direct application of Theorem 6.2.2 shows that $E(z)$ is also a connected subset of $E$: it is therefore *the largest connected subset of $E$ which contains $z$.*

The sets $E(z)$ are called the *components* of $E$. If two components $E(z)$ and $E(w)$ have a point $\zeta$ in common, then, again by Theorem 6.2.2, $E(z) \cup E(w)$ is connected. Thus, say,

$$E(z) \cup E(w) \subset E(z)$$

($E(z)$ being the largest connected subset of $E$ containing $z$) and so $E(w) \subset E(z)$. By symmetry, the reverse inclusion holds and $E(z) = E(w)$.

We have proved that any two components of $E$ are either disjoint or equal. As

$z \in E(z)$, $E$ is the union of its components and we can write

$$E = \bigcup_{\alpha \in A} E_\alpha \tag{6.2.1}$$

where $E_\alpha$ are the components of $E$ and $E_\alpha \cap E_\beta = \emptyset$ if $\alpha \neq \beta$.

The analysis of complex-valued functions is restricted almost entirely to functions defined on an open connected subset of **C**. In view of this the reader is urged to pay particular attention to the following discussion.

*Definition 6.2.2* *A subset of* **C** *is called a domain if and only if it is both open and connected.*

Now let $E$ be a non-empty open set and consider the decomposition (6.2.1) of $E$. We shall show that each component $E_\alpha$ is open and is therefore a domain.

First, if $z \in E$ then, as $E$ is open, there is a positive $\delta$ such that $\mathbf{C}(z, \delta) \subset E$. As $\mathbf{C}(z, \delta)$ is connected we deduce that $\mathbf{C}(z, \delta) \subset E(z)$.

Given any component $E_\alpha$ of $E$, let $z$ be any point in $E_\alpha$. Then $E_\alpha$ and $E(z)$ have the point $z$ in common and so $E_\alpha = E(z)$ and therefore

$$\mathbf{C}(z, \delta) \subset E_\alpha.$$

This shows that $E_\alpha$ is open and we have proved the following result.

*Theorem 6.2.3* *Let E be a non-empty open set. Then the components* $D_\alpha$ *of E are domains and*

$$E = \bigcup_{\alpha \in A} D_\alpha$$

*where* $D_\alpha \cap D_\beta = \emptyset$ if $\alpha \neq \beta$.

We end with two remarks. First, the union of two disjoint non-empty open sets is necessarily disconnected. Indeed, if $A$ and $B$ are open, non-empty, and disjoint, then the function

$$f(x) = \begin{cases} 1 & \text{if} \quad x \in A, \\ 0 & \text{if} \quad x \in B \end{cases}$$

is continuous, integer-valued and not constant on $A \cup B$.

Finally, it should be clear from the definition of components that (in the general situation) any connected subset $K$ of $E$ lies in just one component of $E$. Indeed if $z \in K$, then $K \subset E(z)$.

**Exercise 6.2**

1. By considering $g : \mathbf{R} - \{0\} \to \mathbf{R}$ defined by $g(x) = x / |x|$ show that $\mathbf{R} - \{0\}$ is disconnected.

   Prove that each connected subset of **R** is an interval.

2. Prove that the parabola given by $y = x^2$ (the set of $x + iy$ with $y = x^2$) is connected. Prove that the hyperbola given by $xy = 1$ is disconnected.
3. Let $E_1$ and $E_2$ be sets such that

$$\delta = \inf \{ |z_1 - z_2| : z_1 \in E_1, z_2 \in E_2 \}$$

is positive. Prove that $E_1 \cup E_2$ is disconnected.
4. A set $E$ is *star-shaped* if there is a $z_0$ in $E$ such that for all $z$ in $E$, $[z_0, z] \subset E$. Prove
   (a) a convex set is star-shaped,
   (b) a star-shaped set need not be convex,
   (c) a star-shaped set is connected.
5. Prove that the closure of a connected set is connected. Deduce that each component of a closed set is closed.
6. Prove that each $\mathbf{C}^*(w, r)$, $r > 0$, is connected. Deduce that if $D$ is a domain, then so is $D - \{z_1, \ldots, z_m\}$. [*Hint*: show that if $f$ is continuous and integer-valued on $D - \{z_1, \ldots, z_m\}$, then $f$ extends to a continuous function on $D$.]
7. Let $E$ be any set. We say that $K$ is a *relatively open* subset of $E$ if and only if $K = E \cap A$ for some open set $A$.

   Prove that $E$ is disconnected if and only if $E$ is the disjoint union of two non-empty relatively open subsets of $E$.

## 6.3 LIMITS

In Chapter 2, we introduced the idea of a sequence $z_0, z_1, z_2, \ldots$ as a function defined on the set $\{n \in \mathbf{Z} : n \geqslant 0\}$. It is too restrictive to consider only functions defined on this set; indeed the only essential feature that we require is that the function be defined on an infinite set of integers with a smallest member. This will enable us to discuss, for example, sequences such as $z_0, z_2, z_4, \ldots$.

*Definition 6.3.1* *A sequence s is a function $s : Z_0 \to \mathbf{C}$ where $Z_0$ is an infinite set of integers with a smallest member.*

Given a sequence $s$, we usually denote the image of $n$ in $Z_0$ by $s_n$: for brevity, the reference to $Z_0$ is often omitted. The notations $(s_n : n \in Z_0)$ and $s_{n_1}, s_{n_2}, \ldots$, $Z_0 = \{n_1, n_2, \ldots\}$ for a sequence $s$ are both in common use although the latter clearly presupposes that the general formula is clear. It is essential to distinguish between the sequence $(s_n : n \in Z_0)$ which is a *function* and the *set* $\{s_n : n \in Z_0\}$: for example, if the sequence has each term zero then the function is the constant function zero on $Z_0$ whereas the set is simply $\{0\}$.

With Definition 6.3.1 available, the idea of a subsequence can easily be described. A *subsequence* of $(s_n : n \in Z_0)$ is simply a sequence $(s_n : n \in Z_1)$, where $Z_1$ is an infinite subset of $Z_0$. Loosely speaking, then, the terms of a subsequence occur in the original sequence and in the same order, although some terms of the original sequence may have been omitted.

Observe that if $(s_n : n \in Z_1)$ is a subsequence of $(s_n : n \in Z_0)$, then any subsequence $(s_n : n \in Z_2)$ of $(s_n : n \in Z_1)$ is also a subsequence of $(s_n : n \in Z_0)$. This is simply the statement that if $Z_1 \subset Z_0$ and $Z_2 \subset Z_1$, then $Z_2 \subset Z_0$.

The definition of convergence of sequences is designed to include the case of infinite series discussed in Chapter 2. Given any sequence $z_1, z_2, \ldots$, we write

$$s_n = z_1 + \cdots + z_n; \tag{6.3.1}$$

then

$$\sum_{n=1}^{\infty} z_n = w$$

if and only if for every positive $\epsilon$ there is an integer $n_0$ such that $|s_n - w| < \epsilon$ whenever $n \geqslant n_0$.

We base the general definition of convergence on this.

The sequence $(s_n : n \in Z_0)$ *converges* to $w$ if and only if for every positive $\epsilon$ there is an integer $n_0$ (depending on $\epsilon$) such that $|s_n - w| < \epsilon$ whenever $n \in Z_0$ and $n \geqslant n_0$.

If $s$ converges to $w$ we write $s_n \to w$ as $n \to \infty$ in $Z_0$, $\lim(s) = w$ or sometimes,

$$\lim_{n \to \infty} s_n = w.$$

We say that $s$ is *convergent* (or converges) if it converges to some $w$. If the numbers $s_n$ satisfy a given condition for all but a finite set of $n$ in $Z_0$ we say that $s_n$ satisfies this condition for *almost all* $n$ (in $Z_0$). For example, if $s$ converges to $w$ then $|s_n - w| < 1$ for almost all $n$.

The following facts should now be clear. First, each sequence can converge to at most one $w$. If the sequence $s$ converges to $w$, then so does every subsequence: if each $s_n$ is real, then so is $w$. Next, if $s$ converges to $w$, then it is *bounded*, that is for some positive $r$, $|s_n| \leqslant r$. In fact, if $|s_n - w| < 1$ when $n \geqslant n_0$, then for all $n$ in $Z_0$,

$$|s_n| \leqslant (1 + |w|) + \sup\{|s_m| : m \leqslant n_0, m \in Z_0\}.$$

Finally, if $(s_n : n \in Z_0)$ and $(t_n : n \in Z_0)$ converge to $w_1$ and $w_2$ respectively, then

$$\lim_{n \to \infty} (s_n + t_n) = w_1 + w_2, \qquad \lim_{n \to \infty} (s_n t_n) = w_1 w_2.$$

Let us clarify these remarks with an example.

*Example 6.3.1* Let $(s_n : n \in Z_0)$ be given by

$$s_n = i^n, \qquad Z_0 = \{n \in \mathbf{Z} : n \geqslant 0\}.$$

Then $s$ does not converge although it is bounded ($|s_n| \leqslant 1$). If $Z_1 = \{n! : n \in \mathbf{Z}, n > 0\}$, the subsequence $(s_n : n \in Z_1)$ converges to 1; in fact, $s_n = 1$ for almost all (but not all) $n$ in $Z_1$.

The sequence $s$ is a *real sequence* if for each $n$, $s_n$ is real. The real sequence $s$ is *increasing* if $s_n \geqslant s_m$ whenever $n \geqslant m$: it is *strictly increasing* if $s_n > s_m$ whenever $n > m$. Of course, similar definitions are formulated for (strictly) decreasing sequences.

It is easy to see that if the real sequence $s$ is bounded above ($s_n \leqslant k$ for some $k$) and increasing then it converges to $y$ where

$$y = \sup \{s_n : n \in Z_0\}.$$

Indeed, for every positive $\epsilon$ there is an integer $n_0$ with $s_{n_0} > y - \epsilon$ and so if $n \geqslant n_0$, then

$$y - \epsilon < s_{n_0} \leqslant s_n \leqslant y.$$

Similarly, a bounded decreasing sequence also converges.

A bounded sequence does not in general converge: one can, however, guarantee at least a convergent subsequence.

*Theorem 6.3.1 The Bolzano–Weierstrass Theorem. Every bounded sequence of complex terms has a convergent subsequence.*

*Proof* We first consider any real bounded sequence $(s_n : n \in Z_0)$ with $|s_n| \leqslant r$, $n \in Z_0$. Let

$$E = \{t \in \mathbf{R} : \ s_n \geqslant t \text{ for almost all } n \text{ in } Z_0\}.$$

As $E$ is bounded above by $r$ and as $-r \in E$, we can define $\alpha$ by

$$\alpha = \sup (E).$$

For each positive integer $k$, $\alpha + k^{-1} \notin E$ and so $s_n \leqslant \alpha + k^{-1}$ for infinitely many $n$. Next, $\alpha - k^{-1}$ is not an upper bound of $E$ and so there is some $t$ in $E$ with $\alpha - k^{-1} < t$. Thus $s_n \geqslant \alpha - k^{-1}$ for almost all $n$. Combining these results we find that

$$|s_n - \alpha| < 1/k \tag{6.3.2}$$

for infinitely many $n$ in $Z_0$.

We can now construct a subsequence of $(s_n : n \in Z_0)$ which converges to $\alpha$. Select any $n_1$ in $Z_0$ with $|s_{n_1} - \alpha| < 1$. As (6.3.2) holds with $k = 2$ we can find an integer $n_2$ in $Z_0$ with $n_2 > n_1$ and $|s_{n_2} - \alpha| < ½$. Continuing in this way we can construct (by induction) an increasing sequence of integers $n_1, n_2, \ldots$ in $Z_0$ with

$$|s_{n_k} - \alpha| < 1/k.$$

Obviously, $(s_n : n \in Z_1)$, $Z_1 = \{n_1, n_2, \ldots\}$ converges to $\alpha$.

We have now shown that every real bounded sequence has a convergent subsequence. If $(s_n : n \in Z_0)$ is now any bounded complex sequence we write $x_n = \text{Re } [s_n]$, $y_n = Im \ [s_n]$. Then $(x_n : n \in Z_0)$ is a real bounded sequence and so has a convergent subsequence, say $(x_n : n \in Z_1)$. The sequence $(y_n : n \in Z_1)$ is also real and bounded so it has a convergent subsequence, say $(y_n : n \in Z_2)$. It is now clear that the two sequences $(x_n : n \in Z_2)$ and $(y_n : n \in Z_2)$ converge and as $s_n = x_n + iy_n$, so does $(s_n : n \in Z_2)$.

Theorem 6.3.1 enables us to derive a criterion for the convergence of $(s_n : n \in Z_0)$ which does not depend on guessing (correctly) the value of the limit (which need not exist). This criterion is known both as *Cauchy's Criterion* and as the *General Principle of Convergence.*

*Theorem 6.3.2* *The sequence $(s_n : n \in Z_0)$ is convergent if and only if for each positive $\epsilon$ there is an integer $n_0$ such that $|s_n - s_m| < \epsilon$ whenever $m \geqslant n_0, n \geqslant n_0$.*

A geometric interpretation of this is given in Excercise 6.3.4.

*Proof* If $(s_n : n \in Z_0)$ converges to $w$, say, then $|s_n - w| < \frac{1}{2}\epsilon$ for almost all $n$ and for $m$ and $n$ greater than some $n_0$,

$$|s_n - s_m| \leqslant |s_n - w| + |w - s_m| < \epsilon.$$

Now suppose that the given condition holds. First, this implies that $(s_n : n \in Z_0)$ is bounded and so, by Theorem 6.3.1, there is a subsequence $(s_n : n \in Z_1)$ converging to some $w$. Given any positive $\epsilon$, choose integers $n_0$ and $n_1$ such that

$$|s_n - w| < \tfrac{1}{2}\epsilon \qquad \text{if } n \in Z_1, \quad n \geqslant n_1$$

and

$$|s_n - s_m| < \tfrac{1}{2}\epsilon \qquad \text{if } m, n \in Z_0, \quad m, n \geqslant n_0.$$

If $n$ is in $Z_1$ and exceeds $n_0$ and $n_1$, then $|s_m - w| < \epsilon$ whenever $m \in Z_0$ and $m \geqslant n_0$ and so $(s_n : n \in Z_0)$ converges to $w$.

This result may be rephrased in terms of series using (6.3.1).

*Corollary* *The series $\Sigma_{n=1}^{\infty} z_n$ converges if and only if for each positive $\epsilon$ there is an integer $n_0$ such that*

$$|z_n + \cdots + z_m| < \epsilon$$

*whenever $m \geqslant n \geqslant n_0$.*

It is convenient to extend the notion of convergence to include $\infty$, $+\infty$ and $-\infty$ as limits. First, the *real* sequence $(s_n : n \in Z_0)$ *converges to* $+\infty$ if for each real number $k$, $s_n \geqslant k$ for almost all $n$ in $Z_0$: it *converges to* $-\infty$ if $s_n \leqslant k$ for almost all $n$ in $Z_0$. Note that we are not asserting the existence of numbers $+\infty$ and $-\infty$, less still their properties: we are simply defining, for example, the phrase 'converges to $+\infty$'.

We say that the complex sequence $(s_n : n \in Z_0)$ *converges to* $\infty$ if for each real number $k$, $|s_n| > k$ for almost all $n$ in $Z_0$. If the sequence is real and increasing, then convergence to $+\infty$ is equivalent to convergence to $\infty$.

The purpose behind these definitions is to achieve a greater coherence in the theory rather than to establish new results. For example we can now state that every sequence $(s_n : n \in Z_0)$ has a convergent subsequence. This is because the sequence is either bounded (and the result follows from Theorem 6.3.1) or has a subsequence $s_{n_1}, s_{n_2}, \ldots$ with $|s_{n_k}| > k$, $k = 1, 2, \ldots$. Likewise, every real increasing sequence now converges (possibly to $+\infty$).

It is worth noting, however, that Theorem 6.3.2 applies only in the sense that convergence is to a complex number and not to $\infty$, $+\infty$ or $-\infty$.

Almost all of the preceding discussion can be adapted to apply to the general function rather than just a sequence. For example, let $f$ be defined on $E$ and let $E$ be unbounded. Then we say that *$f(z)$ converges to $w$ as $z$ tends to $\infty$* and write

$$\lim_{z\to\infty} f(z) = w \tag{6.3.3}$$

or $f(z) \to w$ as $z \to \infty$ *if and only if for each positive $\epsilon$ there is a positive $r$ such that*

$$|f(z) - w| < \epsilon$$

*whenever $z \in E$ and $|z| > r$.*

If $E$ is now any set (not necessarily unbounded) with a limit point $\zeta$, we say that *$f$ (defined on $E$) converges to $w$ as $z$ tends to $\zeta$ if and only if for each positive $\epsilon$ there is a positive $\delta$ such that*

$$|f(z) - w| < \epsilon$$

*whenever $z \in E \cap \mathbf{C}^*(\zeta, \delta)$.* If this is so we write $f(z) \to w$ as $z \to \zeta$ or

$$\lim_{z\to\zeta} f(z) = w. \tag{6.3.4}$$

Observe that in this definition we are concerned with $\mathbf{C}^*(\zeta, \delta)$ and not $\mathbf{C}(\zeta, \delta)$. The significance of this is that $\zeta$ need not be in $E$ and even if it is in $E$, the value $f(\zeta)$ has no effect whatsoever on the possible existence of or the value of the limit of $f$ at $\zeta$. This corresponds precisely to the case where a sequence $s$ is defined on $Z_0$ but not at $+\infty$. If the limit of $f$ at $\zeta$ is $f(\zeta)$ then $f$ is continuous at $\zeta$: the converse is true provided that $\zeta$ is a limit point of $E$.

It is clear that many of the remarks made for sequences are equally applicable to the limits (6.3.3) and (6.3.4): the reader should examine these carefully. We mention in particular that Theorem 6.3.2 can be modified to apply to the limits (6.3.3) and (6.3.4) and this will be used later. For example, the limit (6.3.3) exists if and only if for each positive $\epsilon$ there is a positive $r$ such that

$$|f(z) - f(z')| < \epsilon \tag{6.3.5}$$

whenever $z \in E$, $z' \in E$, $|z| > r$ and $|z'| > r$.

Finally, the limits (6.3.3) and (6.3.4) can be defined for the case $w = \infty$ and, if $f$ is real-valued, for $w = +\infty$ and $w = -\infty$: this, too, is left to the reader.

**Exercise 6.3**

1. Let $f : \mathbf{C}^*(w, r) \to \mathbf{C}$, $r > 0$, be any bounded function. Prove that $\lim_{z\to w} f(z)$ exists if and only if there is a function $F : \mathbf{C}(w, r) \to \mathbf{C}$ continuous at $w$ with $f = F$ on $\mathbf{C}^*(w, r)$.

2. Suppose that $(s_n : n \in Z_0)$ converges to $s$ and that $s \neq 0$. Prove that $s_n \neq 0$ for almost all $n$ and that for a suitable $Z_1$, $(1/s_n : n \in Z_1)$ converges to $1/s$.
3. Construct a sequence with the property that for each $x$ in $[0, 1]$ there exists a subsequence converging to $x$. [Consider the sequence

   $$0, 1, 0, \tfrac{1}{2}, 1, 0, \tfrac{1}{4}, \tfrac{1}{2}, \tfrac{3}{4}, 1, \ldots.]$$

   Prove a similar result with $[0, 1]$ replaced by **C**.
4. Let $(s_n : n \in Z_0)$ be any sequence. Prove that the two following properties are equivalent:
   (a) there exists a $w$ such that for every positive $\epsilon$, $s_n \in \mathbf{C}(w, \epsilon)$ for almost all $n$;
   (b) for every positive $\epsilon$, there exists a $w$ such that $s_n \in \mathbf{C}(w, \epsilon)$ for almost all $n$.
   [This is the geometric version of Theorem 6.3.2.]
5. Prove Theorem 6.3.2 as follows. First, construct a sequence $s_{n_1}, s_{n_2}, \ldots$ with

   $$|s_{n_k} - s_{n_{k+1}}| < 2^{-(k+1)}, \qquad k = 1, 2, \ldots.$$

   Next, consider the convergence of the series

   $$\sum_{k=1}^{\infty} (s_{n_k} - s_{n_{k+1}})$$

   and show that the sequence $s_{n_1}, s_{n_2}, \ldots$ converges. Deduce that the original sequence converges.
6. Prove (6.3.5).

## 6.4 COMPACT SETS

In order to discuss compact sets we need the notion of an open cover. The family $\{O_\alpha : \alpha \in A\}$ is an *open cover* of the set $E$ if and only if each $O_\alpha$ is open and $\bigcup_{\alpha \in A} O_\alpha \supset E$. An open cover of $E$ is thus a family of open sets whose union contains $E$. Such families arise naturally in analysis and we shall now study the important and related topic of compactness. First, however, we say that a set $E$ is *bounded* if it is contained in some disc $\mathbf{C}(w, r)$.

*Theorem 6.4.1* *Let E be any subset of* **C**: *the following statements are equivalent*

(a) *E is closed and bounded.*
(b) *Each sequence in E has a subsequence which converges to a point in E.*
(c) *Given any open cover* $\{O_\alpha : \alpha \in A\}$ *of E there is a positive* $\delta$ *such that each disc* $\mathbf{C}(z, \delta)$, $z \in E$, *is contained in some* $O_\alpha$.
(d) *Given any open cover* $\{O_\alpha : \alpha \in A\}$ *of E there are elements* $\alpha_1, \ldots, \alpha_m$ *in A such that* $E \subset O_{\alpha_1} \cup \cdots \cup O_{\alpha_m}$.

A set $E$ is *compact* if and only if it satisfies one (and hence all) of these conditions. The variety of these conditions should give the reader some indication of the usefulness of the concept and we shall consider several important applications later. The implication (a) implies (b) is essentially the *Bolzano–Weierstrass*

*Theorem*: (d) implies (c) is the *Lebesgue Covering Theorem*; (a) implies (d) is the *Heine–Borel Theorem.*

The following simple examples should assist the reader to become familiar with these ideas before embarking on a proof of Theorem 6.4.1. The family $\{\mathbf{C}(0, t) : 0 < t < 1\}$ is an open cover of $\mathbf{C}(0, 1)$ : observe that $\mathbf{C}(0, 1)$ is not in this family nor is it compact. The disc $\bar{\mathbf{C}}(w, r)$ is both closed and bounded and so is compact. Similarly, the rectangle

$$\{x + iy : a \leqslant x \leqslant b, c \leqslant y \leqslant d\}$$

is closed, bounded and hence compact. The set $N = \{n^{-1} : n = 1, 2, \ldots\}$ is not compact; $N \cup \{0\}$ is compact. None of the sets $\mathbf{Z}$, $\mathbf{R}$ and $\mathbf{C}$ are compact.

*Proof of Theorem* 6.4.1 We suppose first that (a) holds and let $s$, $s = (s_n : n \in Z_0)$, be any sequence in $E$. As $E$ is bounded, so is the sequence $s$ and so by Theorem 6.3.1, there is a subsequence, say $(s_n : n \in Z_1)$, converging to $w$. If $w \notin E$, then, as $\mathbf{C} - E$ is open, there is a positive $\epsilon$ with $\mathbf{C}(w, \epsilon) \subset \mathbf{C} - E$. Thus no point of $E$ and so no $s_n, n \in Z_1$, lies in $\mathbf{C}(w, \epsilon)$, and this is a contradiction. We deduce that $w \in E$ and (b) holds .

Now suppose that (b) holds and let $\{O_\alpha : \alpha \in A\}$ be any open cover of $E$. If (c) fails, then for each positive integer $n$ there is a point $z_n$ in $E$ such that $\mathbf{C}(z_n, n^{-1})$ is not contained in any single $O_\alpha$. The sequence $z_1, z_2, \ldots$ is in $E$ and so by (b) has a subsequence $(z_n : n \in Z_1)$ converging to some $w$ in $E$. As $w \in E$, for some $\beta$, $w \in O_\beta$ and so there is a positive $\epsilon$ with

$$\mathbf{C}(w, \epsilon) \subset O_\beta.$$

For almost all $n$ in $Z_1$, we have

$$|z_n - w| + n^{-1} < \epsilon$$

and so

$$\mathbf{C}(z_n, n^{-1}) \subset \mathbf{C}(w, \epsilon) \subset O_\beta.$$

This is a contradiction and so (b) does imply (c).

Next, suppose that (c) holds. We shall first show that $E$ is bounded. For each $w$ satisfying $|w| \geqslant 1$, let $O_w$ be the disc $\mathbf{C}(w, |w|^{-1})$ : this has radius at most 1 and the radius is small if $|w|$ is large. The family $\{O_w : |w| \geqslant 1\}$ together with $\mathbf{C}(0, 1)$ is an open cover of $\mathbf{C}$ and hence of $E$, and according to (c), there is a positive $\delta$ such that if $z \in E$, then $\mathbf{C}(z, \delta)$ is contained either in $\mathbf{C}(0, 1)$ or in some $O_w$. In the first case, $|z| < 1$, while in the second case, $\delta < |w|^{-1}$ and so

$$|z| \leqslant |w| + |w|^{-1} \leqslant \delta^{-1} + 1.$$

In any event, if $z \in E$, then $|z| \leqslant \delta^{-1} + 1$ and $E$ is bounded.

Now let $\{O_\alpha : \alpha \in A\}$ be any open cover of $E$ with $\delta$ given by (c). We subdivide $\mathbf{C}$ (with horizontal and vertical lines) into congruent non-overlapping squares $Q_j$ of diameter $\frac{1}{2}\delta$ and as $E$ is bounded it meets only a finite number of these squares, say $Q_1, \ldots, Q_n$ with $z_j \in E \cap Q_j$. As $Q_j \subset \mathbf{C}(z_j, \delta)$, each $Q_j$ is contained in one single

$O_{\alpha_j}$ and so

$$E \subset \bigcup_{j=1}^{n} O_{\alpha_j},$$

which proves (d). Of course, the subdivision of $\mathbf{C}$ can be expressed quite precisely in algebraic terms and the argument does not depend on geometric intuition.

Finally, we suppose that (d) holds. The family $\{\mathbf{C}(0, n) : n = 1, 2, \ldots\}$ is an open cover of $E$: as (d) holds, $E$ is bounded. If $w \notin E$ we apply (d) to the open cover $\{O_n : n = 1, 2, \ldots\}$ of $E$, where

$$O_n = \{z \in \mathbf{C} : |z - w| > 1/n\},$$

and we find that for some $n$, $E \subset O_n$. This shows that $\mathbf{C}(w, n^{-1}) \subset \mathbf{C} - E$ and so $\mathbf{C} - E$ is open. Thus $E$ is closed and (a) holds.

We shall now give four applications of compactness: two are concerned with sets, two with functions and all are worthy of the title 'Theorem'. In practice, one usually recognizes compactness by (a) and then uses (b), (c) or (d). We have deliberately chosen to illustrate all four conditions in our applications.

*Theorem 6.4.2* *Let $K_1, K_2, \ldots$ be non-empty compact sets such that $K_1 \supset K_2 \supset \cdots$. Then $\bigcap_{n=1}^{\infty} K_n$ is non-empty and compact.*

*Proof* Let $K = \bigcap_{n=1}^{\infty} K_n$. As $K_1$ is bounded so is $K$. As each $K_n$ is closed so is their intersection, namely $K$. Hence $K$ is compact.

Now let $O_n = \mathbf{C} - K_n$. These sets are open, $O_1 \subset O_2 \subset \cdots$ and

$$\bigcup_{n=1}^{\infty} O_n = \mathbf{C} - \bigcap_{n=1}^{\infty} K_n = \mathbf{C} - K.$$

If $K = \emptyset$, then $\{O_n : n = 1, 2, \ldots\}$ is an open cover of $\mathbf{C}$ and hence of $K_1$. Then $K_1$ is contained in a finite number of the $O_n$ and hence in one, say $O_m$. Then $K_m \subset K_1 \subset O_m$, which is false: so $K \neq \emptyset$.

*Theorem 6.4.3* *Let $E$ and $K$ be disjoint non-empty sets and suppose that $E$ is closed and $K$ is compact. Then* dist $(E, K) > 0$, *where*

$$\text{dist}\,(E, K) = \inf\{\,|z - w| : z \in E, w \in K\}.$$

*Proof* The single set $\mathbf{C} - E$ is an open cover of $K$ and so by (c) there is a positive $\delta$ such that if $w \in K$, then $\mathbf{C}(w, \delta) \subset \mathbf{C} - E$. Thus dist $(E, K) \geq \delta > 0$.

*Theorem 6.4.4* *Let $E$ be compact and let $f$ be continuous on $E$. Then $f(E)$ is compact and (if $E \neq \emptyset$) $|f|$ attains its maximum and its minimum value on $E$.*

*Proof* Let $(w_n : n \in Z_0)$ be any sequence in $f(E)$: then we can find $z_n, n \in Z_0$, in $E$ with $f(z_n) = w_n$. The sequence $(z_n : n \in Z_0)$ in $E$ has, by (b), a subsequence

$(z_n : n \in Z_1)$ converging to $z^*$, say. By continuity, $f(z_n) \to f(z^*)$ as $n \to \infty$ in $Z_1$ and so the subsequence $(w_n : n \in Z_1)$ converges to $f(z^*)$ in $f(E)$. This shows that $f(E)$ is compact.

Now let

$$K = \{ |f(z)| : z \in E\}:$$

this is compact as it is the image of the compact set $E$ by the continuous function $|f|$. We deduce first that $K$ is bounded and if $m = \sup(K)$, then $K \subset [0, m]$. If $m \notin K$ then, as $K$ is closed, there is a positive $\delta$ with $K \cap \mathbf{C}(m, \delta) = \emptyset$ and then $K \subset [0, m - \delta]$. This contradicts the definition of $m$: so $m \in K$ and so for some $z$, $|f(z)| = m$. This proves the assertion about the maximum value; a similar argument is used for the minimum value.

The final application is concerned with uniform continuity. We say that a function $f$ *is uniformly continuous on E if given any positive* $\epsilon$ *there is a positive* $\delta$ *such that* $|f(z) - f(w)| < \epsilon$ *whenever z and w are in E with* $|z - w| < \delta$. The crucial issue here is that $\delta$ depends only on $\epsilon$: provided only that $|z - w| < \delta$ (and no matter where these points are located in $E$) we must have $|f(z) - f(w)| < \epsilon$.

Obviously, if $f$ is uniformly continuous on $E$ then it is continuous on $E$. The converse is false: for example $f(x) = x^{-1}$ is continuous but not uniformly continuous on $\{x \in \mathbf{R}: x > 0\}$ (consider $|f(n^{-1}) - f(n^{-2})|$).

*Theorem 6.4.5* *Let E be compact and let f be continuous on E. Then f is uniformly continuous on E.*

*Proof* Let $E$ be compact, $f$ continuous on $E$ and $\epsilon$ any positive number. For each $z$ in $E$ there is a positive $r_z$ such that $|f(w) - f(z)| < \frac{1}{2}\epsilon$ whenever $w \in E \cap \mathbf{C}(z, r_z)$. The family $\{\mathbf{C}(\zeta, r_\zeta) : \zeta \in E\}$ is an open cover of $E$ and so there is an associated $\delta$ given by (c). If $z \in E$, $w \in E$ and $|z - w| < \delta$ then by (c),

$$\{z, w\} \subset \mathbf{C}(z, \delta) \subset \mathbf{C}(\zeta, r_\zeta)$$

for some $\zeta$ and so

$$\begin{aligned} |f(z) - f(w)| &< |f(z) - f(\zeta)| + |f(\zeta) - f(w)| \\ &< \epsilon. \end{aligned}$$

**Exercise 6.4**

1. Let $\{O_\alpha : \alpha \in A\}$ be an open cover of $\{z \in \mathbf{C} : |z| = 1\}$. Prove that for some positive $\epsilon$,

$$\{z \in \mathbf{C} : 1 - \epsilon < |z| < 1 + \epsilon\} \subset \bigcup_{\alpha \in A} O_\alpha.$$

2. Give an example of non-empty closed sets $K_n$, $n = 1, 2, \ldots$, with $K_{n+1} \subset K_n$ and

$$\bigcap_{n=1}^{\infty} K_n = \emptyset.$$

3. Find disjoint closed non-empty sets $E$ and $K$ such that dist $(E, K) = 0$. [Compare this with Theorem 6.4.3]
4. Let $E$ be any set which is non-empty and not compact. Prove that there exists a continuous function $f: E \to \mathbf{C}$ such that $f(E)$ is not compact. [Compare this with Theorem 6.4.4]
5. Let $E$ be a compact subset of $\mathbf{C}$ and let $f: E \to \mathbf{C}$ be continuous. For each $u$ and $v$ in $f(E)$ define

$$\delta(u, v) = \inf\{\,|z - w| : f(z) = u, f(w) = v\}.$$

Obviously, $\delta(u, v) \geqslant 0$: prove that $\delta(u, v) > 0$. Show that this may be false if $E$ is not compact.
6. Let $K$ be non-empty and compact and suppose that $f: K \to K$ satisfies

$$|f(z) - f(w)| < |z - w|, \qquad z, w \in K.$$

Prove that for some $z$ in $K$, $f(z) = z$.

## 6.5 HOMEOMORPHISMS

Topology is the study of open sets, closed sets and continuous functions. One of the primary aims of topology is to identify sets which, as far as continuous functions are concerned, are essentially the same; thus we seek properties which are invariant under continuous functions. For example, if $f$ is continuous on $E$ and if $E$ is connected then so is $f(E)$. Similarly, if $E$ is compact then so is $f(E)$. On the other hand, if $E = (0, 1)$ and $f(x) = x^{-1}$, then $E$ is bounded but $f(E)$ is not.

The process of identification of two sets requires a 1–1 function of one set onto the other. If a continuous function is to respect this identification, then the function and its inverse must be continuous: thus we are led to the fundamental notion of a homeomorphism.

*Definition 6.5.1* *A function $f$ on $X$ is a homeomorphism if $f$ is 1–1 on $X$ and if both $f: X \to f(X)$ and $f^{-1}: f(X) \to X$ are continuous. If there is a homeorphism $f: X \to Y$ with $f(X) = Y$, then $X$ and $Y$ are said to be homeomorphic.*

Clearly, if $f: X \to f(X)$ is a homeomorphism, then so is $f^{-1}: f(X) \to X$. Also, if $X_1$ and $X_2$ are homeomorphic and if $X_2$ and $X_3$ are homeomorphic then so are $X_1$ and $X_3$. A few simple examples should help the reader to appreciate these ideas.

*Example 6.5.1* Any two open intervals, say $(a, b)$ and $(c, d)$, are homeomorphic: we simply take

$$f(x) = c + \left(\frac{d - c}{b - a}\right)(x - a).$$

Further, $(a, b)$ is homeomorphic to $\{x : x > 0\}$ with, say,

$$f(x) = \frac{x - a}{b - x}.$$

*Example 6.5.2* The sets **R** and $\mathbf{R} \cup \{i\}$ are not homeomorphic. Indeed if $X$ and $Y$ are homeomorphic and if $X$ is connected, then so is $Y$. Clearly, **R** is connected but $\mathbf{R} \cup \{i\}$ is not.

*Example 6.5.3* Let $a$ and $b$ be non-zero real numbers. The sets

$$X = \{x + iy : x^2 + y^2 = 1\}, \qquad Y = \left\{x + iy : \left(\frac{x}{a}\right)^2 + \left(\frac{y}{b}\right)^2 = 1\right\}$$

are homeomorphic, for we may take

$$f(x + iy) = ax + iby.$$

*Example 6.5.4* The sets

$$X = \{z : |z| = 1\}, \qquad Y = \{z : |z| = 1, z \neq 1\}$$

are not homeomorphic. If $f$ is a homeomorphism defined on $X$, then $f(X)$ is compact and so cannot be $Y$ (which is not closed).

*Example 6.5.5* The sets

$$\mathbf{R}, \qquad X = \mathbf{R} \cup \{iy : y \geqslant 0\}$$

are not homeomorphic. The easiest proof is by contradiction so we assume that there is a homeomorphism $f$ of $X$ onto **R**. Now let $x = f(0)$; then $f$ is a homeomorphism of $X - \{0\}$ onto $\mathbf{R} - \{x\}$. Clearly there exists a continuous function $g$ mapping $X - \{0\}$ *onto*, say, $\{0, 1, 2\}$ and so $g \circ f^{-1}$ is continuous and maps $\mathbf{R} - \{x\}$ onto $\{0, 1, 2\}$. This, however, is impossible as $\mathbf{R} - \{x\}$ has only two components and $g \circ f^{-1}$ is constant on each component.

We shall content ourselves with one simple but useful criterion for a function to be a homeomorphism.

*Theorem 6.5.1* *Let $E$ be compact and let $f$ be 1–1 and continuous on $E$. Then $f$ is a homeomorphism.*

*Proof* The function $f$ is 1–1 and continuous on $E$, thus $f^{-1} : f(E) \to E$ exists and we need only prove that this is continuous.

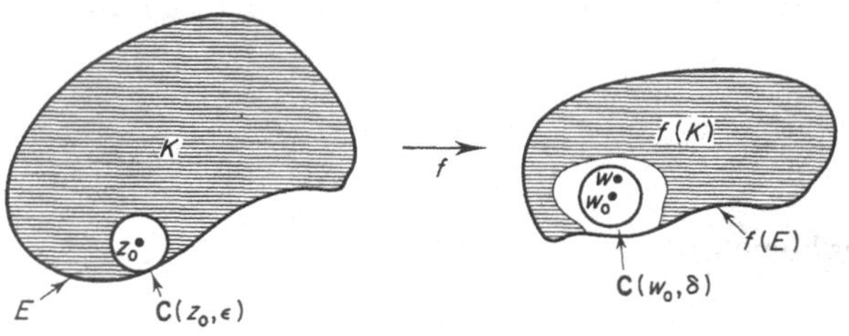


Diagram 6.5.1

Let $w_0$ be in $f(E)$ with, say, $f(z_0) = w_0, z_0 \in E$. For any positive $\epsilon$, $K = E - \mathbf{C}(z_0, \epsilon)$ is closed and bounded and therefore compact. We deduce that $f(K)$ is compact and as $w_0 \in f(K)$ there is a positive $\delta$ with

$$f(K) \cap \mathbf{C}(w_0, \delta) = \emptyset.$$

If $w \in f(E) \cap \mathbf{C}(w_0, \delta)$, then $w \notin f(K)$. Thus $f^{-1}(w) \notin K$ and so $f^{-1}(w) \in \mathbf{C}(z_0, \epsilon)$, that is $|f^{-1}(w) - f^{-1}(w_0)| < \epsilon$. This is illustrated in Diagram 6.5.1.

**Exercise 6.5**

1. Prove that if $X_1$ and $X_2$ are homeomorphic and if $X_2$ and $X_3$ are homeomorphic, then so are $X_1$ and $X_3$.
2. By considering $f(x) = x/(1 + |x|)$, show that **R** and $(-1, 1)$ are homeomorphic.
3. Show that no two of the intervals $(0, 1)$, $[0, 1]$ and

   $$[0, 1) = \{x : 0 \leqslant x < 1\}$$

   are homeomorphic.
4. Prove that if $X$ and $Y$ are homeomorphic then there is a 1–1 correspondence between the components of $X$ and the components of $Y$. [You must construct a 1–1 map of the class of components of $X$ onto the class of components of $Y$. This generalizes Examples 6.5.2 and 6.5.5.]
5. Prove that $X = \{z : |z - i| = 1, z \neq 2i\}$ is homeomorphic to **R**. [Draw a diagram, 'project' $z$ from $2i$ to **R** and let this projection be $f$.]
6. Use Theorem 6.5.1 to give an alternative proof of the existence of a continuous choice of Arg $z$ near the point $z = 1$. [Let $h(z) = z/|z|$ and let $g$: $[-\pi/2, \pi/2] \to \{z : |z| = 1\}$ be defined by $g(t) = e^{it}$. Show that if $g^{-1} \circ h(z) = \theta(z)$, then $\theta$ is a continuous choice of Arg $z$.]

## 6.6 UNIFORM CONVERGENCE

Uniform convergence is concerned with sequences of functions $f_1, f_2, \ldots$, or more generally $(f_n : n \in Z_0)$, each defined on the same set $E$. Among the possible definitions of convergence of $f_n$ to a function $f$, uniform convergence stands out for its relative simplicity and usefulness.

First, if $f$ and $g$ are defined on $E$, we write

$$\| f - g \|_E = \sup \{ |f(z) - g(z)| : z \in E \}$$

(this may be $+\infty$): the suffix $E$ is frequently omitted. We are to regard $\| f - g \|_E$ as a measure of the maximum deviation of $f$ from $g$ on $E$.

*Definition 6.6.1 The sequence of functions* $(f_n : n \in Z_0)$ *converges uniformly to* $f$ *on* $E$ *if and only if*

$$\| f_n - f \|_E \to 0 \qquad \text{as} \qquad n \to +\infty.$$

This can be expressed more fully (by writing down the definition of this limit) in terms of parameters $\epsilon$, $n$, $n_0$ and $z$, but we prefer this more concise formulation. We often write '*$f_n \to f$ uniformly on $E$ as $n \to +\infty$*' to denote the uniform convergence of $f_n$ to $f$.

*Example 6.6.1* Let $f_n(z) = z^n$, $n = 1, 2, \ldots$, and let $f$ denote the zero function so $f(z) = 0$ for all $z$. Then for each $t$, $0 < t < 1$, $f_n \to f$ uniformly on $\bar{C}(0, t)$. This is easily proved, for on $\bar{C}(0, t)$,

$$|f_n(z) - f(z)| = |z^n| \leqslant t^n$$

and so

$$\|f_n - f\| \leqslant t^n.$$

Observe, however, that $f_n$ does not converge uniformly to $f$ on $C(0, 1)$: indeed if $E = C(0, 1)$, then

$$\|f_n - f\|_E = \sup\{|z^n| : |z| < 1\} = 1.$$

Quite generally, if $f_n \to f$ uniformly on each of the sets $E_1, E_2, \ldots$ there is no reason whatsoever to assume that $f_n$ will converge uniformly to $f$ on $\bigcup_{n=1}^{\infty} E_n$. One can say, of course, that for each $z$ in $\bigcup_{n=1}^{\infty} E_n$, the sequence $(f_n(z) : n \in Z_0)$ converges to $f(z)$.

*Theorem 6.6.1* *Let $f_1, f_2, \ldots$ be continuous on $E$ and suppose that $f_n \to f$ uniformly on $E$. Then $f$ is continuous on $E$.*

*Proof* Let $\epsilon$ be any positive number and select an integer $n$ so that $\|f_n - f\| < \frac{1}{3}\epsilon$. Now select any $w$ in $E$. The continuity of $f_n$ at $w$ implies that there is a positive $\delta$ such that if $z \in E \cap C(w, \delta)$, then $|f_n(z) - f_n(w)| < \frac{1}{3}\epsilon$. For these $z$,

$$\begin{aligned} |f(z) - f(w)| &\leqslant |f(z) - f_n(z)| + |f_n(z) - f_n(w)| + |f_n(w) - f(w)| \\ &\leqslant 2\|f - f_n\| + \tfrac{1}{3}\epsilon \\ &< \epsilon \end{aligned}$$

and so $f$ is continuous at $w$.

We shall frequently be concerned with series of functions, say

$$\sum_{n=1}^{\infty} f_n(z). \tag{6.6.1}$$

This series is said to be uniformly convergent on $E$ if the corresponding sequence $F_1, F_2, \ldots$ of partial sums

$$F_n(z) = f_1(z) + \cdots + f_n(z)$$

is uniformly convergent on $E$. If this is so and if each $f_n$ is continuous on $E$, then so is the infinite sum.

There is a very simple test (due to Weierstrass) for the uniform convergence of the series (6.6.1): *if*

$$\sum_{n=1}^{\infty} \| f_n \|_E \tag{6.6.2}$$

*converges then the series* (6.6.1) *is uniformly convergent on E.*

The proof is very easy. For each $z$ in $E$,

$$| f_n(z) | \leqslant \sup \{ | f_n(w) | : w \in E \} = \| f_n \|_E$$

and so by the Comparison Test, (6.6.1) converges to, say, $F(z)$. Then

$$| F_n(z) - F(z) | = \left| \sum_{k=n+1}^{\infty} f_k(z) \right|$$

$$\leqslant \sum_{k=n+1}^{\infty} \| f_k \|_E$$

and because (6.6.2) converges, this is less than any pre-assigned positive $\epsilon$ when $n$ is sufficiently large.

Finally, we have the General Principle of Uniform Convergence: *the functions* $f_n$ *converge uniformly on E to some function f if and only if for every positive* $\epsilon$ *there is an* $n_0$ *such that* $\| f_n - f_m \| < \epsilon$ *whenever* $m, n \geqslant n_0$. By Theorem 6.3.2, if $x \in E$ write $s_n = f_n(x)$; so the sequence $f_n(x)$ converges, say, to $f(x)$. As $| f_n(x) - f_m(x) | < \epsilon$ when $m, n \geqslant n_0$, so (letting $m \to +\infty$) $\| f_n - f \| < \epsilon$ when $n \geqslant n_0$.

**Exercise 6.6**

1. Show that the series

$$\sum_{n=1}^{\infty} \frac{x}{x^2 + n^2}$$

   (a) converges uniformly on each interval $[x_1, x_2]$ and
   (b) does not converge uniformly on **R**.

2. Examine the uniform convergence of

$$f(x) = \begin{cases} \sin(2nx) & \text{if } 0 \leqslant x \leqslant \pi/2n, \\ 0 & \text{if } \pi/2n < x \leqslant 1 \end{cases}$$

   on subsets of [0, 1].

3. Give an example to show that if $f_n \to f$ and $g_n \to g$ uniformly on $E$ then it is not necessarily true that $f_n g_n \to fg$ uniformly on $E$.

4. Show that the series

$$\sum_{n=1}^{\infty} (-1)^n \frac{x^2 + n}{n^2}$$

(a) converges uniformly on each interval $[x_1, x_2]$ and
(b) is not *absolutely* convergent for any real $x$.

5. Let $n_1, n_2, \ldots$ be positive integers. Show that the sequence $f_1, f_2, \ldots$, $f_k(z) = z^{n_k}$ is uniformly convergent on $\{z : |z| = 1\}$ if and only if $n_j = m$, say, for almost all $j$.

# Chapter 7

## 7.1 PLANE CURVES

Our intuitive notion of a curve in **C** is that of a point $z$ moving in a specified manner through certain points of **C**. We wish to distinguish, for example, between the two curves $\gamma_1$ and $\gamma_2$ given by

$$\gamma_1(t) = e^{it}, \qquad 0 \leqslant t \leqslant 2\pi,$$

and

$$\gamma_2(t) = e^{-it}, \qquad \pi \leqslant t \leqslant 4\pi,$$

(the reader is invited to draw these curves). As

$$\{\gamma_1(t) : 0 \leqslant t \leqslant 2\pi\} = \{\gamma_2(t) : \pi \leqslant t \leqslant 4\pi\}$$

we can distinguish between $\gamma_1$ and $\gamma_2$ only by considering them as functions: it is not sufficient simply to consider the set of points through which the moving point passes. We choose, then, to define a curve as a function.

*Definition 7.1.1* *A curve $\gamma$ in* **C** *is a continuous function $\gamma : [a, b] \to$* **C**, *where $[a, b]$ is a bounded real interval.*

The point $\gamma(a)$ is the *initial point* of $\gamma$, $\gamma(b)$ is the *final point* of $\gamma$ and $\gamma$ is said to *join $\gamma(a)$ to $\gamma(b)$*. We say that $\gamma$ is *closed* if $\gamma(a) = \gamma(b)$; that is if the initial point is the same as the final point.

The set of points $\{\gamma(t) : a \leqslant t \leqslant b\}$ is denoted by $[\gamma]$ and we say that $\gamma$ *is in $D$* (or *lies in $D$*) if $[\gamma] \subset D$. Observe that $\gamma$ may be constant on $[a, b]$ : in this case $[\gamma]$ contains only a single point and we call $\gamma$ a *point curve.* Next, $[\gamma] = \gamma([a, b])$ and so $[\gamma]$ is a compact set (Theorem 6.4.4). We deduce that $[\gamma]$ is bounded and so, for example, the real axis is *not* the set $[\gamma]$ for any curve $\gamma$.

Given any curve $\gamma : [a, b] \to$ **C** there is another curve described geometrically by reversing the motion that determines $\gamma$. This curve is denoted by $\gamma^-$ and is defined by

$$(\gamma^-)(t) = \gamma(a + b - t), \qquad a \leqslant t \leqslant b.$$

Observe that the initial and final points of $\gamma^-$ are the final and initial points of $\gamma$, $\gamma^-$ joins $\gamma(b)$ to $\gamma(a)$, $[\gamma] = [\gamma^-]$ and $(\gamma^-)^- = \gamma$.

We shall now state and prove two results which relate the notions of a curve and a domain.

*Theorem 7.1.1* *An open subset $D$ of $\mathbf{C}$ is connected if and only if any two points in $D$ can be joined by a curve in $D$.*

*Theorem 7.1.2* *For any curve $\gamma$, $\mathbf{C} - [\gamma]$ is the disjoint union of domains. Exactly one of these domains contains a set of the form $\{z \in \mathbf{C} : |z| > r\}$.*

*Proof of Theorem* 7.1.1 We suppose first that any two points, say $z$ and $w$ can be joined by a curve $\gamma : [a, b] \to D$ in $D$. Let $f : D \to \mathbf{Z}$ be any continuous function; then $f \circ \gamma : [a, b] \to \mathbf{Z}$ is continuous and hence constant as $[a, b]$ is connected. Thus

$$f(z) = f(\gamma(a)) = f(\gamma(b)) = f(w)$$

and so $f$ is constant. This proves that $D$ is connected.

We now suppose that $D$ is connected and prove a slightly stronger result than is given by Theorem 7.1.1. Let $z_0, z_1, \ldots, z_n$ be complex numbers and define the curve $\gamma : [0, n] \to \mathbf{C}$ by

$$\gamma(t) = z_k + (t - k)(z_{k+1} - z_k), \qquad k \leqslant t \leqslant k + 1, \quad k = 0, 1, \ldots, n - 1.$$

Then $\gamma$ maps $[k, k + 1]$ onto $[z_k, z_{k+1}]$ and so

$$[\gamma] = [z_0, z_1] \cup [z_1, z_2] \cup \cdots \cup [z_{n-1}, z_n].$$

We call $\gamma$ a polygonal curve and we shall prove that any two points in $D$ can be joined by a polygonal curve in $D$.

Select any $w$ in $D$ and define $f : D \to \mathbf{Z}$ by $f(z) = 1$ if $w$ can be joined to $z$ by a polygonal curve in $D$ and $f(z) = 0$ otherwise. If $z \in D$ choose a positive $t$ such that $\mathbf{C}(z, t) \subset D$. Obviously $w$ can be joined by a polygonal curve in $D$ either to each point in $\mathbf{C}(z, t)$ or to no point in $\mathbf{C}(z, t)$. This shows that $f$ is constant on $\mathbf{C}(z, t)$ and so certainly is continuous at $z$. We deduce that $f : D \to \mathbf{Z}$ is continuous and so, as $D$ is connected, $f$ is constant. Finally, $f(w) = 1$. Thus $f(z) = 1$ for all $z$ and this is the desired result.

*Proof of Theorem 7.1.2* As $[\gamma]$ is compact it is both closed and bounded. As $[\gamma]$ is closed, $\mathbf{C} - [\gamma]$ is open and so is a disjoint union of domains (Theorem 6.2.3).

As $[\gamma]$ is bounded there is some $r$ such that $[\gamma] \subset \bar{\mathbf{C}}(0, r)$. By Theorem 7.1.1, $\mathbf{C} - \bar{\mathbf{C}}(0, r)$ is connected and by the remark following Theorem 6.2.3, this lies in one of the components of $\mathbf{C} - [\gamma]$.

**Exercise 7.1**

1. Prove that if $D$ is an open set and if $z$ and $w$ in $D$ can be joined by a curve in $D$, then they can be joined by a polygonal curve in $D$ whose segments are either horizontal or vertical.

2. Show that the set $E$ defined by

$$E = \{x \in \mathbf{R} : x \leqslant 0\} \cup \left\{x + iy : x > 0, y = \frac{1}{x} \sin\left(\frac{1}{x}\right)\right\}$$

is connected but that $-1$ cannot be joined to $1/\pi$ by a curve in $E$.

3. Let $D$ be a domain and let $z_1, z_2, \ldots$ be distinct points in $D$ with the property that for each $w$ in $D$ there is a disc $\mathbf{C}(w, r), r > 0$, which contains only a finite number of the $z_j$. Show that $D - \{z_1, z_2, \ldots\}$ is a domain.

## 7.2 THE INDEX OF A CURVE

Let $\gamma : [a, b] \to \mathbf{C}$ be a curve which does not pass through the origin. As $t$ increases from $a$ to $b$, $\gamma(t)$ moves through $[\gamma]$ and it seems reasonable to suppose that we can select a value $\theta(t)$ of Arg $\gamma(t)$ which varies continuously with $t$. As $\gamma(t) \neq 0$ a choice of $\theta(t)$ is always possible: the problem is to achieve continuity and we now show that this too is possible (see Diagram 7.2.1).

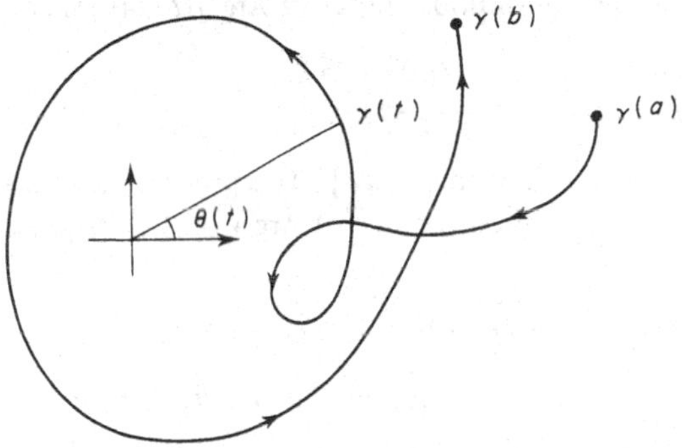


Diagram 7.2.1

First observe that if a curve $\gamma : [a, b] \to \mathbf{C}$ lies in a set $D$ and if there is a continuous choice, say $\theta(z)$, of Arg $z$ for $z$ in $D$ then $\theta(\gamma(t))$ is a choice of Arg $\gamma(t)$ and it is continuous as both $\theta$ and $\gamma$ are. Thus $\theta \circ \gamma$ is a continuous choice of Arg $\gamma$ on $[a, b]$. This is simple and useful: however it is also inadequate. The next result shows that it is not necessary to assume the existence of a continuous choice of Arg $z$ in $D$.

Theorem 7.2.1 *Let* $\gamma : [a, b] \to \mathbf{C}$ *be any curve and suppose that* $0 \notin [\gamma]$. *Then there exists a continuous function* $\theta : [a, b] \to \mathbf{R}$ *such that for each* $t$ *in* $[a, b]$, $\theta(t)$ *is a choice of* Arg $\gamma(t)$.

*Proof* As $\mathbf{C} - [\gamma]$ is an open set which contains zero, there is a positive $\epsilon$ such that

$$|\gamma(t)| \geqslant \epsilon > 0, \qquad t \in [a, b].$$

Next, $\gamma$ is uniformly continuous on $[a, b]$ (Theorem 6.4.5) and so there is a positive $\delta$ such that

$$|\gamma(t) - \gamma(t')| < \epsilon$$

whenever $|t - t'| < \delta$ with $t$ and $t'$ in $[a, b]$.

Now select $t_0, t_1, \ldots, t_n$ satisfying $t_{j+1} - t_j < \delta$ $(j = 0, 1, \ldots, n-1)$ and

$$a = t_0 < t_1 < \cdots < t_n = b$$

and put $z_j = \gamma(t_j)$. The above inequalities show that if $t \in [t_j, t_{j+1}]$, then $|\gamma(t) - z_j| < \epsilon$ and so $\gamma$ maps $[t_j, t_{j+1}]$ into the disc $C(z_j, \epsilon)$. Moreover,

$$|z_j| = |\gamma(t_j)| \geqslant \epsilon$$

and so $C(z_j, \epsilon)$ does not contain the origin. We can now apply Theorem 5.4.2: thus there is a continuous choice, say $\theta_j(t)$ of $\operatorname{Arg} \gamma(t)$ on $[t_j, t_{j+1}]$.

Although $\theta_0$ and $\theta_1$ are defined at $t_1$ they need not be equal there. However $\theta_0(t_1)$ and $\theta_1(t_1)$ are both values of $\operatorname{Arg} z_1$ and so for some integer $n_1$,

$$\theta_0(t_1) = \theta_1(t_1) + 2\pi n_1.$$

As $\theta_1(t) + 2\pi n_1$ is also a continuous choice of $\operatorname{Arg} \gamma(t)$ on $[t_1, t_2]$, the function

$$\theta(t) = \begin{cases} \theta_0(t), & t_0 \leqslant t \leqslant t_1 \\ \theta_1(t) + 2\pi n_1, & t_1 \leqslant t \leqslant t_2 \end{cases}$$

is a continuous choice of $\operatorname{Arg} \gamma$ on $[t_0, t_2]$. This process may clearly be continued. If $\theta$ is defined and is a continuous choice of $\operatorname{Arg} \gamma$ on $[t_0, t_k]$, we extend $\theta$ to $[t_k, t_{k+1}]$ by

$$\theta(t) = \theta_k(t) + [\theta(t_k) - \theta_k(t_k)], \qquad t_k \leqslant t \leqslant t_{k+1}$$

and in this way we construct the required function $\theta$.

Any function $\theta$ having the properties described in Theorem 7.2.1 is called a *branch of* $\operatorname{Arg} \gamma$ *on* $[a, b]$. If $\theta$ and $\phi$ are two such branches, then the function

$$[\theta(t) - \phi(t)]/2\pi$$

is continuous and integer valued on $[a, b]$ and so is constant. Thus

$$\theta(b) - \phi(b) = \theta(a) - \phi(a)$$

or, more significantly,

$$\theta(b) - \theta(a) = \phi(b) - \phi(a).$$

This shows that $\theta(b) - \theta(a)$ depends only on $\gamma$ and not on the particular choice of the branch $\theta$.

The case when $\gamma$ is closed is of supreme importance. In this case $\gamma(a) = \gamma(b)$ and so $\theta(a)$ and $\theta(b)$ are both values of $\operatorname{Arg} z$, $z = \gamma(a) = \gamma(b)$. We conclude that $[\theta(b) - \theta(a)]/2\pi$ is an *integer*.

The importance of these results cannot be overestimated. We generalize a little by considering the function $\gamma - w$, where

$$(\gamma - w)(t) = \gamma(t) - w,$$

rather than $\gamma$ (so that $0 \notin [\gamma - w]$ if and only if $w \notin [\gamma]$), and make the following definition.

*Definition 7.2.1* *Let $\gamma : [a, b] \to \mathbf{C}$ be any curve and suppose that $w \notin [\gamma]$. We define the index $n(\gamma, w)$ of $\gamma$ about $w$ by*

$$n(\gamma, w) = [\theta(b) - \theta(a)]/2\pi,$$

*where $\theta$ is any branch of* Arg $(\gamma - w)$ *on $[a, b]$. If $\gamma$ is closed then $n(\gamma, w)$ is an integer.*

The index $n(\gamma, w)$ is sometimes called the *winding number* of $\gamma$ about $w$, for it represents the number of times that a point $z$ moves around $w$ as it moves from $\gamma(a)$ to $\gamma(b)$ along $\gamma$.

This is an appropriate point to generalize the notion of a branch.

*Definition 7.2.2* *Let $f : E \to \mathbf{C} - \{0\}$ be any function. The function $\theta : E \to \mathbf{R}$ is a branch of* Arg *$f$ on $E$ if and only if $\theta$ is continuous on $E$ and for each $z$ in $E$, $\theta(z)$ is a choice of* Arg *$f(z)$. Similarly, $L$ is a branch of* Log f *in $E$ if and only if $L$ is continuous on $E$ and $L(z)$ is a choice of* Log *$f(z)$, $z \in E$.*

Given a branch $\theta$ (or $L$) on $E$, we may define (for continuous $f$) a branch $L$ (or $\theta$) by

$$L(z) = \log_e |f(z)| + i\theta(z).$$

If $f$ is the identity function, we usually talk of a branch of, say, Arg $z$ on $E$.

The rest of this section consists of examples: these are chosen to illustrate the geometrical significance of the index and the methods used to compute it and they will be used later in this book.

*Example 7.2.1* Let $\gamma(t) = re^{it}$, $0 \leqslant t \leqslant 2\pi$ (this represents a point moving once around the circle $|z| = r$); then

$$n(\gamma, 0) = 1, \qquad n(\gamma, -2r) = 0.$$

First, put $\theta(t) = t$; then $\theta$ is a branch of Arg $\gamma$ on $[0, 2\pi]$ and

$$n(\gamma, 0) = [\theta(2\pi) - \theta(0)]/2\pi = 1.$$

In order to compute $n(\gamma, -2r)$ we need to find a branch of Arg $[\gamma - (-2r)]$ on $[0, 2\pi]$. Observe that as

$$\text{Re}\,[\gamma(t) + 2r] > 0,$$

the curve $\gamma + 2r$ (which is a circle centre $2r$ and radius $r$) lies in the half-plane $H$

given by Re $[z] > 0$. We deduce that

$$\phi(t) = \mathrm{Arg}_{-\pi}\,(\gamma(t) + 2r)$$

is a continuous choice of Arg $(\gamma + 2r)$ on $[0, 2\pi]$ (see Definition 5.4.1) and so

$$\begin{aligned} 2\pi n(\gamma, -2r) &= \phi(2\pi) - \phi(0) \\ &= \mathrm{Arg}_{-\pi}(1 + 2r) - \mathrm{Arg}_{-\pi}(1 + 2r) \\ &= 0. \end{aligned}$$

In fact,

$$n(\gamma, z) = \begin{cases} 1 & \text{if} \quad |z| < r, \\ 0 & \text{if} \quad |z| > r. \end{cases}$$

This could be proved now but we prefer to derive it from the results in the next section.

*Example 7.2.2* Let $\gamma(t) = re^{it}$, $0 \leq t \leq 2\pi p$, let $f(z) = z^q$ and define $\Gamma = f \circ \gamma$ ($p$ and $q$ are positive integers). Thus $\gamma$ represents a circle traversed $p$ times and $\Gamma$ is the image of $\gamma$ under $f$.

Obviously

$$\Gamma(t) = r^q e^{iqt}, \qquad 0 \leq t \leq 2\pi p$$

and taking $\theta(t) = qt$ we find that

$$n(\Gamma, 0) = [\theta(2\pi p) - \theta(0)]/2\pi = pq.$$

If $q = 1$ (so $\gamma = \Gamma$) we obtain $n(\gamma, 0) = p$, confirming our belief that $\gamma$ winds about the origin $p$ times.

By taking $p = 1$ and $q \neq 1$ we can introduce another significant feature of the index. We obtain $n(f \circ \gamma, 0) = q$, and $q$ is exactly the number of zeros of $f$ inside $\gamma$. This is no accident and much of our later work is concerned with a general result of this nature.

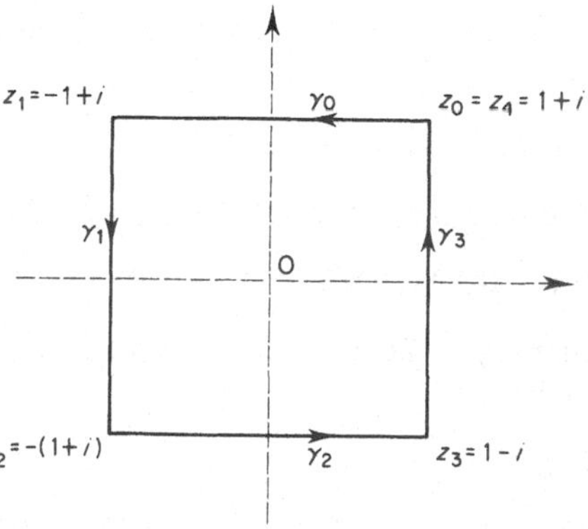


Diagram 7.2.2

*Example 7.2.3* Let $z_0, z_1, z_2$ and $z_3$ be the points $1+i, -1+i, -(1+i), 1-i$ respectively and let $z_4 = z_0$. Define

$$\gamma_k(t) = z_k + (t-k)(z_{k+1} - z_k), \qquad k \leqslant t \leqslant k+1,$$

for $k = 0, 1, 2, 3$ and $\gamma : [0, 4] \to \mathbf{C}$ by $\gamma = \gamma_k$ on $[k, k+1]$ ($\gamma$ is then the boundary of the square with vertices $z_j$).

If $\theta$ is any branch of Arg $\gamma$ on $[0, 4]$ then

$$\begin{aligned} n(\gamma, 0) &= [\theta(4) - \theta(0)]/2\pi \\ &= [\theta(4) - \theta(3) + \theta(3) - \cdots - \theta(0)]/2\pi \\ &= n(\gamma_0, 0) + \cdots + n(\gamma_3, 0). \end{aligned}$$

Each $n(\gamma_j, 0)$ may now be computed independently and we may do this using any convenient branch of Arg $\gamma_j$. For example, $\gamma_3$ maps $[3, 4]$ onto $[z_3, z_0]$, which lies in the half-plane $H$ given in Example 7.2.1. Thus (as in that example)

$$\begin{aligned} n(\gamma_3, 0) &= [\mathrm{Arg}_{-\pi}(z_0) - \mathrm{Arg}_{-\pi}(z_3)]/2\pi \\ &= [\pi/4 - (-\pi/4)]/2\pi \\ &= \tfrac{1}{4}. \end{aligned}$$

Similar computations show that $n(\gamma_j, 0) = \frac{1}{4}$ for each $j$, hence $n(\gamma, 0) = 1$.

*Example 7.2.4* Let $\gamma$ be the semi-circle given by

$$\gamma(t) = \begin{cases} Re^{it}, & 0 \leqslant t \leqslant \pi, \\ t - (\pi + R), & \pi \leqslant t \leqslant \pi + 2R, \end{cases}$$

where $R > 1$; then $n(\gamma, i) = 1$.

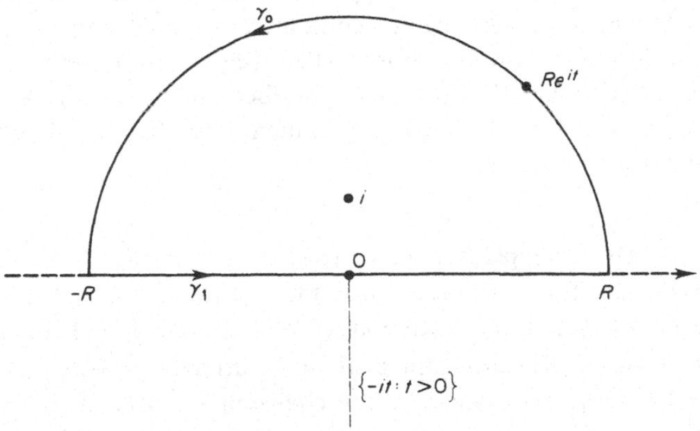


Diagram 7.2.3

Let $\gamma_0$ and $\gamma_1$ be the restriction of $\gamma$ to $[0, \pi]$ and $[\pi, \pi + 2R]$ respectively. Then the curve $\gamma_0 - i$ lies in the complement of the half-line $\{-it : t \geqslant 0\}$ and so

$$\phi(t) = \mathrm{Arg}_{-\pi/2}(\gamma_0(t) - i)$$

is a branch of $\text{Arg}\,(\gamma_0 - i)$ on $[0, \pi]$. Thus

$$2\pi n(\gamma_0, i) = \phi(\pi) - \phi(0).$$

We can compute this but it is easier to note that

$$-\pi/2 < \phi(0) < \phi(\pi) < 3\pi/2$$

and so

$$0 < n(\gamma_0, i) < 1.$$

A similar argument (using $\text{Arg}_{\pi/2}$) shows that

$$0 < n(\gamma_1, i) < 1.$$

As $n(\gamma, i)$ is an integer and as

$$n(\gamma, i) = n(\gamma_0, i) + n(\gamma_1, i)$$

we must have $n(\gamma, i) = 1$.

Obviously this technique is applicable to other reasonably simple curves other than semi-circles.

Given a curve $\gamma : [a, b] \to \mathbf{C}$ with

$$a = t_0 < t_1 < \cdots < t_n = b,$$

then

$$n(\gamma, 0) = \sum_{j=0}^{n-1} n(\gamma_j, 0)$$

where $\gamma_j$ is the restriction of $\gamma$ to $[t_j, t_{j+1}]$. This has been illustrated in the two previous examples, it is implicitly contained in the proof of Theorem 7.1.1, and it is obviously true in the general case. It is worth noting that if $\gamma_j$ lies in an open half-plane bounded by a line through the origin, then $|\,n(\gamma_j, 0)\,| < \frac{1}{2}$. If $\gamma_j$ lies in the complement of a half-line from the origin, then $|\,n(\gamma_j, 0)\,| < 1$. These results will be proved in the next section.

*Example 7.2.5* This example is intended to serve as a warning to the reader. Let $\gamma(t) = e^{it}$, $0 \leqslant t \leqslant 2\pi$. Then there is a continuous choice of $\text{Arg}\,\gamma(t)$ on $[0, 2\pi]$ (Example 7.2.1) but there is no continuous choice of $\text{Arg}\,z$ on $[\gamma]$ (Example 5.4.1).

Despite the apparent similarity, these are quite different problems. According to Theorem 7.2.1, there exists a continuous choice of $\text{Arg}\,\gamma(t)$ on $[0, 2\pi]$ regardless of the general nature of $\gamma$. The fact that in this example, $\gamma(0) = \gamma(2\pi)$ is immaterial: we are concerned with values of $t$ and it is beyond dispute that 0 and $2\pi$ are two such distinct values.

On the other hand, the question of the existence of a continuous choice of $\text{Arg}\,z$ on some set $E$ is a purely topological question and depends (roughly speaking) on whether or not $E$ 'surrounds the point zero': this will be examined in detail in Chapter 11. The fact that in this example, $E = [\gamma]$ is also immaterial: it was chosen this way simply to provide a more striking contrast.

**Exercise 7.2**

1. Let $\gamma : [a, b] \to \mathbf{C}$ be continuous and non-zero and let $h : [c, d] \to [a, b]$ be continuous and 1–1. Prove that

$$n(\gamma \circ h, 0) = \epsilon n(\gamma, 0)$$

where $\epsilon = 1$ if $h$ is increasing and $\epsilon = -1$ if $h$ is decreasing. This shows how the index behaves under a change of parameter for the curve $\gamma$. [It should be clear from the Intermediate Value Theorem that either $h$ is increasing or $h$ is decreasing.]

2. Using Question 1 show that $n(\gamma^-, 0) = -n(\gamma, 0)$.
3. For each positive integer $m$, let

$$\gamma_m(t) = t + i, \qquad -m \leqslant t \leqslant m.$$

Prove that

$$\lim_{m \to \infty} n(\gamma_m, 0) = -\tfrac{1}{2}.$$

4. Let $\gamma$ be the 'figure of eight' curve given by

$$\gamma(t) = \begin{cases} 1 - e^{it}, & 0 \leqslant t \leqslant 2\pi, \\ -1 + e^{-it}, & 2\pi \leqslant t \leqslant 4\pi. \end{cases}$$

Sketch the curve and show that $n(\gamma, 1) = 1$, $n(\gamma, -1) = -1$ and $n(\gamma, i) = 0$.

## 7.3 PROPERTIES OF THE INDEX

In order to be able to use the index effectively we need to develop some of its main properties. There are eight properties which we shall use consistently and these are labelled I1–I8 in the following discussion. The reader should note that it is essential that these are understood as they will frequently be used without being explicitly referred to. We assume throughout that $\gamma$ is defined on $[a, b]$ and that $w \notin [\gamma]$.

I1. *If $\gamma$ is a point curve, then $n(\gamma, w) = 0$.*
Obviously a stationary point cannot wind about any other point. As $\gamma$ is constant we may select a constant value $\theta$ of Arg $(\gamma - w)$ on $[a, b]$. Thus

$$2\pi n(\gamma, w) = \theta - \theta = 0.$$

I2. *If $g(z) = \alpha z + \beta$, $\alpha \neq 0$, then $n(g \circ \gamma, g(w)) = n(\gamma, w)$.*
This shows that the index is invariant under translations ($\alpha = 1$), magnifications ($\alpha > 0, \beta = 0$) and rotations ($|\alpha| = 1, \beta = 0$). We select any branch $\theta$ of Arg $(\gamma - w)$

on $[a, b]$ and any choice $\theta_0$ of Arg $\alpha$. Then

$$g(\gamma(t)) - g(w) = \alpha(\gamma(t) - w)$$

and so $\theta(t) + \theta_0$ is a branch of Arg $[g(\gamma(t)) - g(w)]$ on $[a, b]$. We now have

$$\begin{aligned} n(g \circ \gamma, g(w)) &= [\theta(b) + \theta_0 - (\theta(a) + \theta_0)]/2\pi \\ &= n(\gamma, w). \end{aligned}$$

I3. *Let* $g(x) = \alpha x + \beta$, $\alpha \neq 0$, *map* $[c, d]$ *onto* $[a, b]$. *Then*

$$n(\gamma \circ g, w) = \epsilon n(\gamma, w)$$

*where* $\epsilon = 1$ *if* $g$ *is increasing and* $\epsilon = -1$ *if* $g$ *is decreasing.*

This useful result enables us to regard the curve $\gamma$ as being defined on $[c, d]$ rather than on $[a, b]$. For example, if $s$ is any real number we may define $\gamma^*(t) = \gamma(t - s)$ on $[a + s, b + s]$ and $n(\gamma^*, w) = n(\gamma, w)$. Another important application is when $g(x) = a + b - x$; this gives

$$n(\gamma^-, w) = -n(\gamma, w).$$

The proof of I3 is quite straightforward. Let $\theta$ be any branch of Arg $(\gamma - w)$ on $[a, b]$; then $\theta(g(x))$ is a continuous choice of Arg $(\gamma(g(x)) - w)$ on $[c, d]$ and so

$$2\pi n(\gamma \circ g, w) = \theta(g(d)) - \theta(g(c)).$$

If $g$ is increasing, then $g(c) = a$ and $g(d) = b$: if $g$ is decreasing, then $g(c) = b$ and $g(d) = a$. The result now follows. Observe that $\epsilon = \alpha/|\alpha|$.

I4. *If* $\gamma(t) = w + \gamma_1(t)\gamma_2(t) \cdots \gamma_s(t)$, *then*

$$n(\gamma, w) = n(\gamma_1, 0) + \cdots + n(\gamma_s, 0)$$

As $w \notin [\gamma]$, then $0 \notin [\gamma_j]$ and so we may find a branch $\theta_j$ of Arg $\gamma_j$ on $[a, b]$. Then $\theta_1 + \cdots + \theta_s$ is a branch of Arg $(\gamma - w)$ on $[a, b]$ (Theorem 5.1.2) and so

$$\begin{aligned} 2\pi n(\gamma, w) &= \sum_{j=1}^{s} \theta_j(b) - \sum_{j=1}^{s} \theta_j(a) \\ &= \sum_{j=1}^{s} [\theta_j(b) - \theta_j(a)] \\ &= \sum_{j=1}^{s} 2\pi n(\gamma_j, 0). \end{aligned}$$

I5. *Let D be a domain not containing w and suppose that there exists a continuous choice* $\theta(z)$ *of* Arg $(z - w)$ *in D. Suppose also that* $\gamma$ *is in D, then*

$$n(\gamma, w) = [\theta(\gamma(b)) - \theta(\gamma(a))]/2\pi.$$

The proof is trivial, for $\theta(\gamma(t))$ is a choice of Arg $\gamma(t)$ and it is a continuous function of $t$ as both $\theta$ and $\gamma$ are continuous. The reason for including I5 is that it contains the following useful special case.

If $\gamma$ is contained in the complement $D$ of the half-line $\{te^{i\alpha} : t \geqslant 0\}$, then we may take $w = 0, \theta = \text{Arg}_\alpha$ and so

$$| \theta(\gamma(b)) - \theta(\gamma(a)) | \leqslant (\alpha + 2\pi) - \alpha,$$

which gives

$$| n(\gamma, 0) | < 1.$$

If, in addition, $\gamma$ is closed, then $n(\gamma, 0) = 0$.

Similarly if $\gamma$ is contained in a half-plane given, say, by those $z$ which have some value of Arg $z$ in $(\alpha, \alpha + \pi)$, then $| n(\gamma, 0) | < \frac{1}{2}$.

I6. *Suppose that* $\gamma : [a, b] \to \mathbf{C}$ *and* $\sigma : [a, b] \to \mathbf{C}$ *are both closed and that*

$$| \gamma(t) - \sigma(t) | < | \gamma(t) - w |, \qquad a \leqslant t \leqslant b.$$

*Then*

$$n(\gamma, w) = n(\sigma, w).$$

The geometrical interpretation of the inequality is that for each $t$, $\sigma(t)$ is nearer to $\gamma(t)$ than is $w$. In particular, the segment $[\sigma(t), \gamma(t)]$ does not contain $w$ and we can 'deform' $\sigma$ to $\gamma$ by 'sliding' $\sigma(t)$ along this segment without changing the index (see Diagram 7.3.1).

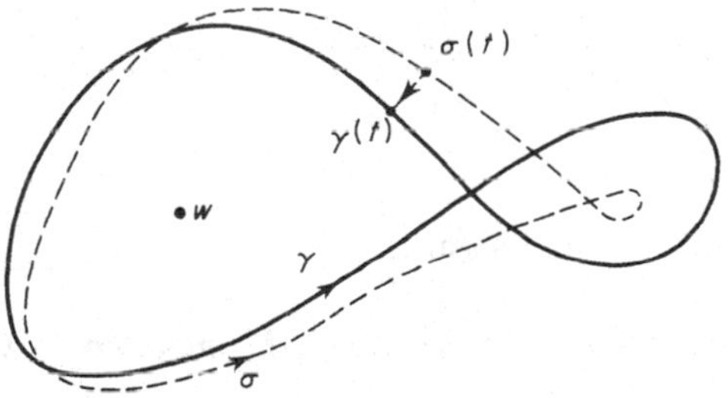


Diagram 7.3.1

We can give a precise description of a simpler but equally important case. As $w \notin [\gamma]$, there is a positive $\epsilon$ such that $| \gamma(t) - w | \geqslant \epsilon, a \leqslant t \leqslant b$. Thus if

$$| \sigma(t) - \gamma(t) | < \epsilon, \qquad a \leqslant t \leqslant b$$

(that is if $\sigma$ is 'uniformly close' to $\gamma$), then $n(\gamma, w) = n(\sigma, w)$.

Finally, there is an alternative form of I6 found by writing $\eta = \sigma - \gamma$. *If* $\eta$ *and* $\gamma$ *are closed and if*

$$| \eta(t) | < | \gamma(t) - w |, \qquad a \leqslant t \leqslant b,$$

*then*

$$n(\gamma, w) = n(\gamma + \eta, w).$$

To prove I6 we define $\Gamma : [a, b] \to \mathbf{C}$ by

$$\Gamma(t) = \frac{\sigma(t) - \gamma(t)}{\gamma(t) - w}.$$

The assumption is that $|\Gamma(t)| < 1$ and so $\Gamma(t) + 1$ lies in the half-plane Re $[z] > 0$. I5 is applicable and we deduce that $n(\Gamma + 1, 0) = 0$. However,

$$\sigma(t) - w = [\Gamma(t) + 1]\,[\gamma(t) - w]$$

and so by I4,

$$\begin{aligned} n(\sigma, w) &= n(\Gamma + 1, 0) + n(\gamma - w, 0) \\ &= n(\gamma, w). \end{aligned}$$

I7. *If $\gamma$ is closed and if $w$ and $z$ are in the same component of* $\mathbf{C} - [\gamma]$, then $n(\gamma, z) = n(\gamma, w)$.

It is only necessary to prove that $n(\gamma, w)$ is a continuous function of $w$ on $\mathbf{C} - [\gamma]$. If this is so, then (as it is integer valued) it is constant on each connected subset of $\mathbf{C} - [\gamma]$ and hence on each component of $\mathbf{C} - [\gamma]$.

Let us now establish continuity. Suppose that $w \in \mathbf{C} - [\gamma]$ and choose a positive $\epsilon$ such that

$$|\gamma(t) - w| \geqslant \epsilon, \qquad a \leqslant t \leqslant b.$$

If $|z - w| < \epsilon$, we put $\eta = w - z$ in I6 and use I2. Then

$$\begin{aligned} n(\gamma, w) &= n(\gamma + w - z, w) \\ &= n(\gamma - z, 0) \\ &= n(\gamma, z) \end{aligned}$$

and so $n(\gamma, z)$ is certainly continuous at $w$.

I8. *If $\gamma$ is closed and if $w$ is in the unbounded component of* $\mathbf{C} - [\gamma]$, *then* $n(\gamma, w) = 0$.

This simply says that $\gamma$ does not wind about any point at a large distance from $\gamma$. As $[\gamma]$ is compact, there is some $r$ such that $[\gamma] \subset \bar{\mathbf{C}}(0, r)$. Then $\gamma + 2r$ lies in the half-plane Re $[z] > 0$ and is closed as $\gamma$ is closed. Thus by I5,

$$n(\gamma, -2r) = n(\gamma + 2r, 0) = 0.$$

Obviously, $-2r$ lies in the unbounded component of $\mathbf{C} - [\gamma]$ and I7 now implies I8.

Let us now review Examples 7.2.1–7.2.4 in the light of these properties. In Example 7.2.1 an easy computation gave $n(\gamma, 0) = 1$. The values of $n(\gamma, z)$ given (but not proved) now follow directly from I7 and I8. In Example 7.2.3 the evaluation of $n(\gamma_j, 0)$ was, of course, a direct application of I5. Using I7 and I8 we may now conclude that $n(\gamma, z)$ is 1 if $z$ is 'inside' the square and zero otherwise. Similar comments apply to Example 7.2.4 where, in addition, we have used I2.

An alternative deduction of the result in Example 7.2.4 may now be given as follows. Put $\Gamma(t) = Re^{it}, 0 \leqslant t \leqslant 2\pi$, so that by Example 7.2.1, $n(\Gamma, i) = 1$. We prefer to write $\Gamma$ as

$$\Gamma(t) = \begin{cases} \Gamma_0(t) = e^{it}, & 0 \leqslant t \leqslant \pi, \\ \Gamma_1(t) = e^{it}, & \pi \leqslant t \leqslant 2\pi, \end{cases}$$

so

$$1 = n(\Gamma, i) = n(\Gamma_0, i) + n(\Gamma_1, i).$$

Next, $i$ lies outside the lower semi-circle given by

$$\gamma^*(t) = \begin{cases} \Gamma_1(t) = e^{it}, & \pi \leqslant t \leqslant 2\pi \\ \sigma_1(t) = R + 2\pi - t, & 2\pi \leqslant t \leqslant 2\pi + 2R \end{cases}$$

so by I8,

$$0 = n(\gamma^*, i) = n(\Gamma_1, i) + n(\sigma_1, i).$$

However,

$$\gamma(t) = \begin{cases} \Gamma_0(t) & 0 \leqslant t \leqslant \pi \\ \sigma_0(t) = t - (\pi + 2R), & \pi \leqslant t \leqslant \pi + 2R \end{cases}$$

and using I3 with $g(x) = 3\pi + 2R - x$, we have

$$n(\sigma_0, i) = -n(\sigma_1, i).$$

Thus

$$\begin{aligned} n(\gamma, i) &= n(\Gamma_0, i) + n(\sigma_0, i) \\ &= n(\Gamma_0, i) + n(\Gamma_1, i) - n(\Gamma_1, i) - n(\sigma_1, i) \\ &= n(\Gamma, i) - n(\gamma^*, i) \\ &= 1. \end{aligned}$$

The index can be used to clarify the difficult question of what is meant by the 'inside' and 'outside' of a closed curve $\gamma$. We shall say

(a) that $z$ *is inside* $\gamma$ if $z \notin [\gamma]$ and $n(\gamma, z) \neq 0$,
(b) that $z$ *is on* $\gamma$ if $z \in [\gamma]$, and
(c) that $z$ *is outside* $\gamma$ if $z \notin [\gamma]$ and $n(\gamma, z) = 0$.

For example, the point $z$ illustrated in Diagram 7.3.2 lies *outside* $\gamma$. The integers given in the components of $\mathbf{C} - [\gamma]$ are the values of $n(\gamma, w)$ for $w$ in that component.

These definitions should take precedence over any geometrical interpretation, as it is this precise use of the index rather than the geometry that consistently and inevitably (unless deliberately suppressed) appears in the theorems of complex analysis. This is not to say that we should ignore our geometric intuition; on the contrary it is essential to a proper understanding of the subject. It is true, however, that it is often surprisingly difficult to make geometrical ideas precise and these ideas are then incompatible with the joint aims of simplicity and rigour.

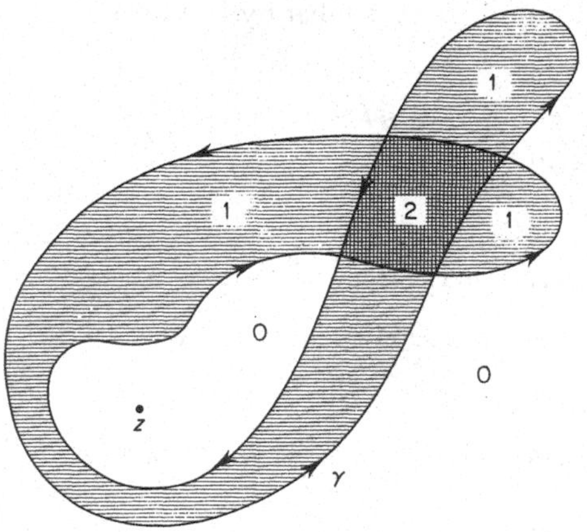


Diagram 7.3.2

We shall use our intuition freely to evaluate $n(\gamma, w)$ in certain specific examples (these will be of the same general character as Examples 7.2.1–7.2.4) and the reader is then invited to supply the formal proofs.

Observe that by I7, the outside of $\gamma$, say $O(\gamma)$, is the union of those components of $\mathbf{C} - [\gamma]$ on which the index is zero. Thus $O(\gamma)$ is an open set. Further, by Theorem 7.2.2 and I8, $O(\gamma)$ contains the complement of some closed disc. If we denote the inside of $\gamma$ by $I(\gamma)$, then

$$C - O(\gamma) = [\gamma] \cup I(\gamma),$$

*and so the set of points which lie inside or on $\gamma$ is a compact set.*

**Exercise 7.3**

1. Suppose that $\gamma : [a, b] \to \mathbf{C}$ is continuous and non-zero (so $\gamma$ is a curve which does not pass through the origin). Prove that

$$n(1/\gamma, 0) = -n(\gamma, 0).$$

Prove also that if neither $w$ nor $1/w$ are on $\gamma$, then

$$n(1/\gamma, w) = n(\gamma, 1/w) - n(\gamma, 0).$$

[*Hint:*

$$\gamma\left(\frac{1}{\gamma} - w\right) = -w\left(\gamma - \frac{1}{w}\right).]$$

Verify this explicitly when $\gamma(t) = e^{it}, 0 \leqslant t \leqslant 2\pi$, and when (a) $w = \frac{1}{2}$, (b) $w = 2$.

2. Let $\gamma_m : [a, b] \to \mathbf{C}$, $m = 1, 2, \ldots$, and $\gamma : [a, b] \to \mathbf{C}$ be curves and suppose that $\gamma_m \to \gamma$ uniformly on $[a, b]$ as $m \to \infty$. Use I6 to show that for all $w$ not in $[\gamma]$

and for all sufficiently large $m$,

$$n(\gamma_m, w) = n(\gamma, w).$$

3. Let

$$p(z) = a(z - z_1) \cdots (z - z_m)$$

where $m \geqslant 1, a \neq 0$ and $|z_j| \neq 1, j = 1, \ldots, m$. Now let $\gamma(t) = e^{it}, 0 \leqslant t \leqslant 2\pi$. Prove that $n(p \circ \gamma, 0)$ is the number of the $z_j$ which lie inside $\gamma$.

4. Let

$$\gamma(t) = \begin{cases} e^{it}, & 0 \leqslant t \leqslant \pi, \\ e^{-it} & \pi \leqslant t \leqslant 2\pi. \end{cases}$$

Prove that there are no points which lie inside $\gamma$.

# Chapter 8

## 8.1 POLYNOMIALS

There are two reasons why it is appropriate to study polynomials now. First, polynomials provide an opportunity to illustrate the use of the index in a less abstract setting and second, they provide us with a preview of the results we might hope to prove in our later study of analytic functions. This section is written with precisely these reasons in mind and should be regarded primarily as an illustration of the use of topological methods in analysis.

A polynomial is either a constant function or is a function $p$ of the form

$$p(z) = a_0 + a_1 z + \cdots + a_d z^d \tag{8.1.1}$$

where $a_d \neq 0$ and the *degree* $d$ of $p$ is a positive integer. We say that $p$ is *monic* if $a_d = 1$: it is *linear* if $d = 1$. *Unless otherwise stated we shall assume that $p$ is not constant.*

We begin by studying the behaviour of $p$ in a neighbourhood of a point $w$. The Binomial Theorem implies that for each $w$, $p(w + z)$ is a polynomial of degree $d$: thus for some $k$,

$$p(w + z) = p(w) + b_k z^k + \cdots + b_d z^d \tag{8.1.2}$$

where $b_k \neq 0$, $b_d = a_d$ and $1 \leqslant k \leqslant d$.

Now let $\gamma$ be the circle defined by

$$\gamma(t) = w + \sigma(t), \qquad \sigma(t) = re^{it}, \qquad 0 \leqslant t \leqslant 2\pi.$$

We substitute $\sigma(t)$ for $z$ in (8.1.2) and obtain

$$p(\gamma(t)) = p(w) + \sigma(t)^k q(\sigma(t)),$$

where

$$q(z) = b_k + b_{k+1} z + \cdots + b_d z^{d-k}.$$

Next, we use I4 and Example 7.2.1: thus

$$\begin{aligned} n(p \circ \gamma, p(w)) &= kn(\sigma, 0) + n(q \circ \sigma, 0) \\ &= k + n(q \circ \sigma, 0). \end{aligned}$$

As $q(z) \to b_k$ as $z \to 0$ and as $b_k \neq 0$ there is a positive $\delta$ such that

$$| q(z) - b_k | < | q(z) |$$

whenever $|z|<\delta$. We deduce that if $0<r<\delta$, then (using I6 and I1)

$$n(q \circ \sigma, 0) = n(b_k, 0) = 0.$$

We conclude that

$$n(p \circ \gamma, p(w)) = k \geqslant 1. \tag{8.1.3}$$

In geometrical terms this means that the image of $\gamma$ under $p$, namely $p \circ \gamma$, winds about $p(w)$ exactly $k$ times. Now let $L$ be any half-line from $p(w)$, that is $L$ is of the form $\{p(w) + t\zeta : t \geqslant 0\}$ for some non-zero $\zeta$. The half-line $L$ is connected and unbounded, thus if $L$ does not meet $p \circ \gamma$, then $L$ lies in the unbounded component of $\mathbf{C} - \lfloor p \circ \gamma \rfloor$ and so by I8, $n(p \circ \gamma, z) = 0$ for all $z$ in $L$. This is false when $z = p(w)$: thus we see that $p \circ \gamma$ meets every half-line $L$ from $p(w)$.

As a special case we consider $\zeta = p(w)$ so that $L$ is of the form $\{tp(w) : t \geqslant 1\}$ (if $p(w) = 0$ any non-zero $\zeta$ will suffice). Then there is a point $z_1$ on $\gamma$ such that $p(z_1) \in L$: thus $|p(z_1)| > |p(w)|$. This proves the Maximum Modulus Theorem: $|p|$ *cannot attain a local maximum at any point of* $\mathbf{C}$.

There is also a Minimum Modulus Theorem: $|p|$ *cannot attain a positive local minimum* in $\mathbf{C}$. The proof is similar: if $p(w) \neq 0$ we can (by continuity) choose $r$ sufficiently small so that if $z$ is on $\gamma$, then

$$|p(z) - p(w)| < |p(w)|.$$

We then take $\zeta = -p(w)$ and observe that if $z_2 \in [\gamma]$ and $p(z_2) \in L$, then $|p(z_2)| < |p(w)|$. The proofs of these results are illustrated (with $k = 2$) in Diagram 8.1.1.

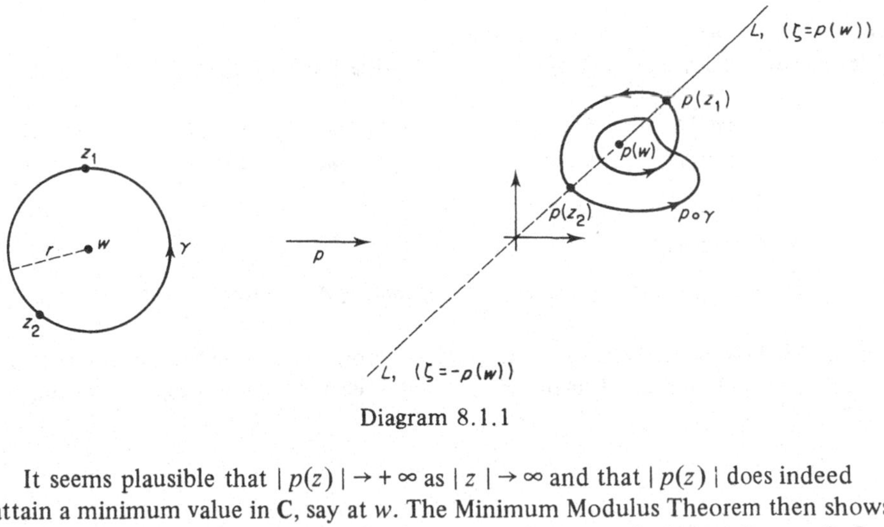


Diagram 8.1.1

It seems plausible that $|p(z)| \to +\infty$ as $|z| \to \infty$ and that $|p(z)|$ does indeed attain a minimum value in $\mathbf{C}$, say at $w$. The Minimum Modulus Theorem then shows that $|p(w)|$ cannot be positive, thus $p(w) = 0$ and we have the basis of a proof of the next result.

*Proposition 8.1.1* *Every non-constant polynomial has at least one zero.*

*Proof* We have seen that we need only establish the existence of some $w$ with the property that for all $z$, $|p(z)| \geq |p(w)|$.

We may clearly assume that $a_d = 1$. If

$$2(|a_0| + \cdots + |a_{d-1}| + 1) = r_0,$$

then for $|z| \geq r_0$,

$$\begin{aligned} |a_0 + a_1 z + \cdots + a_{d-1} z^{d-1}| &\leq (|a_0| + \cdots + |a_{d-1}|)|z|^{d-1} \\ &\leq \tfrac{1}{2}|z|^d, \end{aligned}$$

and therefore

$$\begin{aligned} |p(z)| &\geq |z^d| - |a_0 + a_1 z + \cdots + a_{d-1} z^{d-1}| \\ &\geq \tfrac{1}{2}|z|^d \\ &> 1 + |a_0|. \end{aligned}$$

Now let

$$\mu = \inf\{|p(z)| : |z| \leq r_0\}$$

and observe that $\mu \leq |p(0)| = |a_0|$. Moreover, by Theorem 6.4.4, there is some $w$ in $\mathbf{C}(0, r_0)$ with $|p(w)| = \mu$.

If $z \in \mathbf{C}$, then either $|z| \geq r_0$ and hence

$$|p(z)| > 1 + |a_0| > |p(w)|$$

or $|z| \leq r_0$, in which case

$$|p(z)| \geq \mu = |p(w)|.$$

Thus for all $z$, $|p(w)| \leq |p(z)|$ and so $p(w) = 0$.

Proposition 8.1.1 leads quickly to the following classical result.

*The Fundamental Theorem of Algebra* *Let $p$ be given by* (8.1.1). *Then there are positive integers $m_1, \ldots, m_s$ and distinct complex numbers $z_1, \ldots, z_s$ such that $m_1 + \cdots + m_s = d$ and*

$$p(z) = a_d(z - z_1)^{m_1} \cdots (z - z_s)^{m_s}. \tag{8.1.4}$$

*Further, the set $\{z_1, \ldots, z_s\}$ and the corresponding $m_j$ are unique.*

*Proof* We have proved already that every polynomial has at least one zero. Let $q$ be any monic polynomial of degree $d$, $d \geq 1$, and let $w$ be a zero of $q$. Then using (1.2.2),

$$\begin{aligned} q(z) &= q(z) - q(w) \\ &= \sum_{k=0}^{d} b_k(z^k - w^k) \\ &= (z - w)q_1(z), \end{aligned}$$

where $q_1$ is monic and of degree $d - 1$ and the $b_j$ are the coefficients of $q$. This is the essential step in the proof by induction on $d$ that every monic polynomial of degree $d$ is the product of $d$ linear monic polynomials: the details are omitted.

If $p$ is given by (8.1.1), then $a_d^{-1}p$ is monic and so $p$ can be written in the form (8.1.4).

As $p(z)$ given by (8.1.4) is zero if and only if $z \in \{z_1, \ldots, z_s\}$, this set is uniquely determined by $p$. Next, if $p$ can also be expressed as

$$p(z) = a_d(z - z_1)^{k_1} \cdots (z - z_s)^{k_s},$$

then for each $z_j$,

$$\lim_{z \to z_j} (z - z_1)^{m_1 - k_1} \cdots (z - z_s)^{m_s - k_s} = \lim_{z \to z_j} \frac{p(z)}{p(z)} = 1,$$

and so $m_j = k_j$ ($j = 1, \ldots, s$). This proves the uniqueness.

We call $m_j$ in (8.1.4) the *order* or *multiplicity* of the zero $z_j$ of $p$. When we make an assertion about the number of zeros of $p$ we count a zero of order $m$ as $m$ coincident zeros. With this convention every polynomial of degree $d$ has exactly $d$ zeros. The polynomials $p$ and $p - w$ have the same degree and the zeros of $p - w$ are the solutions of $p(z) = w$. We count the solutions of $p(z) = w$ as the zeros of $p - w$ and we have the following result.

*Corollary* *Each polynomial of degree d maps* **C** *onto itself: each w in* **C** *is the image of exactly d points in* **C**.

We shall now use the factorization (8.1.4) to obtain another geometric interpretation of the index. Let $\gamma$ be any curve not passing through $z_1, \ldots, z_s$. Then by I4,

$$n(p \circ \gamma, 0) = m_1 n(\gamma, z_1) + \cdots + m_s n(\gamma, z_s). \tag{8.1.5}$$

*Now suppose that for each z not on $\gamma$, $n(\gamma, z)$ is either zero or one.* Only the $z_j$ inside $\gamma$ contribute to the sum in (8.1.5) and for these $z_j$, $n(\gamma, z_j) = 1$. Thus we have a special case of the Argument Principle: *$n(p \circ \gamma, 0)$ is the number of zeros of p inside $\gamma$.*

If we can compute $n(p \circ \gamma, 0)$ by some other means we shall then be able to derive information about the location of the zeros of $p$. Let us illustrate this with a specific example (in order to be rigorous we treat this example analytically: it is essential that the reader provides for himself the geometric interpretation).

*Example 8.1.1* Let $p(z) = z^4 + z + 1$: where are the four zeros of $p$?

It is easy to find discs in which $p$ has no zeros. For example, if $p(z) = 0$ then

$$1 = |z^4 + z| \leqslant |z|^4 + |z|.$$

We conclude that if $|z|^4 + |z| < 1$ then $p(z) \neq 0$ and a calculation shows that this is so if $|z| \leqslant 0.72$. This can be (and sometimes is) proved using the Argument Principle but is is completely elementary.

The same procedure can be used at any other point. For example,

$$p(1+z) = z^4 + 4z^3 + 6z^2 + 5z + 3$$

and, for exactly the same reason as above, there are no zeros of $p$ in $\mathbf{C}(1, r)$ if

$$r^4 + 4r^3 + 6r^2 + 5r < 3.$$

This is true if $r = 0.37$.

This random procedure is of little use by itself, but it is obviously helpful when we can use the Argument Principle to assert that there is at least one zero in a certain region. For example, let $\gamma(t) = re^{it}$, $0 \leqslant t \leqslant 2\pi$ and choose $r$ such that $r^4 > 1 + r$: this is true if $r = 1.23$. Then

$$|\gamma(t)|^4 \geqslant 1 + |\gamma(t)|$$

and so by I6,

$$\begin{aligned} n(p \circ \gamma, 0) &= n(\gamma^4 + \gamma + 1, 0) \\ &= n(\gamma^4, 0) \\ &= 4. \end{aligned}$$

We can now assert that all four zeros of $p$ satisfy $|z| < 1.23$ and the earlier results immediately appear more significant. It is clear that with sufficient determination we can use this method to further restrict the location of the zeros within this disc.

We can also obtain more information by a more judicious choice of $\gamma$. Such choices depend in an essential way on the actual form of the polynomial $p$ and are largely a matter of experience. First, we note that

$$\begin{aligned} \operatorname{Re}[p(x+iy)] &= 1 + x + x^4 - 6x^2y^2 + y^4 \\ &= 1 + x - 8x^4 + (y^2 - 3x^2)^2 \end{aligned}$$

and

$$\operatorname{Im}[p(x+iy)] = y + 4xy(x^2 - y^2).$$

In particular, $p(0) = 1$, $\operatorname{Re}[p(x)] > 0$ if $x > 0$ and $\operatorname{Re}[p(iy)] > 0$ if $y > 0$.

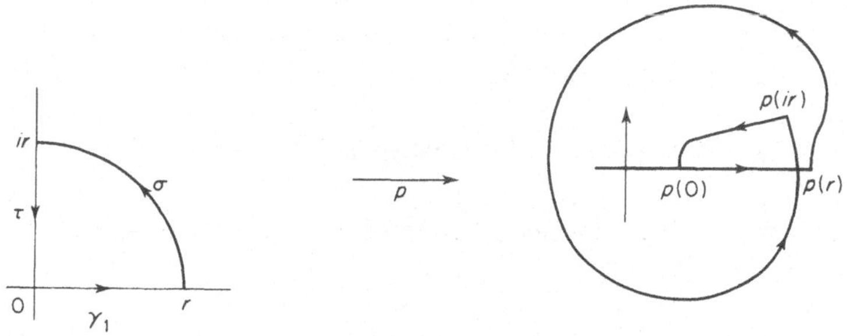


Diagram 8.1.2

It is easy to see that $p$ has no real zeros and as $p$ has real coefficients, the zeros of $p$ are situated symmetrically with respect to the real axis. Further, an examination of $p(iy)$ shows that $p$ has no zeros on the imaginary axis.

Let us now show that $p$ has exactly one zero in each of the open quadrants determined by the co-ordinate axes. Let $\gamma_1$ be the curve described by moving from $r$ ($r = 1.23$) to $ir$ around the shorter arc of $|z| = r$, back along the positive imaginary axis to zero and finally along the real axis to $r$: specifically,

$$\gamma_1(t) = \begin{cases} re^{it}, & 0 \leqslant t \leqslant \dfrac{\pi}{2}; \\[2mm] ir\left(1 + \dfrac{\pi}{2} - t\right), & \dfrac{\pi}{2} \leqslant t \leqslant \dfrac{\pi}{2} + 1; \\[2mm] r\left(t - 1 - \dfrac{\pi}{2}\right), & \dfrac{\pi}{2} + 1 \leqslant t \leqslant \dfrac{\pi}{2} + 2 \end{cases}$$

(this is pictured in Diagram 8.1.2).

Now denote by $\sigma$ and $\tau$ the restriction of $\gamma_1$ to $[0, \pi/2]$ and $[\pi/2, \pi/2 + 2]$ respectively. We have already noted that Re $[p(z)] > 0$ if $z$ is on $\tau$, thus $p \circ \tau$ lies in the right half-plane and so by I5, $|n(p \circ \tau, 0)| < \frac{1}{2}$.

Next, we write

$$p \circ \sigma(t) = \sigma(t)^4 \, [1 + \Sigma(t)]$$

and observe that as $r^4 > 1 + r$,

$$|\Sigma(t)| = \left| \frac{1 + \sigma(t)}{\sigma(t)^4} \right| < 1,$$

and so the curve $1 + \Sigma$ also lies in the right half-plane. Thus $|n(1 + \Sigma, 0)| < \frac{1}{2}$.

Obviously, $n(\sigma, 0) = \frac{1}{4}$ and so if we use I4 we find that

$$\begin{aligned} |n(p \circ \gamma_1, 0) - 1| &= |n(p \circ \sigma, 0) + n(p \circ \tau) - 1| \\ &\leqslant |4n(\sigma, 0) - 1| + |n(1 + \Sigma, 0)| + |n(p \circ \tau, 0)| \\ &< 1. \end{aligned}$$

We conclude that $n(p \circ \gamma_1, 0) = 1$ and so $p$ has exactly one zero inside $\gamma_1$. As all zeros of $p$ lie inside $|z| = r$ we conclude that $p$ has exactly one zero in each open quadrant.

Let us now concentrate our attention on the single zero in the first quadrant. By constructing a curve $\gamma_2$ similar to $\gamma_1$ but using the circle $|z| = 1$ instead of $|z| = r$ we can show that this zero satisfies $|z| > 1$. Indeed, if $0 \leqslant t \leqslant \pi/2$, then

$$\begin{aligned} \text{Re } [p(e^{it})] &= 1 + \cos t + \cos 4t \\ &= (1 + \cos 4t) + \cos t \\ &> 0, \end{aligned}$$

and so the entire curve $p \circ \gamma_2$ lies in the right half-plane. Using I5 we deduce that $n(p \circ \gamma_2, 0) = 0$ and so $p$ has no zeros in the region

$$\{x + iy \in \mathbf{C} : x > 0, y > 0, x^2 + y^2 \leqslant 1\}.$$

Next, if $z$ lies in the region

$$D = \{x + iy \in \mathbf{C} : 0 \leqslant y \leqslant x\}$$

we find that either Im $[p(z)] > 0$ or $y = 0$ in which case Re $[p(z)] > 0$. Again, we use I5 and deduce that for any circle $\gamma_3$ in $D$, $n(p \circ \gamma_3, 0) = 0$. Thus $p$ has no zeros inside $\gamma_3$ and hence no zeros in $D$.

We shall consider one more curve and we ask the reader to provide the detailed argument for himself. We take a fixed positive value of $x$ (an appropriate value will be seen to be 0.67) and construct the curve $\gamma_4$ (which the reader should draw) given by

$$\gamma_4(t) = \begin{cases} \sigma(t) = re^{it}, & -t_0 \leqslant t \leqslant t_0, \\ \tau(t) = z_1 + (t - t_0)(z_2 - z_1), & t_0 \leqslant t \leqslant t_0 + 1, \end{cases}$$

where $r > 2$, $z_1 = x + i(r^2 - x^2)^{½}$, $z_2 = x - i(r^2 - x^2)^{½}$ and $t_0$ is the unique solution $t$ in $[0, \pi/2]$ of $\cos t = x/r$.

Exactly as in our discussion of $\gamma_1$ we find that

$$\begin{aligned} n(p \circ \sigma, 0) &= 4n(\sigma, 0) + n(1 + \Sigma, 0) \\ &= (4/2\pi)(t_0 - (-t_0)) + n(1 + \Sigma, 0) \\ &= 4t_0/\pi + n(1 + \Sigma, 0). \end{aligned}$$

It is clear that as $r \to +\infty$, $t_0 \to \pi/2$. Further, as $|\Sigma(t)| \leqslant 2/r^3$ the curve $1 + \Sigma$ lies in the disc $\mathbf{C}(1, 2/r^3)$ and so

$$n(1 + \Sigma, 0) \to 0 \qquad \text{as} \qquad r \to +\infty.$$

Thus

$$n(p \circ \sigma, 0) \to 2 \qquad \text{as} \qquad r \to +\infty.$$

Next, if $1 + x > 8x^4$ (and this is so if $x = 0.67$), then Re $[p(z)] > 0$ where $z = x + iy$. We deduce that $p \circ \tau$ lies in the right half-plane and so $|n(p \circ \tau, 0)| < ½$.

It is now apparent that for sufficiently large $r$,

$$\begin{aligned} n(p \circ \gamma_4, 0) &= n(p \circ \sigma, 0) + n(p \circ \tau, 0) \\ &= 2 \end{aligned}$$

and we conclude that $p$ has exactly two zeros satisfying Re $[z] > 0.67$. We have now shown that the single zero of $p$ in the first quadrant lies in the shaded region in Diagram 8.1.3.

We end this discussion of polynomials with a brief (and rather informal) examination of the relationship between the coefficients and the zeros of a polynomial. If polynomials $p$ and $q$ have different coefficients then $p - q$ is a non-

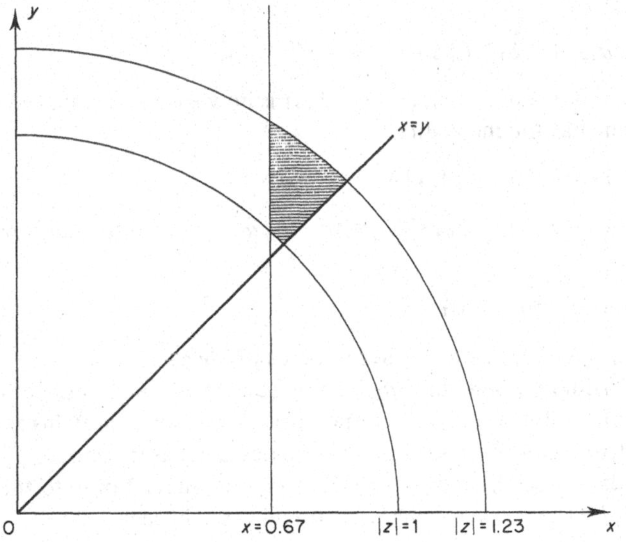


Diagram 8.1.3

constant polynomial and for some $z, p(z) \neq q(z)$: thus if two polynomials are equal (for all $z$) we may 'equate coefficients'. For simplicity we shall assume that $a_d = 1$.

If $z_1, \ldots, z_d$ are the zeros of $p$ (not necessarily distinct) then

$$a_0 + a_1 z + \cdots + a_{d-1} z^{d-1} + z^d = (z - z_1) \cdots (z - z_d),$$

and so

$$a_{d-j} = (-1)^j \Sigma z_{n_1} \ldots z_{n_j}, \tag{8.1.6}$$

the sum being over all $j$-tuples $(n_1, \ldots, n_j)$ of distinct integers chosen from $1, \ldots, d$. For example,

$$a_0 = (-1)^d z_1 \ldots z_d, \qquad a_{d-1} = -(z_1 + \cdots + z_d).$$

It is clear from (8.1.6) that a small change in each $z_j$ yields only a small change in each $a_n$ (the details are omitted but see Exercise 8.1.7). We express this by saying that *each coefficient* $a_n$ *depends continuously on the zeros* $z_1, \ldots, z_d$.

It is true (but less obvious) that *each zero* $z_j$ *depends continuously on the coefficients* $a_n$ and this is meant in the following sense.

Let $\gamma$ be any circle, say $\gamma(t) = w + re^{it}$, $0 \leqslant t \leqslant 2\pi$, and suppose that $p$ is not zero on $\gamma$. Then

$$\epsilon = \inf\{\, |p(z)| : z \in [\gamma]\,\} > 0.$$

Now let

$$q(z) = b_0 + b_1 z + \cdots + b_{d-1} z^{d-1} + z^d$$

and note that

$$|p(z) - q(z)| \leqslant \delta(1 + |z| + \cdots + |z^{d-1}|)$$

whenever $|a_j - b_j| < \delta, j = 0, \ldots, d-1$. It is now easy to see that there is a positive $\delta$ such that the inequality

$$|q(z) - p(z)| < \epsilon \leqslant |p(z)|) \tag{8.1.7}$$

holds on $\gamma$ whenever $|a_j - b_j| < \delta, j = 0, \ldots, d-1$. In conjunction with I6, this implies that

$$n(p \circ \gamma, 0) = n(q \circ \gamma, 0)$$

and so $p$ and $q$ have the same number of zeros inside $\gamma$.

We now consider a polynomial $p$ given by (8.1.4) and we construct circles $\gamma_1, \ldots, \gamma_s$ with centres $z_1, \ldots, z_s$ respectively such that $\gamma_j$ contains $z_j$ and no other $z_n$. The (positive) radii of these circles may be as small as we choose. For a suitable $\delta$, (8.1.7) holds on each $\gamma_j$ and so the polynomial $q$, subject only to the restriction $|a_j - b_j| < \delta, j = 0, \ldots, d-1$, has exactly $m_j$ zeros in each $\gamma_j$.

## Exercise 8.1

1. Prove the Minimum Modulus Theorem for a polynomial as follows. Suppose that $p(w) \neq 0$ and consider

$$p(w + re^{i\alpha}) = p(w) + b_s(re^{i\alpha})^s + \cdots + b_d(re^{i\alpha})^d$$

$$= p(w) + b_s(re^{i\alpha})^s \left[1 + \cdots + \left(\frac{b_d}{b_s}\right)(re^{i\alpha})^{d-s}\right].$$

Show that for a suitable choice of $r$ and $\alpha$,

$$|p(w + re^{i\alpha})| < |p(w)|.$$

[*Hint:* $|z_1 + z_2| < |z_1|$ if $z_2/z_1 \in C(-1, 1)$.]

2. Given (8.1.4), use the Argument Principle to show that $m_j$ is uniquely determined by $p$ and $z_j$.
3. Let

$$p_m(z) = 1 + z + \frac{z^2}{2!} + \cdots + \frac{z^m}{m!} \qquad (m = 1, 2 \ldots),$$

and let $\delta_m$ be the largest positive $r$ such that $p_m$ is non-zero on $C(0, r)$. Prove that $\delta_m \to +\infty$ as $m \to \infty$.

4. Let $\gamma$ be a closed curve such that $n(\gamma, z)$ is either zero or one. Let $p$ and $q$ be polynomials which are both non-zero on $[\gamma]$ and let $r(z) = p(z)/q(z)$. Show that $n(r \circ \gamma, 0)$ is the number of zeros of $p$ inside $\gamma$ minus the number of zeros of $q$ inside $\gamma$.
5. Let

$$p(z) = z^5 + 13z^2 + 15.$$

Show that $p$ has two zeros satisfying $1 < |z| < 2$ and three satisfying $2 < |z| < 5/2$.

6. Let

$$p(z) = z^5 + 11z + 9.$$

How many zeros has $p$ in each of the domains (a) $\frac{3}{4} < |z| < 1$, (b) $1 < |z| < 2$, (c) $2 < |z| < 3$, (d) $x > 0, y > 0$ and (e) $x < 0, y > 0$?

7. Let $a_{d-j}$ be defined by (8.1.6) and let $b_{d-j}$ be defined similarly in terms of $w_j$. Show that if $|z_j - w_j| < \delta < 1$, then

$$|a_{d-j} - b_{d-j}| \leqslant \delta M,$$

where $M$ depends only on the $z_j$.

## 8.2. POWER SERIES

A *power series* is a function of the form

$$f(z) = \sum_{n=0}^{\infty} a_n (z - z_0)^n \tag{8.2.1}$$

and this is defined at each $z$ at which the series converges. Before we can discuss the properties of $f$ we must examine the nature of the set on which $f$ is defined: this is our first task.

For brevity we shall rewrite (8.2.1) in the form

$$f(z_0 + z) = \sum_{n=0}^{\infty} a_n z^n \tag{8.2.2}$$

and we shall denote by $S$ the set of $z$ at which this series converges. Obviously $0 \in S$ and so we may define

$$\rho = \sup \{|z| : z \in S\}$$

with the convention that $\rho = +\infty$ if $S$ is unbounded. The definition of $\rho$ implies that if $|z| > \rho$ then $z \notin S$ and so the series in (8.2.2) diverges.

Now suppose that $|z| < \rho$: the definition of $\rho$ does not directly imply that $z \in S$ but we shall prove that this is so. As $|z| < \rho$ there is a $w$ in $S$ with $|z| < |w| \leqslant \rho$. As $w \in S$ the series

$$\sum_{n=0}^{\infty} a_n w^n$$

converges and so there is a constant $m$ such that

$$|a_n w^n| \leqslant m \qquad (n = 1, 2, \ldots). \tag{8.2.3}$$

As $|z| < |w|$, we have

$$|a_n z^n| \leqslant m(|z| / |w|)^n \qquad (n = 0, 1, 2, \ldots)$$

and so using the Comparison Test we see that the series in (8.2.2) is absolutely convergent.

We rewrite these results in terms of (8.2.1) and restate them as our next theorem.

*Theorem 8.2.1* *Let f be the power series given by* (8.2.1), *let S be the set of z at which this converges and let*

$$\rho = \sup \{ |z - z_0| : z \in S \}.$$

*The series in* (8.2.1) *is absolutely convergent if* $|z - z_0| < \rho$ *and is divergent if* $|z - z_0| > \rho$.

Observe that if (8.2.3) holds (whether or not $w \in S$), then (8.2.2) converges for each $z$ satisfying $|z| < |w|$; thus $\rho \geqslant |w|$. The question of convergence on the circle $|z - z_0| = \rho$ requires more delicate analysis and there is no need to study this here.

We shall use the following terminology. The series in (8.2.1) converges at each point in the *disc of convergence* $\mathbf{C}(z_0, \rho)$: $\rho$ is the *radius of convergence* and $z_0$ the *centre of convergence* of the series. If $\rho = +\infty$ we interpret $\mathbf{C}(z_0, \rho)$ as $\mathbf{C}$.

As simple examples, the radii of convergence of the series

$$\sum_{n=0}^{\infty} \frac{z^n}{n!}, \qquad \sum_{n=0}^{\infty} z^n, \qquad \sum_{n=0}^{\infty} (nz)^n$$

are $+\infty$, 1 and 0 respectively (the latter is zero because if $z \neq 0$ then $(nz)^n \to \infty$ as $n \to +\infty$).

It is easy to see that the power series (8.2.1) is continuous on its disc of convergence $\mathbf{C}(z_0, \rho)$. First, we may assume that $z_0 = 0$. Next, consider any $z$ in $\mathbf{C}(0, \rho)$ and select a positive $\delta$ so that $|z| + 2\delta < \rho$. Then if $|h| < \delta$, we use (1.2.2) and then (1.2.1) to derive the inequalities

$$\begin{aligned} |f(z+h) - f(z)| &\leqslant \sum_{n=1}^{\infty} |a_n| |(z+h)^n - z^n| \\ &\leqslant \sum_{n=1}^{\infty} |h| n |a_n| (|z| + \delta)^{n-1} \\ &\leqslant (|h|/\delta) \sum_{n=1}^{\infty} |a_n| (|z| + 2\delta)^n. \end{aligned}$$

This upper bound tends to zero as $h \to 0$; thus $f$ is continuous at $z$ and hence on $\mathbf{C}(0, \rho)$. For future reference, we state this as a theorem.

*Theorem 8.2.2* *The series* (8.2.1) *is continuous in its disc of convergence.*

This result enables us to 'equate coefficients' of equal power series.

*Corollary* *Suppose that the series*

$$\sum_{n=0}^{\infty} a_n (z - z_0)^n, \qquad \sum_{n=0}^{\infty} b_n (z - z_0)^n$$

*are convergent and are equal in some disc* $C(z_0, \delta)$, $\delta > 0$. *Then* $a_n = b_n$, $n = 0, 1, \ldots$.

*Proof* As the series are equal at $z_0$, $a_0 = b_0$. Now suppose that $a_j = b_j$ when $j = 0, \ldots, m-1$. Then

$$\sum_{n=m}^{\infty} a_n(z - z_0)^n = \sum_{n=m}^{\infty} b_n(z - z_0)^n$$

in $C(z_0, \delta)$ and so if $0 < |z - z_0| < \delta$, then

$$\sum_{n=m}^{\infty} a_n(z - z_0)^{n-m} = \sum_{n=m}^{\infty} b_n(z - z_0)^{n-m}.$$

Each side of this equation is a power series which converges if $0 < |z - z_0| < \delta$: thus each side must converge on $C(z_0, \delta)$. The two series are continuous on $C(z_0, \delta)$ and equal on $C^*(z_0, \delta)$; hence they are equal on $C(z_0, \delta)$ and so at $z_0$. This gives $a_m = b_m$ and the proof is completed by induction.

We shall see later that a power series which is not identically zero can only have a countable number of zeros.

Finally, if $f$ is given by (8.2.1) and if $|z - z_0| < t < \rho$, then

$$|a_n(z - z_0)^n| \leqslant |a_n| t^n.$$

As

$$\sum_{n=0}^{\infty} |a_n| t^n$$

converges (Theorem 8.2.1) we find from (6.6.2) that $f$ converges uniformly on $\bar{C}(z_0, t)$. We can now deduce from Theorem 6.6.1 that $f$ is continuous on $\bar{C}(z_0, t)$. Better still, if $z \in C(z_0, \rho)$, choose $t$ so that $|z - z_0| < t < \rho$: then $f$ is continuous on $C(z, \delta)$, where

$$\delta = t - |z - z_0| > 0.$$

This shows that $f$ is continuous at $z$ (an arbitrary point of $C(z_0, \rho)$) and hence is continuous on $C(z_0, \rho)$. This, then, is an alternative proof of Theorem 8.2.2.

**Exercise 8.2**

1. Prove that if (8.2.1) has radius of convergence $\rho$ then so do the series

$$\sum_{n=1}^{\infty} n a_n z^{n-1}, \qquad \sum_{n=0}^{\infty} a_n z^{n+1}/(n+1)$$

2. Let $\epsilon$ be any positive number. Prove that if $|w| = 1$ there is a $z$ satisfying $|z - w| < \epsilon$, $|z| = 1$ and such that

$$\sum_{n=1}^{\infty} \frac{(-1)^n z^{n!}}{n}$$

converges. [This series has disc of convergence $\mathbf{C}(0, 1)$ and converges on a dense set of points on the circumference.]

3. Prove that

$$f(z) = \sum_{n=1}^{\infty} z^n/n^2$$

(a) converges if and only if $|z| \leqslant 1$ and
(b) is continuous on $\bar{\mathbf{C}}(0, 1)$.
[This shows that a power series may be continuous (and hence bounded) on the closure of the disc of convergence.]

4. Find the radius of convergence of $\Sigma_{n=1}^{\infty} a_n z^n$, where for each positive integer $n$, $n = 2^k p$ ($k$ and $p$ are non-negative integers with $p$ odd) and $a_n = k$.

5. Let $f$ be given by (8.2.1) and be convergent in $\mathbf{C}(z_0, \rho)$ and suppose that $0 < t < \rho$. Prove that

$$\sup \{ |f(z)| : |z - z_0| \leqslant t\} = \sup \{ |f(z)| : |z - z_0| = t\}.$$

[*Hint*: consider polynomials converging uniformly to $f$.]

## 8.3 ANALYTIC FUNCTIONS

The class of polynomials is contained in the class of power series. We now introduce a still larger class of functions, namely the class of those functions which can be expressed locally as a power series: this is the class of analytic functions.

As an introduction to the idea of analyticity let us consider the following simple example.

*Example 8.3.1* Let $f(z) = z^{-1}, z \neq 0$. This is certainly not a power series yet if $z_0 \neq 0$ then

$$f(z) = z_0^{-1} \sum_{n=0}^{\infty} \left[\frac{-(z - z_0)}{z_0}\right]^n$$

whenever $|z - z_0| < |z_0|$. Thus the restriction of $f$ to the disc $\mathbf{C}(z_0, |z_0|)$ is a power series in that disc.

*Definition 8.3.1* *A function $f$ is said to be analytic in a domain $D$ if and only if for each $z_0$ in $D$ there exist complex numbers $a_0, a_1, \ldots$ and a positive $\rho$ such that*

$$f(z) = \sum_{n=0}^{\infty} a_n (z - z_0)^n$$

*in the disc* $\mathbf{C}(z_0, \rho)$.

Of course the numbers $a_0, a_1, \ldots$ depend on $z_0$ and when we have reason to examine this dependence we shall write

$$f(z) = \sum_{n=0}^{\infty} a_n(z_0)(z - z_0)^n.$$

The reader should note that we are *not* assuming that $\mathbf{C}(z_0, \rho)$ is necessarily the largest disc in $D$ with centre $z_0$. This is so in Example 8.3.1 and it is true in general: we shall prove this later and it is not necessary to assume it now.

An immediate consequence of Definition 8.3.1 is that an analytic function $f$ inherits all of the *local* properties of power series: for example, if $f$ is analytic in $D$ then it is continuous in $D$.

The object of this section is simply to familiarize the reader with the concept of analyticity, and we shall content ourselves with some illustrative examples and a few elementary remarks.

First, let $f_1$ and $f_2$ be power series converging in $\mathbf{C}(z_0, \rho_1)$ and $\mathbf{C}(z_0, \rho_2)$ respectively and suppose that $\rho, \rho = \min\{\rho_1, \rho_2\}$, is positive. Then $kf_1$ ($k$ is any constant), $f_1 + f_2$ and $f_1 f_2$ are all power series converging in $\mathbf{C}(z_0, \rho)$ (Theorems 2.1.1 and 2.3.1). We conclude that if $f$ and $g$ are analytic in a domain $D$ then so are $kf$, $f+g$ and $fg$: more generally, finite sums, finite products and finite linear combinations of analytic functions are analytic.

*Example 8.3.2* Every polynomial is analytic in $\mathbf{C}$: this is just (8.1.2). Alternatively, a linear polynomial is clearly analytic in $\mathbf{C}$ and the general polynomial is obtained as a finite product of linear polynomials.

*Example 8.3.3* The functions exp, sin and cos are analytic in $\mathbf{C}$. First, for any $z_0$,

$$\exp(z) = \exp(z_0)\exp(z - z_0)$$

$$= \sum_{n=0}^{\infty} \exp(z_0)\frac{(z - z_0)^n}{n!}$$

and this shows that exp is analytic in $\mathbf{C}$.

Next

$$\sin z = \sin(z - z_0)\cos z_0 + \cos(z - z_0)\sin z_0$$

and this may be written as a power series with centre of convergence $z_0$. Thus sin is analytic in $\mathbf{C}$: the proof for cos is similar.

*Example 8.3.4* A rational function $r$ is defined to be the quotient $p/q$ of two polynomials $p$ and $q$ (with $q$ not identically zero). We show that $r$ is analytic in $\mathbf{C} - Q$, where $Q$ is the set of zeros of $q$.

If $z$ and $z_0$ are different from $a$, then

$$\frac{1}{z - a} = \frac{1}{a - z_0} \sum_{n=0}^{\infty} \left(\frac{z - z_0}{a - z_0}\right)^n \qquad (z_0 \neq a)$$

and so $(z - a)^{-1}$ is analytic in $\mathbf{C} - \{a\}$. Taking products, we deduce that $1/q$, and hence $p/q$, are analytic in $\mathbf{C} - Q$.

*Example 8.3.5* Let $f$ be analytic in $D$, then so is

$$F(z) = \begin{cases} \dfrac{f(z) - f(z_0)}{z - z_0} & \text{if } z \neq z_0, \\ a_1(z_0) & \text{if } z = z_0. \end{cases}$$

Note first that $f(z) - f(z_0)$ and $(z - z_0)^{-1}$ are both analytic in $D - \{z_0\}$, thus $F$ is analytic in $D - \{z_0\}$. This means that $F$ is expressible as a power series about any point of $D$ except possibly $z_0$. We now examine this possible exception.

If

$$f(z) = \sum_{n=0}^{\infty} a_n(z - z_0)^n \tag{8.3.1}$$

say, in $\mathbf{C}(z_0, \rho)$, then (by considering $z = z_0$ and $z \neq z_0$ separately)

$$F(z) = \sum_{n=1}^{\infty} a_n(z - z_0)^{n-1}$$

in $\mathbf{C}(z_0, \rho)$. Thus $F$ is expressible as a power series with centre of convergence $z_0$ and so $F$ is analytic in $D$.

More generally, if $f$ is analytic in $D$ and if (8.3.1) holds, then

$$g(z) = \begin{cases} \dfrac{f(z) - [a_0 + a_1(z - z_0) + \cdots + a_n(z - z_0)^n}{(z - z_0)^{n+1}} & \text{if } z \neq z_0; \\ a_{n+1}(z_0) & \text{if } z = z_0 \end{cases}$$

is also analytic in $D$.

*Example 8.3.6* Let $f$ be analytic in $D$ and suppose that

$$f(z_1) = f(z_2) = \cdots = f(z_n) = w$$

for distinct points $z_1, \ldots, z_n$ in $D$. Then

$$F(z) = \frac{f(z) - w}{(z - z_1) \cdots (z - z_n)}$$

is analytic in $D$ when suitably defined at $z_1, \ldots, z_n$.

Obviously $F$ is analytic in $D - \{z_1, \ldots, z_n\}$. Moreover, both $[f(z) - w]/[z - z_1]$ and $[(z - z_2) \cdots (z - z_n)]^{-1}$ are analytic in $D - \{z_2, \ldots, z_n\}$ and so is their product, namely $F$. This shows that $F$ can be expressed as a power series with centre of convergence $z_1$ and a similar argument holds for $z_2, \ldots, z_n$.

We began this section with the assertion that the class of analytic functions contains all power series. The last result in this section justifies this claim: this could be left until much later and its only purpose is to provide now yet more examples of analytic functions.

*Theorem 8.3.1* *Let f be a power series with disc of convergence* $\mathbf{C}(z_0, \rho)$, $\rho > 0$. *Then f is analytic in* $\mathbf{C}(z_0, \rho)$.

*Proof* First, we may take $z_0 = 0$. The proof is based on a discussion of the functions

$$f^{(k)}(z) = \sum_{n=k}^{\infty} n(n-1)\cdots(n-k+1)a_n z^{n-k} \tag{8.3.2}$$

where $k = 0, 1, \ldots$ and $f^{(0)} = f$. Of course, for the readers who have met the derivative, the function $f^{(k)}$ is obtained from $f$ by differentiating each term of $f$ exactly $k$ times. This is of no importance now, we only need the convergence of $f^{(k)}$ in $\mathbf{C}(0, \rho)$.

Given any $z$ in $\mathbf{C}(0, \rho)$, select a positive $\delta$ so that $|z| + \delta < \rho$. Then

$$\begin{aligned} n(n-1)\cdots(n-k+1)\,|z^{n-k}| &= \frac{k!}{\delta^k}\binom{n}{k}|z^{n-k}|\,\delta^k \\ &\leqslant \frac{k!(|z| + \delta)^n}{\delta^k}, \end{aligned}$$

and as the series

$$\sum_{n=0}^{\infty} a_n(|z| + \delta)^n$$

converges so does the series (8.3.2). This shows that $f^{(k)}$ has radius of convergence at least $\rho$: in fact, the radius is $\rho$ and this is elementary but unimportant.

For $z$ and $z + h$ in $\mathbf{C}(0, \rho)$ we have the identity

$$\begin{aligned} f(z+h) - &\sum_{n=0}^{p} \frac{h^n}{n!} f^{(n)}(z) \\ &= \sum_{n=p+1}^{\infty} a_n\left[(z+h)^n - \left(z^n + nhz^{n-1} + \cdots + \binom{n}{p} h^p z^{n-p}\right)\right]. \end{aligned}$$

This is obvious for $p = 0$ and the general case follows easily by induction on $p$. We conclude that

$$\begin{aligned} \Big|f(z+h) - &\sum_{n=0}^{p} \frac{h^n}{n!} f^{(n)}(z)\Big| \\ &\leqslant \sum_{n=p+1}^{\infty} |a_n|\left[(|z| + |h|)^n + |z|^n + \cdots + \binom{n}{p}|h^p z^{n-p}|\right] \\ &\leqslant 2 \sum_{n=p+1} |a_n|(|z| + |h|)^n, \end{aligned}$$

and this upper bound tends to zero as $p \to +\infty$. Thus

$$f(z+h) = \sum_{n=0}^{\infty} \frac{h^n}{n!} f^{(n)}(z)$$

and $f$ is analytic in $\mathbf{C}(0, \rho)$.

**Exercise 8.3**

1. Prove that $f(z) = |z|^2$ and $g(z) = \text{Re}\,[z]$ are not analytic in any domain.
2. Let $f$ be analytic in $D$ and let

   $$D^* = \{z : \bar{z} \in D\}$$

   ($D^*$ is the reflection of $D$ in the real axis). Define $f^*$ on $D^*$ by $f^*(z) = \overline{f(\bar{z})}$: prove that $f^*$ is analytic in $D^*$.
3. Let $f$ be analytic and non-zero in a domain $D$. Prove that $1/f$ is analytic in $D$. [Proceed as follows.

   Suppose that $z_0$ is in $D$ and let

   $$f(z) = \sum_{n=0}^{\infty} a_n(z - z_0)^n$$

   in $\mathbf{C}(z_0, \rho)$, $\rho > 0$. Define $b_0, b_1, \ldots$ inductively by $b_0 = 1/a_0$ and

   $$a_0 b_n + a_1 b_{n-1} + \cdots + a_n b_0 = 0 \qquad (n = 1, 2, \ldots)$$

   and let

   $$g(z) = \sum_{n=0}^{\infty} b_n(z - z_0)^n.$$

   Choose $r$, $0 < r < \rho$, so that

   $$\sum_{n=1}^{\infty} |a_n| r^n \leqslant |a_0|.$$

   (a) Prove by induction that $|b_n| r^n \leqslant |b_0|$.
   (b) Deduce that $g$ converges in $\mathbf{C}(z_0, r)$.
   (c) Show that $f(z)g(z) = 1$ in $\mathbf{C}(z_0, r)$; thus

   $$(1/f)(z) = \sum_{n=0}^{\infty} b_n(z - z_0)^n.]$$
4. Prove that $z^{-2}(1 - \cos(z))$ is analytic in $\mathbf{C}$.

## 8.4 INEQUALITIES

We observed in §8.1 that given a non-constant polynomial $p$ and any $w$ in $\mathbf{C}$ we may write

$$p(w+z) = p(w) + b_s z^s + \cdots + b_d z^d, \quad z \in \mathbf{C},$$

where $b_s \neq 0$. This simple fact was an essential part of the proof of the Maximum Modulus Theorem. In this section we shall establish the corresponding results for analytic functions: these are then used to derive several important and classical inequalities.

*Theorem 8.4.1* *Let f be analytic and not constant in a domain D, let $z_0$ be in D and let*

$$f(z) = f(z_0) + \sum_{n=1}^{\infty} a_n(z - z_0)^n$$

*in some disc $\mathbf{C}(z_0, \delta)$. Then there exists a positive integer s such that $a_s \neq 0$.*

Assuming the validity of Theorem 8.4.1 we can write

$$\begin{aligned} f(z) &= f(z_0) + (z - z_0)^s [a_s + a_{s+1}(z - z_0) + \cdots] \\ &= f(z_0) + (z - z_0)^s g(z), \end{aligned}$$

say, where $g(z) \to a_s$ as $z \to z_0$ and $a_s \neq 0$. The proof of the Maximum Modulus Theorem now follows exactly as for polynomials.

*The Maximum Modulus Theorem. Let f be analytic and not constant in a domain D. Then $|f|$ cannot have a local maximum in D.*

If the disc $\bar{\mathbf{C}}(z_0, \delta)$ is contained in $D$, then $|f|$ actually attains its maximum value on $\bar{\mathbf{C}}(z_0, \delta)$ (Theorem 6.4.4). The Maximum Modulus Theorem shows that this maximum cannot be attained in $\mathbf{C}(z_0, \delta)$: thus

$$\sup\{\,|f(z)| : |z - z_0| \leqslant \delta\,\} = \sup\{\,|f(z)| : |z - z_0| = \delta\,\} \tag{8.4.1}$$

Observe that this reasoning only requires that $f$ be analytic in $\mathbf{C}(z_0, \delta)$ and continuous in $\bar{\mathbf{C}}(z_0, \delta)$.

The proof of Theorem 8.4.1 involves a discussion of the zeros of an analytic function. We shall say that $\zeta$ in $D$ is an *isolated zero* of $f$ if $f(\zeta) = 0$ and if there is a positive $\delta$ such that $\mathbf{C}(\zeta, \delta) \subset D$ and $f(z) \neq 0$ when $0 < |z - \zeta| < \delta$. Obviously each zero of a non-constant polynomial is an isolated zero. Next, we say that $\zeta$ is an *interior zero* of $f$ if there is a positive $\delta$ such that $\mathbf{C}(\zeta, \delta) \subset D$ and $f(z) = 0$ when $|z - \zeta| < \delta$.

Apparently, there may exist zeros which are neither isolated nor interior zeros: for example the (non-analytic) function $g : \mathbf{C} \to \mathbf{C}$ defined by $g(x + iy) = \max\{x, 0\}$ has a zero at the origin and this is neither an isolated nor an interior zero. It is easy to see, however, that this situation cannot arise for analytic functions.

*Theorem 8.4.2* *Let f be analytic in a domain D: then every zero of f is either an isolated zero or an interior zero. If f has an interior zero then f is identically zero in D.*

This remarkable result is entirely topological in character for it says that the set $f^{-1}\{0\}$ is either $D$ or consists of a set of isolated points in $D$. In analytic terms it says that $f$ is uniquely determined throughout $D$ by its values on an arbitrary small open disc $\Delta$ in $D$. Indeed, if $f$ and $g$ are analytic in $D$ and if $f = g$ on $\Delta$, then $f - g$ is analytic in $D$ and has an interior zero, so $f = g$ throughout $D$. Better still, if $f - g$ has a non-isolated zero then $f = g$ throughout $D$.

We come now to the proofs of these results. We shall prove Theorem 8.4.2 first and Theorem 8.4.1 follows easily from this. As we have already seen, The Maximum Modulus Theorem follows from Theorem 8.4.1.

*Proof of Theorem 8.4.2* We suppose that $\zeta \in D$ and $f(\zeta) = 0$ and write

$$f(z) = \sum_{n=1}^{\infty} a_n(z - \zeta)^n$$

in some $\mathbf{C}(\zeta, \delta)$. Then either (a) $a_n = 0, n = 1, 2, \ldots$ or (b)

$$f(z) = (z - \zeta)^s [a_s + a_{s+1}(z - \zeta) + \cdots]$$

in $\mathbf{C}(\zeta, \delta)$, where $a_s \neq 0$. If (a) occurs, then $f$ is identically zero in $\mathbf{C}(\zeta, \delta)$ and $\zeta$ is an interior zero of $f$. If (b) occurs, we may find a positive $r$ such that

$$a_s + a_{s+1}(z - \zeta) + \cdots \neq 0$$

in $\mathbf{C}(\zeta, r)$ (this series is continuous and non-zero at $\zeta$) and so $\zeta$ is an isolated zero of $f$. This proves the first statement.

Now let $D_1$ be the set of interior zeros of $f$ and suppose that $D_1 \neq \emptyset$. As $D_1$ is the union of open discs on which $f$ is identically zero, $D_1$ is an open subset of $D$.

If $w \in D - D_1$, then either $f(w) \neq 0$ or $w$ is an isolated zero of $f$. In each case there is a disc $\mathbf{C}(w, \rho)$ on which $f$ is never zero except possibly at $w$. Thus $\mathbf{C}(w, \rho) \cap D_1 = \emptyset$ and so $D - D_1$ is also an open subset of $D$.

As $D$ is connected it cannot be the disjoint union of two non-empty open sets. As $D_1 \neq \emptyset$ we conclude that $D - D_1 = \emptyset$ and so $D = D_1$. This proves that $f$ is identically zero on $D$.

*Proof of Theorem 8.4.1* Let $\zeta$ be in $D$ and suppose that

$$f(z) = f(\zeta) + \sum_{n=1}^{\infty} a_n(z - \zeta)^n$$

in some $\mathbf{C}(\zeta, \delta)$ where $a_n = 0$ for $n = 1, 2, \ldots$. This implies that the function $F$, $F(z) = f(z) - f(\zeta)$, is analytic in $D$ and that $\zeta$ is an interior zero of $F$. Theorem 8.4.1 implies that $F$ and hence $f$ is constant in $D$.

We have now proved the three results stated above. One of the most useful applications of the Maximum Modulus Theorem is to identify functions by their boundary values. For example, if $f$ is analytic in $\mathbf{C}(0, 1)$, continuous in $\bar{\mathbf{C}}(0, 1)$ and if $f(z) = \exp(z)$ when $|z| = 1$, then we may apply (8.4.1) to $f(z) - \exp(z)$ and conclude that $f(z) = \exp(z)$ throughout $\bar{\mathbf{C}}(0, 1)$.

We end this section with three classical results: all are direct consequences of the Maximum Modulus Theorem.

*Liouville's Theorem* *Let f be analytic and bounded in* **C**. *Then f is constant.*

*Proof* The function $[f(z) - f(0)]/z$ is analytic in **C** (Example 8.5.3). By (8.4.1), if $0 < |z| < t$, then

$$\left|\frac{f(z) - f(0)}{z}\right| \leqslant \sup\left\{\left|\frac{f(w) - f(0)}{w}\right| : |w| = t\right\}$$

$$\leqslant \frac{M + |f(0)|}{t},$$

where $|f(w)| \leqslant M$ for $w$ in **C**. This inequality holds for each $z$ and $t$ satisfying $0 < |z| < t$. Thus choosing any $z$ and letting $t$ tend to $+\infty$ we find that $f(z) = f(0)$ and $f$ is constant.

*Schwarz's Lemma* *Let f be analytic and satisfy* $|f(z)| \leqslant M$ *in* $\mathbf{C}(0, r)$ *and suppose also that* $f(0) = 0$. *Then*

$$|f(z)| \leqslant M|z|/r, \qquad z \in \mathbf{C}(0, r).$$

*If equality holds for one z in* $\mathbf{C}(0, r)$, *then* $f(z) = \alpha z$ *for some constant* $\alpha$.

*Proof* Exactly as in the previous proof we can show that if $0 < |z| < t < r$, then

$$|f(z)/z| \leqslant M/t$$

and we then let $t$ tend to $r$.

If equality holds for some $z$ in $\mathbf{C}(0, r)$, then the analytic function $f(z)/z$ does attain a local (indeed a global) maximum modulus and according to the Maximum Modulus Theorem it is therefore constant.

The inequality in Schwarz's Lemma gives no useful information when $z = 0$. However, if

$$f(z) = \sum_{n=1}^{\infty} a_n z^n \tag{8.4.2}$$

in some $\mathbf{C}(0, \delta)$, then as $f(z)/z \to a_1$ as $z \to 0$, we have

$$|a_1| \leqslant M/r:$$

this is the appropriate form of the inequality when $z = 0$. A similar inequality is, in fact, satisfied by all the coefficients $a_n$ as we shall now see.

*Proposition 8.4.1* *Let f be analytic in* $\mathbf{C}(0, r)$ *and suppose that* (8.4.2) *holds in some* $\mathbf{C}(0, \delta)$. *If* $0 < t < r$, *then*

$$|a_n| t^n \leqslant \sup\{|f(z)| : |z| = t\}, \qquad n = 1, 2, \ldots. \tag{8.4.3}$$

*Proof* Let $n$ be a positive integer, put $\omega = \exp(2\pi i/n)$ and

$$F(z) = [f(z) + f(\omega z) + \cdots + f(\omega^{n-1}z)]/n.$$

This function is analytic in $C(0, r)$ and for large $n$, $F(z)$ approximates the average value of $f$ on the circle $\{\zeta : |\zeta| = |z|\}$. Further, as $|\omega^k z| = |z|$,

$$\sup\{|F(z)| : |z| = t\} \leqslant \sup\{|f(z)| : |z| = t\}.$$

In terms of power series, we see that

$$F(z) = n^{-1} \sum_{k=1}^{\infty} a_k z^k [1 + \omega^k + \cdots + \omega^{k(n-1)}]$$

$$= a_n z^n + b_{n+1} z^{n+1} + \cdots$$

for some sequence $b_k$ as the geometric series

$$1 + \omega^k + \cdots + \omega^{k(n-1)} = \begin{cases} 0 \text{ if } k = 1, 2, \ldots, n-1, \\ n \text{ if } k = n. \end{cases}$$

We deduce that $F(z)/z^n$ is analytic in $C(0, r)$, its value at $z = 0$ being $a_n$ (see Example 8.3.5). Thus by (8.4.1), if $0 < t < r$, then

$$|a_n| \leqslant \sup\{|F(z)/z^n| : |z| = t\}$$

$$\leqslant t^{-n} \sup\{|f(z)| : |z| = t\},$$

and this is (8.4.3).

The inequalities (8.4.3) are valid without the assumption that $f(0) = 0$ and are then known as *Cauchy's Inequalities* (Exercise 8.4.9) : we shall use (8.4.3) to establish the next result, which was first mentioned in §8.3.

*Theorem 8.4.3* *Let $f$ be analytic in $D$, let $z_0$ be in $D$ and let*

$$f(z) = \sum_{n=0}^{\infty} a_n (z - z_0)^n$$

*in some disc $C(z_0, \delta)$. Then this series converges and equals $f$ in the largest open disc $C(z_0, r)$ which is contained in $D$.*

*Proof* First, we apply Proposition 8.4.1 to the function $f(z_0 + z) - a_0$: then (8.4.3) becomes

$$|a_n| t^n \leqslant \sup\{|f(w)| + |a_0| : |w - z_0| = t\}$$

$$= M(t),$$

say.

Next, the function

$$g(z) = \frac{f(z_0 + z) - [a_0 + a_1 z + \cdots + a_n z^n]}{z^{n+1}}$$

is analytic in $C(0, r)$. Given any $z$, $0<|z|<r$, we choose any $t$ satisfying $|z|<t<r$: then by (8.4.1),

$$|g(z)| \leqslant [M(t) + |a_1 t| + \cdots + |a_n t^n|]/t^{n+1}$$
$$\leqslant (n+1)M(t)t^{-(n+1)}.$$

In terms of $f$, this becomes

$$|f(z_0+z) - [a_0 + a_1 z + \cdots + a_n z^n]| \leqslant M(t)[(n+1)(|z|/t)^{n+1}]$$

and the result follows as this upper bound tends to zero as $n \to \infty$ (see Example 2.3.1).

**Exercise 8.4**

1. Prove that if $f$ and $g$ are analytic in a domain $D$ and if $\text{Re}\,[f(z)] = \text{Re}\,[g(z)]$ throughout $D$, then $f = g$ on $D$.
   Find all functions $f$ which are analytic in $\mathbf{C}$ and for which
   $$\text{Re}\,[f(z)] = \sin(x)\cosh(y) + x^2 - y^2.$$
2. Let $f$ be analytic and not constant in a domain $D$. Prove that $\text{Re}\,[f]$ and $\text{Im}\,[f]$ cannot attain a local maximum or a local minimum in $D$.
3. (a) Let $f$ be analytic in a domain $D$ and let $K$ be a compact subset of $D$. Prove that there exists a point $w$ in $\partial K$ such that for all $z$ in $K$, $|f(z)| \leqslant |f(w)|$.
   (b) Let $f$ be analytic in a domain $D$, let $\bar{D}$ be compact and suppose that $f$ is continuous on $\bar{D}$. Prove that $|f|$ attains its maximum on $\bar{D}$ at a point of $\partial D$.
4. Let $f$ be analytic and not constant in $\mathbf{C}$ and define
   $$M(r) = \sup\{|f(z)| : |z| \leqslant r\},$$
   $$m(r) = \inf\{|f(z)| : |z| \leqslant r\}.$$
   Prove that (a) $M(r)$ is continuous, strictly increasing and tends to $+\infty$ as $r \to +\infty$; (b) $m(r)$ is continuous, decreasing (not necessarily strictly) and tends to zero as $r \to +\infty$.
5. Let $f, f^*, D, D^*$ be as in Exercise 8.3.2. Suppose, in addition, that $D$ meets the real axis in an interval $I$ and that $f$ is real-valued on $I$. Prove that $f = f^*$ on some open subset of $D \cup D^*$ which contains $I$.
   Give an example to show that $f$ need not equal $f^*$ on $D \cup D^*$.
6. (a) Let $f$ be analytic in the square
   $$S = \{x + iy : |x| < 1, |y| < 1\},$$
   continuous in $\bar{S}$, and let $|f|$ be bounded by $m_1, \ldots, m_4$ respectively on the four sides $\gamma_1, \ldots, \gamma_4$ of $S$. Prove that
   $$|f(0)| \leqslant |m_1 m_2 m_3 m_4|^{1/4}.$$
   [*Hint*: consider $f(z)f(iz)f(-z)f(-iz)$.]
   (b) Let $f$ be analytic in $C(0, 1)$, continuous on $\bar{C}(0, 1)$ and let $f(e^{i\theta}) = 0$ for $\theta_1 < \theta < \theta_2$. Prove that $f = 0$ on $\bar{C}(0, 1)$.

7. Extend Liouville's Theorem as follows.
   (a) Prove that if $f$ is analytic in $\mathbf{C}$ and satisfies $|f(z)| \leqslant M|z|^s$ ($M$ and $s$ are non-negative constants) for $|z| > r$, say, then $f$ is a polynomial of degree $d$ where $d \leqslant s$. [Liouville's Theorem is the case $s = 0$.]
   (b) Prove that if $f$ is analytic and not constant in $\mathbf{C}$, then for every disc $\mathbf{C}(w, \delta)$, $\delta > 0$, $f$ assumes values in $\mathbf{C}(w, \delta)$.
   [*Hint*: consider $(f(z) - w)^{-1}$. Liouville's Theorem says that $f$ assumes values in every set $\{z : |z| > r\}$.]
8. Let $f$ be analytic and satisfy $|f(z)| \leqslant M$ in $\mathbf{C}(0, r)$. Suppose also that $f(z_j) = 0$ for distinct points $z_1, \ldots, z_n$ in $\bar{\mathbf{C}}(0, \lambda r)$, $0 \leqslant \lambda < 1$. Prove that in $\mathbf{C}(0, r)$,

$$|f(z)| \leqslant Mr^{-n}(1 - \lambda)^{-n} |(z - z_1) \cdots (z - z_s)| .$$

   [Schwarz's Lemma is the case $n = 1, z_1 = 0$ and $\lambda = 0$.]
9. Let $f$ be analytic in $\mathbf{C}(0, r)$ and suppose that

$$f(z) = \sum_{m=0}^{\infty} a_m z^m, \qquad |z| < r.$$

   Define $\omega = \exp(i\pi/n)$ and $F$ by

$$2nF(z) = \sum_{k=0}^{2n-1} (-1)^k f(z\omega^k).$$

   Prove that $F(0) = 0$ and that

$$F(z) = a_n z^n + b_{n+1} z^{n+1} + \cdots .$$

   Deduce that if $0 < t < r$, then

$$|a_n| \leqslant t^{-n} \sup\{|f(z)| : |z| = t\}.$$

   [Proposition 8.4.1 is based on (8.4.2) which assumes that $f(0) = 0$: here we derive the same inequalities (Cauchy's Inequalities) but do not assume that $f(0) = 0$.]

## 8.5 THE ZEROS OF ANALYTIC FUNCTIONS

We continue with our programme of generalizing (in as far as is possible) the results of §8.1 on polynomials to analytic functions. In this section we focus our attention on the two results concerned with the zeros of polynomials, namely the Fundamental Theorem of Algebra and the Argument Principle.

We have already made considerable progress in understanding the nature of the zeros of analytic functions by proving Theorem 8.4.2. The following examples help to show both the strength and the limitation of this result.

*Example 8.5.1* Let $f$ be analytic in $\mathbf{C}$ and suppose, for example, that $f(z) = \sin z$ whenever $z = 1/n, n = 1, 2, \ldots$ . The function $g, g(z) = f(z) - \sin z$, has zeros at the points $1/n$ and, by continuity, $g(0) = 0$. It is evident that the point zero is not an isolated zero of $g$: thus by Theorem 8.4.2, $g(z) = 0$ and $f(z) = \sin z$ throughout $\mathbf{C}$.

Observe that this means that $f$ is determined throughout **C** by its values at the points $1/n$. For a general result of this nature, see Exercise 8.5.1.

*Example 8.5.2* In contrast to the preceding Example, let us consider the function $s(z) = \sin(\pi/z)$ (we shall see in Chapter 9 that $s$ is analytic in $\mathbf{C}^*$, $\mathbf{C}^* = \{z : z \neq 0\}$). In this case $s(1/n) = 0, n = 1, 2, \ldots,$ and yet $s$ is not identically zero in $\mathbf{C}^*$. The explanation of this is, of course, that each point $1/n$ is an isolated zero of $s$ whereas the point zero is not in $\mathbf{C}^*$. Theorem 8.4.2 relates only to zeros in $D$: in this Example $D = C^*$ and $s$ cannot even be defined at zero so as to be continuous there.

We now return to the general situation of a function $f$ analytic and not constant in a domain $D$. Each zero $\zeta$ of $f$ is an isolated zero and so

$$f(z) = (z - \zeta)^s[a_s + a_{s+1}(z - \zeta) + \cdots]$$

in some $\mathbf{C}(\zeta, \delta)$ where $a_s \neq 0$. Exactly as for polynomials, we say that $\zeta$ is a zero of *order* or *multiplicity* $s$ of $f$ and we shall always consider $\zeta$ as $s$ coincident zeros of $f$. Even so, $\zeta$ is still an isolated zero of $f$.

There is no reason, however, to give special consideration to the value zero. Quite generally, if $\zeta \in D$, we have

$$f(z) = f(\zeta) + (z - \zeta)^s[a_s + a_{s+1}(z - \zeta) + \cdots]$$

where $a_s \neq 0$. We then say that $f$ has *valency* $s$ at $\zeta$ and we use $v_f(\zeta)$ to denote the valency of $f$ at $\zeta$. We have already used the fact that if

$$\gamma(t) = \zeta + re^{it}, \qquad 0 \leqslant t \leqslant 2\pi,$$

and if $r$ is sufficiently small, then

$$n(f \circ \gamma, f(\zeta)) = s = v_f(\zeta)$$

and this gives an intrinsic definition of the valency in terms of the index. The reader should now compare this with (8.1.3).

We shall now crystallize the above remarks and come as near as we are able to an analogue of the Fundamental Theorem of Algebra for analytic functions.

*Theorem 8.5.1 The Principle of Isolated Zeros. Let $f$ be analytic and not constant in a domain D, let K be any compact subset of D and let w be any complex number. Then $f$ assumes the value w at only a finite number of points $z_1, \ldots, z_s$ in K and we may write*

$$f(z) = w + (z - z_1)^{v_1} \cdots (z - z_s)^{v_s} g(z), \tag{8.5.1}$$

*where $v_j = v_f(z_j), j = 1, \ldots, s$, and g is analytic in D and not zero in K.*

If $f$ does not assume the value $w$ in $K$, then (8.5.1) is to be interpreted as $s = 0$ and $g(z) = f(z) - w$.

*Example 8.5.3* Using Theorem 8.5.1, we find that for each positive integer $p$,

$$\sin \pi z = z(1 - z^2)\left(1 - \frac{z^2}{2^2}\right) \cdots \left(1 - \frac{z^2}{p^2}\right) g(z),$$

where $g$ is analytic in $C$ and not zero when $|z| < p + 1$.

*Proof of Theorem 8.5.1* By considering $f(z) - w$ instead of $f$, we may assume that $w = 0$. As $f$ is not constant, each zero of $f$ is isolated and so each $z$ in $D$ is the centre of some open disc which contains only a finite number of zeros of $f$. These discs form an open cover of the compact set $K$ and so a finite number of these discs cover $K$. Thus $f$ assumes the value zero at only a finite number of distinct points, say $z_1, \ldots, z_s$, in $K$.

We now define $g$ by

$$g(z) = f(z)(z - z_1)^{-v_1} \cdots (z - z_s)^{-v_s}, \qquad v_j = v_f(z_j)$$

this is certainly defined and analytic in $D - \{z_1, \ldots, z_s\}$. However in some $\mathbf{C}(z_1, \delta)$, for example, we have

$$f(z) = (z - z_1)^{v_1} \sum_{n=v_1}^{\infty} a_n (z - z_1)^{n - v_1}$$

where $a_{v_1} \neq 0$. Thus $f(z)(z - z_1)^{-v_1}$, and hence $g$ is analytic in $\mathbf{C}(z_1, \delta)$ provided, of course, that we define

$$g(z_1) = a_{v_1} / (z_1 - z_2)^{v_2} \cdots (z_1 - z_s)^{v_s}.$$

A similar argument holds for $z_2, \ldots, z_s$ and we conclude that $g$ is analytic in $D$. By definition, $g(z_j) \neq 0$: if $z \neq z_j$ and $z \in K$, then $f(z) \neq 0$ and so $g(z) \neq 0$, thus $g$ is not zero in $K$.

We shall now attempt to prove the Argument Principle for analytic functions. The difficulties which arise in the general case are quite formidable and at this stage we shall only be partially successful. Nevertheless, it is most instructive to see the exact nature of the difficulties.

We begin with a function $f$ analytic and not constant in a domain $D$ and any closed curve $\gamma$ in $D$ on which $f$ is not zero. The first difficulty that arises is that $f$ may have infinitely many zeros inside $\gamma$: indeed, the inside of $\gamma$ need not be entirely contained in $D$. It is natural to impose the condition that $\gamma$ and the points inside it lie in $D$: equivalently, if $z$ lies outside $D$, then $z$ lies outside $\gamma$. The precise formulation of this assumption involves the index and is

(A): *if* $z \notin D$, *then* $n(\gamma, z) = 0$.

The reader should recall that we have defined $z$ to be outside $\gamma$ if and only if $n(\gamma, z) = 0$ and the interpretation of (A) is that $\gamma$ does not wind about any point in $\mathbf{C} - D$. Observe that (A) is satisfied if $D$ is a disc (Example 7.2.1) or if $D = \mathbf{C}$. It is worth noting that (A) will play a crucial part in Cauchy's Theorem (Chapter 9).

We now denote by $K$ the set of points which are either inside or on $\gamma$. Then $K$ is a compact set (see § 7.3) and (A) is precisely the statement that $K \subset D$. Thus $K$ is a compact subset of $D$, Theorem 8.5.1 is applicable with $w = 0$ and (8.5.1) holds. We now substitute $\gamma(t)$ for $z$ in (8.5.1) and in the usual way we obtain

$$n(f \circ \gamma, 0) = \sum_{j=1}^{s} v_f(z_j) n(\gamma, z_j) + n(g \circ \gamma, 0). \tag{8.5.2}$$

It remains to prove that (A) (which does not involve $f$ or $g$) implies that $n(g \circ \gamma, 0) = 0$, for if this is true we obtain from (8.5.2) the natural generalization of (8.1.5). This implication is, however, a substantial exercise in plane topology and it does not depend on analyticity in any essential way. There are a variety of special cases available (of differing degrees of generality and corresponding difficulty) and we shall content ourselves now with the simplest case of all, namely when $D$ is a disc (or **C**). The general case will be proved directly (and without any intermediate cases) in the next chapter.

*The Argument Principle (for a disc)* *Let $D$ be an open disc or* **C**, *let $\gamma$ be any closed curve in $D$ and let $f$ be analytic in $D$ and non-zero on $\gamma$. Then*

$$n(f \circ \gamma, 0) = \sum_{j=1}^{s} v_f(z_j) n(\gamma, z_j),$$

*the sum being over the distinct zeros of $f$ inside $\gamma$.*

Observe that if $n(\gamma, z) = 0$ or $1$ for each $z$ not on $\gamma$, we may calculate that $n(f \circ \gamma, 0)$ is simply the number of zeros of $f$ inside $\gamma$. The reader should now look again at Example 7.2.2.

*Proof* As (A) holds when $D$ is a disc, the preceding discussion shows that $f$ is zero at only a finite set of points $z_1, \ldots, z_s$ inside $\gamma$ and further, that (8.5.2) holds. As we have already remarked, it only remains to prove that $n(g \circ \gamma, 0) = 0$.

The proof is based on the ideas of uniform convergence, however it is just as easy to be quite explicit. As $g$ is analytic in $D$ and as $D$ is a disc, say $D = \mathbf{C}(z_0, r)$, we may write

$$g(z) = \sum_{n=0}^{\infty} a_n (z - z_0)^n$$

where, by Theorem 8.4.3, this is valid throughout $D$. The set $K$ of points inside and on $\gamma$ is a compact subset of $D$ and $g$ is continuous and non-zero on $K$: thus we may write

$$\epsilon = \inf \{ |g(z)| : z \in K \} > 0.$$

Next, as $K$ is a compact subset of $D$, $K$ lies in some disc $\bar{\mathbf{C}}(z_0, t)$ where $t < r$ (Theorem 6.4.1) and as $t < r$, the series

$$\sum_{n=0}^{\infty} |a_n| t^n$$

converges. We conclude that for some positive integer $k$ and for all $z$ in $K$,

$$\begin{aligned} |g(z) - [a_0 + a_1(z - z_0) + \cdots + a_k(z - z_0)^k]| &\leqslant \sum_{n=k+1}^{\infty} |a_n| t^n \\ &< \epsilon \\ &\leqslant |g(z)|. \end{aligned} \tag{8.5.3}$$

We now substitute $\gamma(t)$ for $z$ and use I6; thus

$$n(g \circ \gamma, 0) = n(p \circ \gamma, 0),$$

where $p$ is the polynomial in (8.5.3). Note that if $p(z) = 0$ for some $z$ in $K$, then by (8.5.3), $|g(z)| < |g(z)|$ which is, of course, false. Thus $p$ has no zeros in $K$ and hence no zeros inside $\gamma$. The Argument Principle for polynomials now shows that $n(p \circ \gamma, 0) = 0$; thus $n(g \circ \gamma, 0) = 0$ and the proof is complete.

Although this version of the Argument Principle is only a very special case of the general result it is quite adequate for most purposes, in particular it is adequate for the analysis of local properties of analytic functions. We shall illustrate its use now with one example and one more result and we shall return to it later.

*Example 8.5.4* Let $S$ be the strip in **C** given by

$$S = \{x + iy : |x| < \pi/2, y > 0\}.$$

If $x + iy \in S$, then

$$\sin(x + iy) = \sin x \cosh y + i \cos x \sinh y$$

and so sin maps $S$ into the upper half plane $H$ given by $\operatorname{Im}[z] > 0$. We shall show that sin is actually a *one-to-one* map of $S$ *onto* $H$.

Let $w$ be in $H$ and choose a positive $t$ with $\sinh t > |w|$. Let $\gamma$ be the boundary of the rectangle given in Diagram 8.5.1 and let $\Gamma$ be the image curve (so $\Gamma(t) = \sin \gamma(t)$). The Argument Principle (with $D = \mathbf{C}$) applied to $\sin z - w$ shows directly that $n(\Gamma, w)$ is the number of solutions of $\sin z = w$ inside $\gamma$.

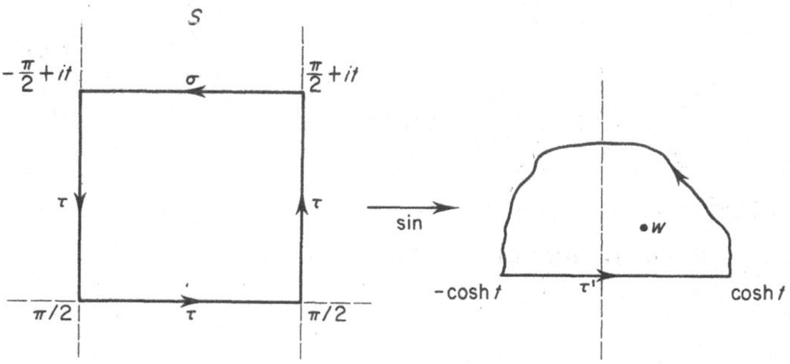


Diagram 8.5.1

Now let $\tau$ be the three segments of $\gamma$ joining $-\pi/2 + it$ to $\pi/2 + it$ and let $\sigma$ be the top segment of $\gamma$. The image $\tau'$ of $\tau$ is the segment of the real axis from $-\cosh t$ to $\cosh t$ and so $0 < n(\tau', w) < \frac{1}{2}$. If $z$ is on $\sigma$ we have $\operatorname{Im}[\sin z] \geqslant 0$ and

$$|\sin z|^2 = \sin^2 x + \sinh^2 t > |w|^2$$

and so $0 < n(\sigma', w) < 1$. Thus

$$n(\Gamma, w) = n(\sigma', w) + n(\tau', w) = 1$$

and the given result follows.

Our last result is a considerable strengthening of the Maximum Modulus Theorem.

*Theorem 8.5.2* *Let $f$ be analytic and not constant in a domain $D$. If $\Delta$ is any open subset of $D$, then $f(\Delta)$ is an open set.*

As $D$ is connected so is $f(D)$: if we take $\Delta = D$ we obtain the following result.

*Corollary* *If $f$ is analytic and not constant in a domain $D$, the $f(D)$ is a domain.*

It is clear that $|w|$ cannot attain a maximum value as $w$ varies in the *open* set $f(D)$. Each such $w$ is of the form $f(z)$, $z \in D$, and this is the Maximum Modulus Theorem.

*Proof of Theorem 8.5.2* We select any point $f(\zeta)$ in $f(\Delta)$ where $\zeta \in \Delta$ and choose a positive $r$ which is sufficiently small so that

(a) $\bar{C}(\zeta, r) \subset \Delta$ and
(b) $n(f \circ \gamma, f(\zeta)) = v_f(\zeta)$, where $\gamma(t) = \zeta + re^{it}$, $0 \leqslant t \leqslant 2\pi$.

Now let $\Delta_\zeta$ be the component of $\mathbf{C} - [f \circ \gamma]$ that contains $f(\zeta)$ and select any $w$ in $\Delta_\zeta$. By I7 and (b),

$$\begin{aligned} n(f \circ \gamma, w) &= n(f \circ \gamma, f(\zeta)) \\ &= v_f(\zeta), \end{aligned}$$

and so by the Argument Principle (applied to $f(z) - w$), $f(z) = w$ has exactly $v_f(\zeta)$ solutions inside $\gamma$. This, combined with (a), shows that $w \in f(\Delta)$ and so $\Delta_\zeta \subset f(\Delta)$.

We have now shown that each point $f(\zeta)$ of $f(\Delta)$ lies in some open subset $\Delta_\zeta$ of $f(\Delta)$ and this proves that $f(\Delta)$ is open. (The proof is illustrated in the case $v_f(\zeta) = 1$ in Diagram 8.5.2).

Now suppose that

$$f(z) = f(\zeta) + \sum_{n=1}^{\infty} a_n (z - \zeta)^n.$$

If $a_1 \neq 0$ then $v_f(\zeta) = 1$ and for each $w$ in $\Delta_\zeta$, $f(z) = w$ has exactly one solution inside $C(\zeta, r)$. Of course, points inside $C(\zeta, r)$ may map to points outside $\Delta_\zeta$, but by decreasing $r$ we can ignore this possibility.

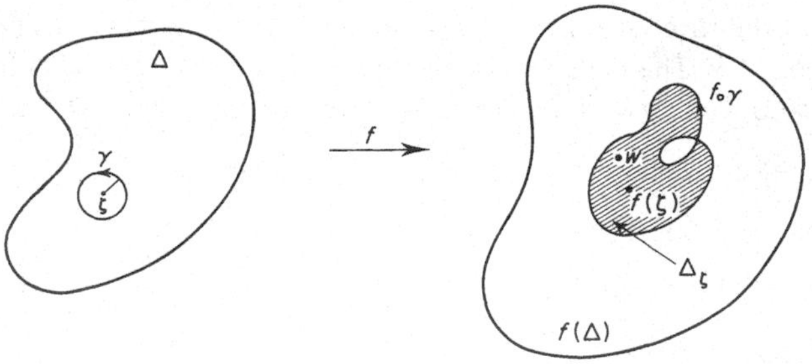


Diagram 8.5.2

To see this, let

$$\sigma(t) = \zeta + \delta e^{it}, \qquad 0 \leqslant t \leqslant 2\pi, 0 < \delta < r,$$

where $\delta$ is chosen so small that (by continuity)

$$f : \bar{C}(\zeta, \delta) \to \Delta_\zeta.$$

Now $f$ is 1–1 on $\bar{C}(\zeta, \delta)$, for if $w \in \Delta_\zeta$, then $f(z) = w$ has only one solution in $C(\zeta, r)$ and hence at most one solution in $\bar{C}(\zeta, \delta)$. For brevity, write $\Sigma = C(\zeta, \delta)$. Then $\Sigma$ and $f(\Sigma)$ are domains and $f : \Sigma \to f(\Sigma)$ is 1–1 and analytic.

We can use Theorem 6.1.1 (applied to $f^{-1}$) to show that $f^{-1} : f(\Sigma) \to \Sigma$ is also continuous. If $A$ is any open set then (writing $g$ for $f^{-1}$)

$$g^{-1}(A) = g^{-1}(A \cap \Sigma) = f(A \cap \Sigma),$$

and this too is an open set (Theorem 8.5.2).

We have now established the following corollary (which we shall discuss again in Chapter 10).

*Corollary* *Let $f$ be analytic in $D$ and suppose, as above, that $\zeta \in D$ and $a_1 \neq 0$. Then there is an open disc $\Sigma$ containing $\zeta$ such that*

(a) *$f(\Sigma)$ is a domain containing $f(\zeta)$ and*
(b) *$f : \bar{\Sigma} \to f(\bar{\Sigma})$ is a homeomorphism.*

We have achieved our aim of generalizing the results of §8.1 to analytic functions. In a certain sense, analytic functions are completely characterized by the purely topological conclusions of Theorem 8.4.2 and 8.5.2 [Whyburn, p. 103]. As the Argument Principle contains both of these results as well as a method for counting the zeros of analytic functions, we should clearly regard the Argument Principle as our most important result so far. In view of this we end Chapter 8 with a pictorial example of the Argument Principle. In Diagram 8.5.3, $f$ maps the closed disc bounded by $\gamma$ onto $f \circ \gamma$ and the shaded regions $\Delta_1$ and $\Delta_2$. For each $w$ in $\Delta_1$ there is exactly one solution of $f(z) = w$ inside $\gamma$; for $w$ in $\Delta_2$ there are exactly two solutions. Geometrically speaking (and as the diagram itself suggests) $\Delta_1$ is covered once and $\Delta_2$ twice by the image of the disc.

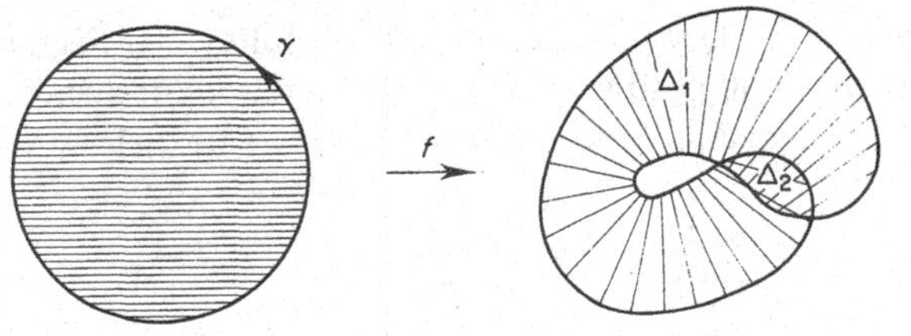


Diagram 8.5.3

**Exercise 8.5**

1. Let $f$ and $g$ be analytic in a domain $D$. Suppose that there are infinitely many distinct points $w_1, w_2, \ldots$ in a compact subset of $D$ such that $f(w_j) = g(w_j)$, $j = 1, 2, \ldots$ . Prove that $f = g$ on $D$.
2. Let $D$ be any domain and define

$$D_m = \{z \in D : \mathbf{C}(z, 1/m) \subset D\}, \qquad m = 1, 2, \ldots .$$

Prove that $D_m$ is a closed set: hence

$$D_m^* = D_m \cap \bar{\mathbf{C}}(0, m)$$

is a compact subset of $D$. Prove that $D_m^* \subset D_{m+1}^*$ and that $D = \cup_{m=1}^{\infty} D_m^*$. This shows that $D$ is the union of an increasing sequence of compact sets.

Now let $f$ be analytic and not constant in $D$. Show that $f$ has only a finite number of zeros in each $D_m^*$. Deduce that $f$ has only a countable number of zeros in $D$.
3. Prove Rouché's Theorem: let $f$ and $g$ be analytic in a domain $D$, let $\gamma$ be a curve in $D$ whose interior lies in $D$ and is such that $n(\gamma, z) = 0$ or $1$. Prove that if $|f| > |g|$ on $[\gamma]$ then $f$ and $f + g$ have the same number of zeros inside $\gamma$. [This is I6 together with the Argument Principle: see Exercise 8.5.5.]
4. Use Exercise 8.5.3 to show that there are $k$ solutions of $3z^k = \exp(z)$ in $\mathbf{C}(0, 1)$.
5. Generalize Rouché's Theorem (in Exercise 8.5.3) as follows. Let $D, \gamma, f, g$ be as in Exercise 8.5.3. Define

$$\Lambda = \{\lambda \in \mathbf{C} : f(z) + \lambda g(z) \neq 0 \quad \text{when} \quad z \in [\gamma]\}.$$

Prove that $\Lambda$ is open: thus $\Lambda = \cup_j \Lambda_j$ where the $\Lambda_j$ are domains.

Prove that the function $N$,

$$N(\lambda) = n((f + \lambda g) \circ \gamma, 0)$$

is integer valued and continuous on $\Lambda$: so *$N$ is constant on each $\Lambda_j$.*

To recapture Rouché's Theorem show that if $|f| > |g|$ on $[\gamma]$ then $\bar{\mathbf{C}}(0, 1) \subset \Lambda_j$, some $j$. Thus $N(0) = N(1)$: this is Rouché's Theorem.

Prove also that if $|f| + |g| > |f - g|$ on $[\gamma]$ then

$$\{\lambda : \lambda \geqslant 0\} \subset \Lambda_j$$

for some $j$ and so $f$ and $f + g$ have the same number of zeros inside $\gamma$.

6. Derive the second Corollary of Theorem 8.5.2 as follows. If $f$ is not 1–1 in some $C(\zeta, r)$, then for all positive $r$ there are distinct points $z$ and $w$ in $C(\zeta, r)$ with $f(z) = f(w)$. Using a power series representation for $f$ about $\zeta$ show that

$$|a_1| = \left| \frac{f(z) - f(w)}{(z - \zeta) - (w - \zeta)} - a_1 \right| \leqslant \sum_{n=2}^{\infty} |a_n| n r^{n-1}.$$

This is false for small $r$.

7. Prove Hurwitz's Theorem. Let $f, f_n$ ($n = 1, 2, \ldots$) be analytic and not constant in a domain $D$ and let $f_n \to f$ uniformly in $D$. Suppose that $\bar{C}(z, r) \subset D$ and let $\gamma(t) = z + re^{it}$, $0 \leqslant t \leqslant 2\pi$. Then for all sufficiently large $n$, $f$ and $f_n$ have the same number of zeros inside $\gamma$. [See the remarks at the end of §8.1.]

# Chapter 9

## 9.1 DERIVATIVES

The function $f$ is continuous at $w$ if, roughly speaking, $f(z)$ is approximately $f(w)$ when $z$ is near $w$. The differentiability of $f$ is concerned with finding a more accurate approximation of $f$ by using a polynomial of degree one rather than the constant $f(w)$.

Let $f$ be any function defined on $E$ and let $w$ be both in $E$ and a limit point of $E$.

*Definition 9.1.1* *The function $f$ is differentiable at $w$ if and only if there is a constant $\alpha$ and a function $\eta : E \to \mathbf{C}$ satisfying*

$$f(z) = f(w) + \alpha(z - w) + (z - w)\eta(z) \tag{9.1.1}$$

*on $E$, where $\eta(z) \to 0$ as $z \to w$ in $E$.*

The identity (9.1.1) holds for all $z$ in $E$. We shall not use a great variety of sets $E$ and the definition is given in this form only so as to include the cases of real or complex-valued functions of a real or complex-valued variable.

Note that $f$ is differentiable at $w$ if and only if

$$\frac{f(z) - f(w)}{z - w} \to \alpha$$

as $z \to w$ in $E$. Definition 9.1.1 is preferable to this formulation for several reasons: for example, it does not involve division and because of this it generalizes to maps between Euclidean spaces of any dimension. However, the alternative form does show that $\alpha$ is uniquely determined by $f$ and $w$. If $f$ is differentiable at each point in a set $E$ we say that $f$ is *differentiable in E.* The function which maps $w$ to the corresponding value $\alpha$ is the *derivative* of $f$: it is denoted by $f^{(1)}$ and so $f^{(1)}(w) = \alpha$.

Suppose now that every point in $E$ is a limit point of $E$ (so $E$ has no isolated points) and let $f$ be differentiable. If so, we denote its derivative by $f^{(2)}$. In general, if $f^{(n)}$ is defined on $E$ then $f^{(n+1)}$ denotes the derivative (when it exists) of $f^{(n)}$. Thus

$$(f^{(n)})^{(1)} = f^{(n+1)};$$

for convenience we write $f = f^{(0)}$.

The reader may recall that we have already used the notation $f^{(n)}$ for power series in §8.3: there is no ambiguity here and it will be proved shortly that the two apparently different definitions of $f^{(n)}$ do agree.

There are several immediate consequences of Definition 9.1.1 which are worth mentioning. First, if $f$ is differentiable at $w$ then it is continuous at $w$. Next, if $f$ and $g$ are differentiable at $w$ then so are the sum $f+g$ and the product $fg$ and

$$(f+g)^{(1)}(w) = f^{(1)}(w) + g^{(1)}(w), \tag{9.1.2}$$

$$(fg)^{(1)}(w) = f^{(1)}(w)g(w) + f(w)g^{(1)}(w). \tag{9.1.3}$$

Another direct consequence of Definition 9.1.1 is the

*Chain Rule: if $g : E_1 \to E_2$ is differentiable at $w$ and if $f : E_2 \to \mathbf{C}$ is differentiable at $g(w)$ then the composite function $f \circ g$ is differentiable at $w$ and*

$$\begin{aligned}(f \circ g)^{(1)}(w) &= f^{(1)}(g(w))g^{(1)}(w)\\ &= (f^{(1)} \circ g)(w)g^{(1)}(w)\end{aligned} \tag{9.1.4}$$

Note the two equivalent forms of this derivative. The proof is simply to replace $z$ and $w$ in (9.1.1) by $g(z)$ and $g(w)$ respectively and then to use the differentiability of $g$, say

$$g(z) = g(w) + (z-w)g^{(1)}(w) + (z-w)\eta^*(z),$$

to compute $f(g(z))$ in terms of $(z-w)$. This gives (9.1.4) provided that $\eta(g(z)) \to 0$ as $z \to w$ and this is so as $g$ is continuous at $w$. For an alternative proof, see Exercise 9.1.1.

*Example 9.1.1* Let $h(z) = z^{-1}, z \neq 0$: then $h^{(1)}(w) = -1/w^2$. This is a direct consequence of the identity

$$\frac{1}{z} = \frac{1}{w} - \frac{(z-w)}{w^2} + \frac{(z-w)^2}{zw^2},$$

which is already in the required form.

If $g(w) \neq 0$ we may regard $f/g$ as the product of $f$ and $h \circ g$ and we obtain the usual formula

$$(f/g)^{(1)}(w) = [f^{(1)}(w)g(w) - f(w)g^{(1)}(w)]/[g(w)]^2 \tag{9.1.5}$$

Of course, one should first show that $w$ is a limit point of the set $\{z \in E \colon g(z) \neq 0\}$ on which $f/g$ is defined.

*Example 9.1.2* We show that $\exp^{(1)} = \exp$. By Theorem 4.1.1,

$$\begin{aligned}\exp(z) &= \exp(z-w)\exp(w)\\ &= \exp(w) + (z-w)\exp(w) + (z-w)\exp(w)\,\eta(z),\end{aligned}$$

where

$$\eta(z) = \sum_{n=2}^{\infty} \frac{(z-w)^{n-1}}{n!}.$$

This series is continuous on $\mathbf{C}$ and is zero when $z$ is $w$: the result follows.

*Example 9.1.3* Let $D$ be any domain not containing zero and suppose that there exists a continuous choice $L(z)$ of Log $(z)$, $z \in D$. Then $L$ is differentiable in $D$ and $L^{(1)}(z) = 1/z$.

As exp is differentiable at $L(w)$,

$$\exp(L(z)) = \exp(L(w)) + [L(z) - L(w)]\,[\exp(L(w)) + \eta(L(z))]$$

where $\eta(\zeta) \to 0$ as $\zeta \to L(w)$. The continuity of $L$ shows that $\eta(L(z)) \to 0$ as $z \to w$ and as, for example, $\exp(L(z)) = z$ we obtain the required form

$$\begin{aligned} L(z) - L(w) &= \frac{(z-w)}{w + \eta(L(z))} \\ &= \frac{(z-w)}{w} - \frac{(z-w)\eta(L(z))}{w^2 + w\eta(L(z))}. \end{aligned}$$

Of course, this is valid only if $\eta(L(z)) \neq -w$, but this holds in some $\mathbf{C}(w, \delta)$, $\delta > 0$.

Finally, we must remove the ambiguity of the notation $f^{(n)}$.

*Theorem 9.1.1* *Let $f$ be the power series* (8.2.1) *with positive radius of convergence $\rho$. Then $f$ is differentiable in $\mathbf{C}(z_0, \rho)$ with derivative*

$$\sum_{n=1}^{\infty} na_n(z - z_0)^{n-1}. \tag{9.1.6}$$

This means that the function (9.1.6) is equal to both $f^{(1)}$ as defined in §8.3 and to the derivative $f^{(1)}$ as defined above. Repeated applications of Theorem 9.1.1 show that the $n$th derivative of $f$ exists and is indeed $f^{(n)}$ in $\mathbf{C}(z_0, \rho)$ and there is no longer any ambiguity.

*Proof* This can be based on uniform convergence but we prefer to use a direct and elementary estimate.

First we may assume that $z_0 = 0$. Then using (1.2.1) and (1.2.2) we have

$$\begin{aligned} |(z+h)^n - z^n - nhz^{n-1}| &= \left| \sum_{k=2}^{n} \binom{n}{k} h^k z^{n-k} \right| \\ &\leqslant \sum_{k=2}^{n} \binom{n}{k} |h|^k |z|^{n-k} \\ &= (|z| + |h|)^n - |z|^n - n|h||z|^{n-1} \\ &= |h| \left[ \sum_{k=0}^{n-1} (|z| + |h|)^k |z|^{n-1-k} \right] - n|h||z|^{n-1} \\ &\leqslant |h| \left[ \sum_{k=0}^{n-1} (|z| + |h|)^{n-1} \right] - n|h||z|^{n-1} \\ &\leqslant n|h|[(|z| + |h|)^{n-1} - |z|^{n-1}] \\ &\leqslant n(n-1)|h|^2 (|z| + |h|)^{n-2}. \end{aligned}$$

Now suppose that $|z| + 2\delta < \rho$ and $|h| < \delta$. Then using $g(z)$ to denote (9.1.6) with $z_0 = 0$,

$$|f(z+h)-f(z)-hg(z)| = \left| \sum_{n=2}^{\infty} a_n [(z+h)^n - z^n - nhz^{n-1}] \right|$$

$$\leqslant |h|^2 \sum_{n=2}^{\infty} n(n-1)|a_n|(|z|+|h|)^{n-2}$$

$$\leqslant 2|h|^2 \delta^{-2} \sum_{n=2}^{\infty} |a_n|(|z|+2\delta)^n.$$

This shows that

$$\lim_{h\to 0} \frac{f(z+h)-f(z)}{h} = g(z)$$

as required.

An immediate consequence of Theorem 9.1.1 is that if $f$ is analytic in $D$, then it is differentiable in $D$; in fact, repeated applications of Theorem 9.1.1 shows that all derivatives $f^{(n)}$ exist in $D$. It is one of the truly remarkable results of complex analysis that the converse is also true: if $f$ is differentiable in $D$, then it is analytic in $D$ and so all derivatives of $f$ necessarily exist. We state this as a theorem now although the proof (which depends on integration) is not given until §9.6: a proof free of integration is sketched in the Appendix.

*Theorem 9.1.2* *A function $f$ is analytic in a domain $D$ if and only if it is differentiable in $D$.*

We repeat, if $f$ is differentiable in $D$, then all derivatives $f^{(n)}$ exist in $D$ and even more, $f$ is analytic in $D$. We now see that *the results of Chapter 8 are applicable to functions which are merely differentiable in $D$.* It hardly needs saying that this exhibits a striking difference between the theory of differentiation in real analysis and that in complex analysis.

Despite the fact that we are delaying the proof of Theorem 9.1.2 let us give one particular application now. We suppose that $f$ is differentiable in $\mathbf{C}^*(w, r)$, $r > 0$, and that

$$\lim_{z\to w} (z-w)f(z) = 0. \tag{9.1.7}$$

This holds if, for example, $f$ is bounded in $\mathbf{C}^*(w, r)$. The function

$$F(z) = \begin{cases} (z-w)^2 f(z) & \text{if } z \neq w; \\ 0 & \text{if } z = w \end{cases}$$

is easily seen to be differentiable in $\mathbf{C}(w, r)$ (one need only check that $F^{(1)}(w)$ exists) and $F^{(1)}(w) = 0$. By Theorem 9.1.2, $F$ is analytic in $\mathbf{C}(w, r)$ and so we can write

$$F(z) = \sum_{n=0}^{\infty} a_n(z - w)^n, \qquad z \in \mathbf{C}(w, r).$$

As $F(w) = F^{(1)}(w) = 0$, we have $a_0 = a_1 = 0$ and so for $0 < |z - w| < r$,

$$f(z) = \frac{F(z)}{(z - w)^2} = \sum_{n=0}^{\infty} a_{n+2}(z - w)^n.$$

If we now *define* $f$ to be $a_2$ at $w$ we see that $f$ is actually a power series in $\mathbf{C}(w, r)$ and hence is analytic there.

We say that $w$ is a *removable singularity* of $f$ if $f$ is differentiable in $\mathbf{C}^*(w, r)$ and satisfies (9.1.7): *then with a suitable definition of $f(w)$, $f$ is analytic in* $\mathbf{C}(w, r)$.

Of course, (9.1.7) holds if $f$ is continuous in $\mathbf{C}(w, r)$ or even bounded there. Finally, if $f$ is differentiable in $\mathbf{C}(w, r)$, then $w$ is a removable singularity of

$$g(z) = \frac{f(z) - f(w)}{z - w} \tag{9.1.8}$$

as this is bounded near $w$.

**Exercise 9.1**

1. Prove the Chain Rule (9.1.4) as follows. Suppose first that there exists a positive $r$ such that

   $$g(z) \neq g(w) \qquad \text{if } z \in E_1 \cap \mathbf{C}^*(w, r).$$

   By considering

   $$\frac{f(g(z)) - f(g(w))}{g(z) - g(w)} \frac{g(z) - g(w)}{z - w}$$

   as $z \to w$, derive (9.1.4).

   If the above condition on $g$ fails, then there is a sequence $z_1, z_2, \ldots$ in $E_1 - \{w\}$ with $g(z_n) = g(w)$ and $z_n \to w$. Show that in this case (9.1.4) holds with both sides equal to zero.
2. Let $f : \mathbf{R} \to \mathbf{R}$ be defined by

   $$f(x) = x(x + |x|).$$

   Show that $f$ is differentiable in $\mathbf{R}$ (even though $|x|$ is not). Where does $f^{(2)}$ exist?
3. Prove that

   $$\sin^{(1)} = \cos, \qquad \sinh^{(1)} = \cosh.$$
4. Generalize Example 9.1.3 to prove the following result. If $f$ is 1–1 and analytic in a domain $D$, then

   (a) $f^{-1} : f(D) \to D$ is continuous;
   (b) $f^{(1)}(z) \neq 0, z \in D$;

(c) $f^{-1}$ is differentiable in $f(D)$ with

$$(f^{-1})^{(1)}(w) = \frac{1}{f^{(1)}(f^{-1}(w))}.$$

5. Let $f$ be differentiable in some disc $\mathbf{C}(0, r)$, $r > 0$, and suppose that $f(0) = 1$ and

$$f(z) = f^{(1)}(2z), \qquad |z| < \tfrac{1}{2}r.$$

   Assuming the validity of Theorem 9.1.2, find $f$ explicitly.
   Where is $f$ analytic?
6. Let $f$ be analytic in $\mathbf{C}$. Prove that $f$ is a non-constant polynomial if and only if $f(z) \to \infty$ as $z \to \infty$.
   [*Hint:* consider $g(z) = [f(1/z)]^{-1}$ on some $\mathbf{C}^*(0, \delta)$ and use the idea of a removable singularity.]

## 9.2 LINE INTEGRALS

We shall assume that the reader is familiar with the definition and basic properties of the Riemann integral

$$\int_a^b f, \qquad \int_a^b f(x)\, dx \tag{9.2.1}$$

(we use both of these notations) of a real-valued continuous function $f$ defined on a real interval $[a, b]$. The extension to complex-valued continuous functions $f: [a, b] \to \mathbf{C}$ is achieved by the definition

$$\int_a^b f = \int_a^b \mathrm{Re}\,[f] + i \int_a^b \mathrm{Im}\,[f]. \tag{9.2.2}$$

This implies, for example, that

$$\mathrm{Re}\left[\int_a^b f\right] = \int_a^b \mathrm{Re}\,[f].$$

We now define the integral of a function $f$ on a curve $\gamma$, and to avoid unnecessary analytic difficulties we assume at first that $f$ is continuous and $\gamma$ is continuously differentiable. Geometrically, this means that $\gamma$ has a continuously varying tangent.

*Definition 9.2.1* *A curve $\gamma : [a, b] \to \mathbf{C}$ is said to be a path if $\gamma^{(1)}$ exists and is continuous on $[a, b]$.*

*If $\gamma$ is a path and if $f$ is continuous on $[\gamma]$ we define the integral of $f$ on $\gamma$ by*

$$\int_\gamma f = \int_\gamma f(z)\, dz = \int_a^b f(\gamma(t))\gamma^{(1)}(t)\, dt. \tag{9.2.3}$$

By assumption, $f(\gamma(t))\gamma^{(1)}(t)$ is continuous on $[a, b]$ and the right-hand side of

(9.2.3) is defined in terms of (9.2.2). This definition is, of course, motivated by the familiar rule for the change of variable ($z = \gamma(t)$) in an integral. We adopt it here in preference to other definitions because it is readily available, easy to use and entirely adequate for our purposes. In this section $\gamma$ will always be used to denote a path defined on $[a, b]$.

There is no need to seek general conditions under which we may change the variable of integration in (9.2.3). Although this is of interest in its own right we shall only need the following trivial case. If $h(x) = \alpha x + \beta$, $\alpha \neq 0$, maps $[c, d]$ onto $[a, b]$, then for every continuous $f$,

$$\int_{h(c)}^{h(d)} f(\gamma(t))\gamma^{(1)}(t)\,\mathrm{d}t = \int_c^d f(\gamma(\alpha x + \beta))\gamma^{(1)}(\alpha x + \beta)\alpha\,\mathrm{d}x$$

or, more briefly,

$$\epsilon \int_\gamma f = \int_{\gamma \circ h} f,$$

where $\epsilon = 1$ or $-1$ as $h$ is increasing or decreasing respectively. The significance of this is that we may regard $\gamma$ as being defined on a pre-assigned interval $[c, d]$ rather than the given interval $[a, b]$.

An important special case is when $h(x) = a + b - x$, for then $\gamma \circ h = \gamma^-$, $\epsilon = -1$ and

$$\int_{\gamma^-} f = -\int_\gamma f.$$

If $f$ and $g$ are continuous on $[\gamma]$ and if $\lambda$ and $\mu$ are any complex numbers, then

$$\int_\gamma [\lambda f + \mu g] = \lambda \int_\gamma f + \mu \int_\gamma g:$$

this is true for real $f$, $g$, $\lambda$ and $\mu$ in the case (9.2.1) and it extends without difficulty to (9.2.3).

Another basic result which is easily generalized is the Fundamental Theorem of Calculus: if $f: [a, b] \to \mathbf{R}$ has a continuous derivative, then

$$\int_a^b f^{(1)} = f(b) - f(a). \qquad (9.2.4)$$

*Theorem 9.2.1* *Let $D$ be any domain and suppose that $f: D \to \mathbf{C}$ has a continuous derivative in $D$. Then for any path $\gamma$ in $D$,*

$$\int_\gamma f^{(1)} = f(\gamma(b)) - f(\gamma(a)). \qquad (9.2.5)$$

*Proof* Observe first that if $g: [a, b] \to \mathbf{C}$ is differentiable on the *real* interval $[a, b]$ and if $u = \operatorname{Re}[g]$, then $\operatorname{Re}[g^{(1)}(t)] = u^{(1)}(t)$. The proof is easy but we must appreciate that the two terms are defined in different ways and we must make essential use of the fact that $t$ is real.

We now write $g = f \circ \gamma$, $u = \text{Re}\,[g]$ and apply (9.2.4) to $u$: thus

$$\text{Re}\,[f(\gamma(b))] - \text{Re}\,[f(\gamma(a))] = \int_a^b \text{Re}\,[(f \circ \gamma)^{(1)}].$$

A similar expression holds for $\text{Im}\,[f \circ \gamma]$ and the result follows.

Theorem 9.2.1 has immediate applications. If $\gamma$ is closed, then

$$\int_\gamma f^{(1)} = 0. \tag{9.2.6}$$

This holds, for example, for all closed paths when $f$ is a polynomial. By virtue of Theorem 9.1.1, it also holds for all closed paths inside the disc of convergence of a power series $f$.

As specific examples of (9.2.5), let $\gamma$ be *any* path joining $z_1$ to $z_2$. Then

$$\int_\gamma z^k \, dz = (z_2^{k+1} - z_1^{k+1})/(k+1) \qquad (k \text{ an integer}, \neq -1)$$

while

$$\int_\gamma \cos z \, dz = \sin z_2 - \sin z_1.$$

Finally, Theorem 9.2.1 shows that if $f^{(1)} = 0$ throughout $D$ then $f$ is constant on each path in $D$. In particular, $f$ is then constant on each segment $[z_1, z_2]$ in $D$. This combines with the proof of Theorem 7.1.1 to yield a proof of the following result.

*Corollary* *If $f^{(1)} = 0$ throughout $D$, then $f$ is constant in $D$.*

**Exercise 9.2**

1. Prove (under suitable assumptions which should be stated) that

$$\int_{f \circ \gamma} g(w) \, dw = \int_\gamma (g \circ f)(z) f^{(1)}(z) \, dz.$$

2. Let $D = \{x + iy : x > 0\}$ and for $z = re^{i\theta}$ in $D$ let $f(z) = \sqrt{r}\, e^{i\theta/2}$, where $|\theta| < \pi/2$. Prove that $f$ is differentiable in $D$ and that

$$f^{(1)}(z) = \tfrac{1}{2}(\sqrt{r}\, e^{i\theta/2})^{-1}.$$

[If $f$ is differentiable, this must be the derivative as we can use $(f(z))^2 = z$ and differentiate.]

3. Let $L(z)$ be the continuous choice of $\text{Log}\,(1 - z)$ in $\mathbf{C}(0, 1)$ with $L(0) = 0$.

Apply the Corollary to Theorem 9.2.1 to the function

$$L(z) + \sum_{n=1}^{\infty} \frac{z^n}{n}$$

and deduce that

$$L(z) = -\sum_{n=1}^{\infty} \frac{z^n}{n}, \qquad |z| < 1.$$

Show that if $L_1(z)$ is any continuous choice of Log $(1 - z)$ in $\mathbf{C}(0, 1)$, then

$$L_1(z) = 2m\pi - \sum_{n=1}^{\infty} \frac{z^n}{n}$$

for some integer $m$.

## 9.3 INEQUALITIES

We shall frequently be confronted with integrals which, it seems, cannot be evaluated directly. In many cases it is enough to estimate the value of the integral and we now tackle the problem of obtaining such estimates.

We shall recall that if $f: [a, b] \to \mathbf{R}$ is continuous, then $f \leqslant |f|$, $-f \leqslant |f|$ and so

$$\begin{aligned} \left| \int_a^b f \right| &= \max\left\{ \int_a^b f, -\int_a^b f \right\} \\ &= \max\left\{ \int_a^b f, \int_a^b (-f) \right\} \\ &\leqslant \int_a^b |f|. \end{aligned} \tag{9.3.1}$$

The inequality remains true, although this proof is no longer valid, when $f$ is complex valued. This section contains a proof of this inequality for complex $f$ and a brief discussion of its consequences.

*Theorem 9.3.1* *Let $f: [a, b] \to \mathbf{C}$ be continuous. Then*

$$\left| \int_a^b f \right| \leqslant \int_a^b |f|.$$

*Proof* The result is obviously true if the integral of $f$ is zero. If not we define $\lambda$ by

$$\lambda \int_a^b f = \left| \int_a^b f \right|$$

and observe that $|\lambda| = 1$.

The required inequality now follows for using (9.3.1) (with $f$ replaced by Re $[\lambda f]$),

$$\left|\int_a^b f\right| = \text{Re}\left[\left|\int_a^b f\right|\right]$$

$$= \text{Re}\left[\lambda\int_a^b f\right]$$

$$= \int_a^b \text{Re}\,[\lambda f]$$

$$\leqslant \int_a^b |\lambda f|$$

$$= \int_a^b |f|.$$

The applications now follow. First, we define the *length* $L(\gamma)$ of a path $\gamma : [a, b] \to \mathbf{C}$ by

$$L(\gamma) = \int_a^b |\gamma^{(1)}|$$

(this is finite as $\gamma^{(1)}$ is continuous). For example, if $\gamma(t) = r\,e^{it}$, $0 \leqslant t \leqslant 2\pi$, then $L(\gamma) = 2\pi r$: if $\sigma(t) = \alpha t + \beta$, $a \leqslant t \leqslant b$, then $L(\sigma) = |\alpha|(b - a)$.

Theorems 9.2.1 and 9.3.1 imply that for any path $\gamma : [a, b] \to \mathbf{C}$ joining $z_1$ to $z_2$,

$$|z_2 - z_1| = \left|\int_a^b \gamma^{(1)}(t)\,\mathrm{d}t\right|$$

$$\leqslant L(\gamma).$$

As equality holds for some choices of $\gamma$, we find that the Euclidean distance between $z_1$ and $z_2$ is indeed the infimum of the lengths of all paths joining $z_1$ to $z_2$.

We can now give a simple (but usually effective) estimate of a general integral. For any path $\gamma$ and any continuous $f$,

$$\left|\int_\gamma f\right| = \left|\int_a^b f(\gamma(t))\gamma^{(1)}(t)\,\mathrm{d}t\right|$$

$$\leqslant \int_a^b |f(\gamma(t))|\,|\gamma^{(1)}(t)|\,\mathrm{d}t$$

$$\leqslant L(\gamma)\sup\{|f(z)| : z \in [\gamma]\}. \qquad (9.3.2)$$

This estimate will be used repeatedly.

An important yet simple application of this estimate is to a sequence of functions $f_n$ which converges uniformly to $f$ on $[\gamma]$. If the functions $f_n$ are continuous on $[\gamma]$, then so is $f$ (Theorem 6.6.1) and so

$$\left| \int_\gamma (f_n - f) \right| \leqslant L(\gamma) \sup \{ |f_n(z) - f(z)| : z \in [\gamma] \},$$

where the upper bound tends to zero as $n \to \infty$ as $L(\gamma)$ is finite. We state this result as a theorem.

*Theorem 9.3.2* *Let $\gamma$ be any path and suppose that the sequence of continuous functions $f_n$ converges uniformly to $f$ on $[\gamma]$. Then*

$$\lim_{n\to\infty} \int_\gamma f_n = \int_\gamma f.$$

In terms of uniformly convergent series, we write $f_n = g_1 + \cdots + g_n$ and this becomes

$$\sum_{n=0}^{\infty} \int_\gamma g_n = \int_\gamma \left( \sum_{n=0}^{\infty} g_n \right).$$

Theorem 9.3.2 provides us with a large class of examples of analytic functions.

*Corollary* *Let $\gamma$ be any path and let $g$ be continuous on $\gamma$. Then*

$$G(z) = \int_\gamma \frac{g(w)}{w - z}\, dw$$

*is analytic in each component of* $\mathbf{C} - [\gamma]$.

*Proof* If $z_0 \notin [\gamma]$ we choose a positive $r$ such that $C(z_0, 3r)$ does not meet $\gamma$. If $|z - z_0| < r$ and if $w \in [\gamma]$, then $|(z - z_0)/(w - z_0)| < \frac{1}{2}$ and so the series

$$\frac{g(w)}{w - z} = \frac{g(w)}{w - z_0} \sum_{n=0}^{\infty} \left( \frac{z - z_0}{w - z_0} \right)^n$$

viewed as a function of $w$ (with fixed $z$ and $z_0$) converges uniformly on $[\gamma]$. Thus

$$G(z) = \int_\gamma \frac{g(w)}{w - z_0} \sum_{n=0}^{\infty} \left( \frac{z - z_0}{w - z_0} \right)^n dw$$

$$= \sum_{n=0}^{\infty} \left[ \int_\gamma \frac{g(w)}{(w - z_0)^{n+1}}\, dw \right] (z - z_0)^n.$$

This shows that $G$ may be expressed as a power series with centre of convergence $z_0$ and this implies that $G$ is analytic in each component of $\mathbf{C} - [\gamma]$.

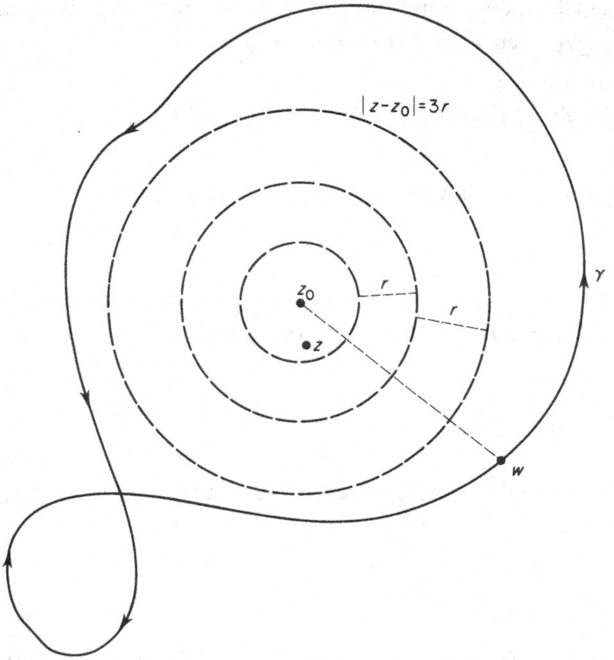


Diagram 9.3.1

**Exercise 9.3**

1. Prove that if $f$ is a power series with disc of convergence $\mathbf{C}(0, \rho)$, $\rho > 0$, and if $\gamma$ is any closed path in $\mathbf{C}(0, \rho)$, then

$$\int_\gamma f = 0.$$

2. Let $\gamma(t) = e^{it}$, $0 \leqslant t \leqslant 2\pi$: evaluate

$$\int_\gamma z^{-1}\,dz.$$

3. Let $G$ be as in the Corollary to Theorem 9.3.2. Prove that for each positive integer $m$,

$$G^{(m)}(z) = m! \int_\gamma \frac{g(w)}{(w-z)^{m+1}}\,dw.$$

4. Let $\gamma$ be any closed path and define $f$ on $\mathbf{C} - [\gamma]$ by

$$f(z) = \frac{1}{2\pi i}\int_\gamma \frac{dw}{w-z}.$$

Use Exercise 9.3.3 to show that $f^{(1)}(z) = 0$ for all $z$ in $\mathbf{C} - [\gamma]$ and deduce that $f$ is constant on each component of the open set $\mathbf{C} - [\gamma]$.

Use Exercise 9.3.2 to show that $f$ need not be constant on $\mathbf{C} - [\gamma]$.

5. As in Example 8.1.1, let $p(z) = z^4 + z + 1$. Show that if $\zeta$ is the zero of $p$ in the first quadrant and if $z$ is in this quadrant and satisfies $|z| < 1.23$, then

$$|p(z)| \leqslant 9|z - \zeta|.$$

[Given $z$, this excludes $\zeta$ from the disc $\mathbf{C}(z, |p(z)|/9)$.]

6. Let $f(z) = \Sigma_{n=0}^{\infty} a_n z^n$ have disc of convergence $\mathbf{C}(0, \rho)$, $\rho > 0$. Prove that

$$\int_0^{2\pi} |f(r e^{i\theta})|^2 \, d\theta = 2\pi \sum_{n=0}^{\infty} |a_n|^2 r^{2n}.$$

[Use the uniform convergence of the partial sums of $f$: this avoids the use of double series.]

Deduce that if $|f(z)| \leqslant |f(0)|$ on some $\mathbf{C}(0, \delta)$, then $f$ is constant.

## 9.4 CHAINS AND CYCLES

It is too restrictive to only consider integration on paths: for example we shall want to integrate around the boundary of a square and this is not a path. Fortunately, we need only consider integration over a finite collection of paths. This generalization is simple enough but because we wish to emphasize the logical structure of the discussion we prefer a formal definition.

*Definition 9.4.1* *A chain is a finite sequence* $(\gamma_1, \ldots, \gamma_s)$ *of paths. The chain* $(\gamma_1, \ldots, \gamma_s)$ *is said to be closed if the sequence of initial points* $(z_1, \ldots, z_s)$ *is a permutation of the sequence of final points* $(w_1, \ldots, w_s)$. *A closed chain will be called a cycle.*

Of course, a single path $\gamma$ is itself a chain and, as a chain, it is closed if and only if $\gamma$ is a closed path. As another example, consider the curves $\gamma$ and $\gamma_j$ defined in Example 7.2.3. The curve $\gamma$ is not a path, the chain $(\gamma_0, \ldots, \gamma_3)$ is a cycle and we distinguish between the curve $\gamma$ and the chain $(\gamma_0, \ldots, \gamma_3)$.

From a geometric viewpoint a cycle $\Gamma$, $\Gamma = (\gamma_1, \ldots, \gamma_s)$, may be regarded as a collection of closed curves. The curve $\gamma_1$ starts at $z_1$ and ends at $w_1$. If $z_1 = w_1$, then $\gamma_1$ is closed. If not, then $w_1 = z_j$, say, and we may relabel $\gamma_2, \ldots, \gamma_s$ so that $w_1 = z_2$. Thus $\gamma_2$ starts at $w_1$ and ends at $w_2$. This process can be continued and eventually we find some $\gamma_k$ which ends at $z_1$. In this sense we may regard $(\gamma_1, \ldots, \gamma_k)$ as a closed curve. The same argument applies to $(\gamma_{k+1}, \ldots, \gamma_s)$ and we can view $\Gamma$ as a collection of closed curves: see Diagram 9.4.1.

We can be more precise (but perhaps less clear). If $\gamma_j$ is defined on $[a_j, b_j]$ we can find increasing linear functions $h_j$ which map $[j, j+1]$ onto $[a_j, b_j]$ and we define $\gamma_j^* = \gamma_j \circ h_j$. Thus $\gamma : [1, k+1] \to \mathbf{C}$ defined by

$$\gamma(t) = \gamma_j^*(t), \qquad j \leqslant t \leqslant j+1, \quad j = 1, \ldots, k$$

is a curve and $\gamma$ is a realization of $(\gamma_1, \ldots, \gamma_k)$ as a single curve.

A cycle consisting (in the above sense) of a single (often closed) curve is often

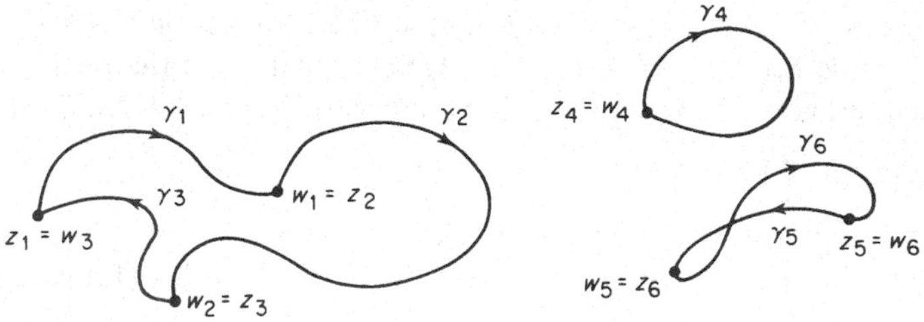


Diagram 9.4.1

called a *contour*. We repeatedly need to consider cycles which are not contours, however, and there seems little point in retaining the additional terminology.

The definition of the integral of $f$ over a chain is the natural one. If $\Gamma$ is the chain $(\gamma_1, \ldots, \gamma_s)$ we write

$$[\Gamma] = [\gamma_1] \cup \cdots \cup [\gamma_s],$$

$$n(\Gamma, z) = n(\gamma_1, z) + \cdots + n(\gamma_s, z)$$

and, provided only that $f$ is continuous on $[\Gamma]$, we define

$$\int_{(\gamma_1, \ldots, \gamma_s)} f = \sum_{j=1}^{s} \int_{\gamma_j} f.$$

This is not a self-evident fact: it is a definition.

In contrast to this, if $\gamma : [a, b] \to \mathbf{C}$ is a path and if

$$a = t_0 < t_1 < \cdots < t_n = b,$$

then the restriction $\gamma_j$ of $\gamma$ to $[t_j, t_{j+1}]$ is a path and

$$\int_\gamma f = \sum_{j=0}^{n-1} \int_{\gamma_j} f.$$

This is not a definition: it is a straightforward result in integration theory and does not depend in any way on the idea of a chain.

It is convenient to have both notations

$$\int_\Gamma f, \quad \int_{(\gamma_1, \ldots, \gamma_s)} f$$

available for the integral of $f$ over the chain $\Gamma$, $\Gamma = (\gamma_1, \ldots, \gamma_s)$.

Theorem 9.2.1 and its Corollary extend to chains and cycles: if $f : D \to \mathbf{C}$ has a continuous derivative in $D$, then

$$\int_\Gamma f^{(1)} = 0 \tag{9.4.1}$$

for every cycle $\Gamma$ in $D$. Indeed, let $\Gamma = (\gamma_1, \ldots, \gamma_s)$, where $\gamma_j$ has initial point $z_j$ and final point $w_j$. Then

$$\begin{aligned}\int_\Gamma f^{(1)} &= \sum_{j=1}^{s} \int_{\gamma_j} f^{(1)}\\ &= \sum_{j=1}^{s} [f(w_j) - f(z_j)]\\ &= \sum_{j=1}^{s} f(w_j) - \sum_{j=1}^{s} f(z_j)\\ &= 0\end{aligned}$$

as $(w_1, \ldots, w_s)$ is a permutation of $(z_1, \ldots, z_s)$.

The special case of a chain which consists only of linear segments deserves special mention. Let $z_0, z_1$ be any distinct complex numbers and let $\sigma$ be the curve

$$\sigma(t) = z_0 + t(z_1 - z_0), \qquad 0 \leqslant t \leqslant 1.$$

If $f$ is continuous on $[z_0, z_1]$ $(=[\sigma])$ we write

$$\int_{[z_0, z_1]} f = \int_\sigma f$$

in the sense that this *defines* the first of these integrals. Obviously,

$$\int_{[z_0, z_1]} f = -\int_{[z_1, z_0]} f. \tag{9.4.2}$$

More generally, we consider any sequence $z_0, z_1, \ldots, z_n$ of complex numbers and let $\gamma$ be the polygonal curve defined on $[0, n]$ by

$$\gamma(t) = \gamma_j(t), \qquad j \leqslant t \leqslant j+1, \quad j = 0, 1, \ldots, n-1,$$

where

$$\gamma_j(t) = z_j + (t-j)(z_{j+1} - z_j), \qquad j \leqslant t \leqslant j+1.$$

We naturally regard the polygonal curve as a chain and we write

$$\int_{[z_0, z_1, \ldots, z_n]} f = \sum_{j=0}^{n-1} \int_{\gamma_j} f.$$

This gives the useful formula

$$\int_{[z_0, z_1, \ldots, z_n]} f = \int_{[z_0, z_1]} f + \cdots + \int_{[z_{n-1}, z_n]} f. \tag{9.4.3}$$

An obvious application is suggested by Diagram 9.4.2. As, for example,

$$\int_{[z_2, z_3]} f = \int_{[z_2, w_1]} f + \int_{[w_1, z_3]} f$$

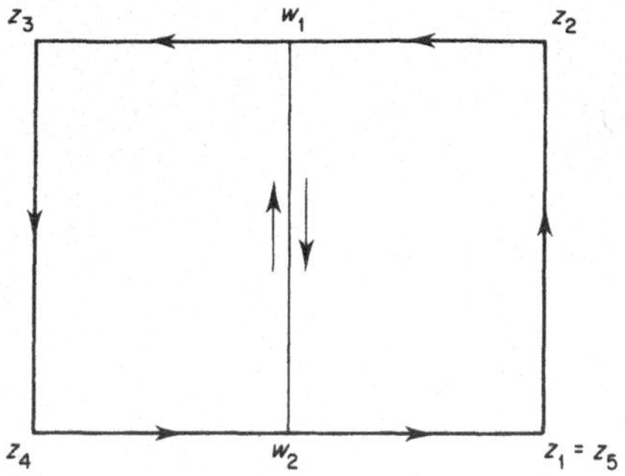


Diagram 9.4.2

and as (9.4.3) holds, we have

$$\int_{[z_1,\ldots,z_5]} f = \int_{[z_1,z_2,w_1,w_2,z_5]} f + \int_{[w_1,z_3,z_4,w_2,w_1]} f.$$

In general, when we speak of a rectangle $Q$ we shall only mean a set of the form

$$\{x+iy\colon a \leqslant x \leqslant b, c \leqslant y \leqslant d\}. \tag{9.4.4}$$

The 'perimeter' of $Q$ is of course the boundary $\partial Q$ as described in §6.1. It is obviously convenient to use $\partial Q$ also to denote the cycle consisting of the four sides of $Q$. We shall do this and shall always assume that $\partial Q$ is oriented so that $n(\partial Q, w) = 1$ if $w$ is inside $\partial Q$. Henceforth when we use such expressions as, for example,

$$\int_{\partial Q} f$$

we refer only to this unique parametrization and orientation of $\partial Q$. These ideas can be made precise (see Exercise 9.4.1).

**Exercise 9.4**

1. Let $Q$ be the set given by (9.4.4). Prove that $\partial Q$ (as defined in §6.1) is the set

$$\{x+iy \in Q : (x-a)(x-b)(y-c)(y-d) = 0\}.$$

   Show that the four sides of $Q$ can be represented as a cycle, which we now denote by $\partial Q$, such that $n(\partial Q, z)$ is 0 or 1.
2. Prove (9.4.2) and (9.4.3).

## 9.5 EVALUATION OF INTEGRALS

One of the standard techniques for evaluating real integrals is to use the partial fraction expansion of rational functions. As every complex polynomial factorizes completely, the complex partial fraction expansion of rational functions is much more satisfactory than in the real case. Moreover, the same techniques are applicable to functions other than rational functions and this gives a powerful, though limited method of integration. We shall now explore these ideas further, although at present we restrict ourselves to situations which are free of topological difficulties.

We begin by considering the partial-fraction expansion of a rational function $r$ where $r = p/q, p$ and $q$ are polynomials and $q$ is not identically zero. If $q$ has no zeros, then $r$ is a polynomial.

Otherwise, $q$ has, say, exactly $k$ distinct zeros $z_1, \ldots, z_k$ and we may write

$$q(z) = (z - z_1)^s q_1(z), \qquad s = v_q(z_1),$$

where $q_1(z_1) \neq 0$. The function $p/q_1$ is analytic in some disc containing $z_1$ and so for some constants $b_j$,

$$r(z) = (z - z_1)^{-s} \sum_{n=0}^{\infty} b_n (z - z_1)^n.$$

We prefer to write this as

$$r(z) = \frac{c_{-s}}{(z - z_1)^s} + \cdots + \frac{c_{-1}}{(z - z_1)} + \sum_{n=0}^{\infty} c_n (z - z_1)^n,$$

where $c_j = b_{j+s}$ or, better still, as

$$r(z) = P_1\left(\frac{1}{z - z_1}\right) + \sum_{n=0}^{\infty} c_n (z - z_1)^n,$$

where $P_1$ is the polynomial

$$P_1(z) = c_{-s} z^s + \cdots + c_{-1} z.$$

The function

$$r(z) - P_1\left(\frac{1}{z - z_1}\right)$$

(which is defined to be $c_0$ at $z_1$) is analytic in $\mathbf{C} - \{z_2, \ldots, z_k\}$ and is clearly a rational function.

The above argument may be repeated and after $k - 1$ repetitions we obtain a rational function

$$r(z) - P_1\left(\frac{1}{z - z_1}\right) - \cdots - P_k\left(\frac{1}{z - z_k}\right)$$

(where the $P_j$ are polynomials) which is analytic in $\mathbf{C}$. As polynomials are the only rational functions which are analytic in $\mathbf{C}$ we have proved the following result.

*Theorem 9.5.1* *Let $r$ be a rational function. Then there are polynomials $P_0, P_1, \ldots, P_k$ with $P_j(0) = 0, j = 1, \ldots, k$ and complex numbers $z_1, \ldots, z_k$ such that*

$$r(z) = P_1\left(\frac{1}{z - z_1}\right) + \cdots + P_k\left(\frac{1}{z - z_k}\right) + P_0(z), \qquad z \in \mathbf{C}.$$

We now see that each rational function is a finite linear combination of terms of the form $(z - w)^m$ and this means that we can evaluate integrals of rational functions as soon as we can integrate functions of the form $(z - w)^m, m \in \mathbf{Z}$. Fortunately, this is easy.

*Theorem 9.5.2* *Let w be any complex number and* $\Gamma$ *any cycle with* $w \notin [\Gamma]$. *Then*

$$\int_\Gamma (z - w)^m \, dz = \begin{cases} 0 & \text{if } m \neq -1, m \in \mathbf{Z}; \\ 2\pi i n(\Gamma, w) & \text{if } m = -1. \end{cases}$$

*Proof* If $m \neq -1$, then $(z - w)^m$ is the derivative of

$$(z - w)^{m+1}/(m + 1)$$

and so by (9.2.6) the integral is zero.

The case $m = -1$ requires more effort. As $\Gamma$ is compact there is a positive $\epsilon$ such that $\mathbf{C}(w, \epsilon)$ does not meet $[\Gamma]$. Now let $\gamma : [a, b] \to \mathbf{C}$ be any path in the cycle $\Gamma$. Then $\gamma$ is uniformly continuous on $[a, b]$ and so we can find a partition $t_0$, $t_1, \ldots, t_h$ of $[a, b]$ which satisfies

$$a = t_0 < t_1 < \cdots < t_h = b$$

and is such that the image under $\gamma$ of each interval $[t_j, t_{j+1}]$ lies in some disc $\Delta_j$ of radius $\epsilon$. This implies that $w \notin \Delta_j$ and so there is a continuous choice, say $L_j(z)$, of Log $(z - w)$ on $\Delta_j$. From Example 9.1.3 we deduce that $L_j$ is differentiable and that $L_j^{(1)}(z) = (z - w)^{-1}$.

We now denote the restriction of $\gamma$ to $[t_j, t_{j+1}]$ by $\gamma_j$, and we then find from Theorem 9.2.1 that

$$\begin{aligned} \int_{\gamma_j} \frac{dz}{z - w} &= \int_{\gamma_j} L_j^{(1)}(z) \, dz \\ &= L_j(\gamma(t_{j+1}) - w) - L_j(\gamma(t_j) - w) \\ &= \log_e | \gamma(t_{j+1}) - w | - \log_e | \gamma(t_j) - w | + 2\pi i n(\gamma_j, w). \end{aligned}$$

After summing over $j$ we find that

$$\int_\gamma \frac{dz}{z - w} = \log_e | \gamma(b) - w | - \log_e | \gamma(a) - w | + 2\pi i n(\gamma, w)$$

and this is true for each $\gamma$ in the cycle $\Gamma$.

Let $\Gamma$ be the cycle $(\gamma_1, \ldots, \gamma_m)$ and let $\gamma_j$ have initial point $w_j$ and final point $w_j'$: then

$$\int_\Gamma \frac{dz}{z - w} = \sum_{j=1}^m \log_e | w_j' - w | - \sum_{j=1}^m \log_e | w_j - w | + 2\pi i \sum_{j=1}^m n(\gamma_j, w).$$

The first two sums are rearrangements of each other and so their joint contribution is zero. As

$$n(\Gamma, w) = \sum_{j=1}^m n(\gamma_j, w),$$

the proof is complete.

These two theorems may now be combined. Observe first from the proof of Theorem 9.5.1 that the coefficient of $(z - z_1)^{-1}$ in the given partial fraction expansion is $c_{-1}$ and this is $P_1^{(1)}(0)$. Thus we obtain the general formula

$$\int_\Gamma \frac{p(z)}{q(z)}\,dz = 2\pi i \sum_{j=1}^{k} n(\Gamma, z_j)P_j^{(1)}(0). \tag{9.5.1}$$

Let us now illustrate this with two examples.

*Example 9.5.1* Let $p$ and $q$ be polynomials such that $|q(z)| \neq 0$ when $|z| = 1$, let $r = p/q$ and let $\gamma(t) = e^{it}$, $0 \leqslant t \leqslant 2\pi$. Then, by definition,

$$\int_\gamma \frac{r(z)}{iz}\,dz = \int_0^{2\pi} r(\exp(it)\,dt.$$

The first integral can be computed using Theorems 9.5.1 and 9.5.2: the second integral is a rational function of cos $(t)$ and sin $(t)$.

We show, for example, that if $a > 1$, then

$$I = \int_0^{2\pi} \frac{dt}{a + \sin(t)} = \frac{2\pi}{\sqrt{(a^2 - 1)}}.$$

We choose $r$ so that $r(e^{it})$ is the integrand in $I$; thus

$$r(z) = \left[a + \frac{1}{2i}\left(z - \frac{1}{z}\right)\right]^{-1}.$$

This yields

$$\frac{r(z)}{iz} = \frac{2}{(z_1 - z_2)}\left[\frac{1}{z - z_1} - \frac{1}{z - z_2}\right],$$

where $z_1 = -i(a + A)$, $z_2 = -i(a - A)$ and $A = \sqrt{(a^2 - 1)}$. As $|z_1| > 1$ and $|z_2| < 1$, (9.5.1) yields

$$I = \frac{-4\pi i}{z_1 - z_2},$$

and this is the required result.

*Example 9.5.2* We shall show that

$$\int_{-\infty}^{+\infty} \frac{dt}{(1 + t^2)^2} = \pi/2.$$

To begin with we take

$$\begin{aligned} r(z) &= (1 + z^2)^{-2} \\ &= (z - i)^{-2}(z + i)^{-2} \end{aligned}$$

and let $z_1 = i$, $z_2 = -i$. The cycle $\gamma$ given in Example 7.2.4 is a semi-circle in the upper half-plane with diameter $[-R, R]$ and

$$n(\gamma, i) = 1, \qquad n(\gamma, -i) = 0.$$

Together with (9.5.1), this implies that

$$\int_\gamma r(z)\ dz = 2\pi i P_1^{(1)}(0),$$

and the following simple computation gives $P_1^{(1)}(0)$. Writing $z = i + \zeta$ we have

$$\begin{aligned} r(i+\zeta) &= \frac{1}{\zeta^2(\zeta+2i)^2} \\ &= \frac{-1}{4\zeta^2(1+\zeta/2i)^2} \\ &= \frac{-1}{4\zeta^2}\sum_{n=1}^{\infty} n\left(\frac{-\zeta}{2i}\right)^{n-1}, \end{aligned}$$

the last line being derived from Example 2.3.1 (with $z = -\zeta/2i$).
Thus

$$P_1(z) = (-z^2/4) + (z/4i)$$

and so

$$\int_\gamma r(z)\ dz = \pi/2.$$

Now let $\gamma_1$ be the arc of $\gamma$ which lies on the circle $|z| = R$. Then

$$\int_\gamma r(z)\ dz = \int_{-R}^{R} \frac{dt}{(1+t^2)^2} + \int_{\gamma_1} r(z)\ dz$$

and so

$$\left| \int_{-R}^{R} \frac{dt}{(1+t^2)^2} - \pi/2 \right| = \left| \int_{\gamma_1} r(z)\ dz \right| \leqslant \pi R/(R^2-1)^2$$

as $|z^2+1| \geqslant |z|^2 - 1$ on $\gamma_1$. This shows that

$$\lim_{R\to+\infty} \int_{-R}^{R} \frac{dt}{(1+t^2)^2} = \pi/2.$$

Perhaps the most surprising feature of these examples is that we are using complex line integrals to evaluate *real* integrals. In fact, this technique is far more general than we have seen so far. First, the preceding discussion involving the rational function $p/q$ does not demand that $p$ be a polynomial. Exactly the same argument shows that if $f = p/q$, where $p$ is analytic in a domain $D$ and where $q$ is a polynominal with distinct zeros $z_1, \ldots, z_k$, then there are polynomials $P_1, \ldots, P_k$

with $P_j(0) = 0$ and a function $g$ analytic in $D$ such that if $z$ is in $D$, then

$$f(z) = P_1\left(\frac{1}{z - z_1}\right) + \cdots + P_k\left(\frac{1}{z - z_k}\right) + g(z). \tag{9.5.2}$$

The reader is urged to read this section again with this in mind.

We shall say that such functions $f$ are *meromorphic* in $D$ (this definition will be slightly modified later). The points $z_1, \ldots, z_k$ are called the *poles* of $f$ : $z_j$ is said to be a pole of *order* $v$ if it is a zero of order $v$ of $q$ and $p(z_j) \neq 0$. The function $P_j(1/z - z_j)$ is called the *principal part* of $f$ at $z_j$ and the constant $P_j^{(1)}(0)$ is called the *residue* of $f$ at $z_j$. It is convenient to introduce a notation to handle the residues of $f$ without the explicit mention of the $P_j$ and we now denote the residue of $f$ at $z_j$ by Res $(f, z_j)$.

If $f$ is meromorphic in $D$ and if $\Gamma$ is a cycle in $D$ with no pole of $f$ on $\Gamma$, we may apply Theorem 9.5.2 and deduce from (9.5.2) that

$$\int_\Gamma f = 2\pi i \sum_{j=1}^{k} n(\Gamma, z_j) \operatorname{Res}(f, z_j) + \int_\Gamma g. \tag{9.5.3}$$

The general question of when this integral of $g$ is zero is one of the major themes in this subject and it is discussed in detail in the next two sections. For the moment we simply observe that if $D$ is a disc with centre $z_0$ then $g$ may be expressed as a power series with centre of convergence $z_0$, and hence by Theorem 8.4.3 this series converges throughout $D$. As we have seen in §9.2, this implies that the integral of $g$ is zero and we obtain the following generalization of (9.5.1): if $f$ is meromorphic in a disc $D$ with poles $z_1, \ldots, z_k$, then

$$\int_\Gamma f = 2\pi i \sum_{j=1}^{k} n(\Gamma, z_j) \operatorname{Res}(f, z_j). \tag{9.5.4}$$

Observe that if $z_j$ is a pole of order *one* then we may write

$$f(z) = \frac{\alpha}{z - z_j} + f_1(z),$$

where $f_1$ is analytic in some disc $\mathbf{C}(z_j, \epsilon)$. This shows that

$$\operatorname{Res}(f, z_j) = \alpha = \lim_{z \to z_j} (z - z_j) f(z) \tag{9.5.5}$$

and this provides a useful and easy way to compute certain residues.

We end this section with two more examples: these are of a more substantial nature than Examples 9.5.1 and 9.5.2.

*Example 9.5.3* We first prove that

$$\int_{-\infty}^{+\infty} \frac{\cos(t)}{1 + t^2}\, dt = \frac{\pi}{e}.$$

Contrary, perhaps, to expectation we are going to consider the function

$f(z) = \exp(iz)/(1+z^2)$ rather than the function $\cos(z)/(1+z^2)$. The reason is that cos is not bounded on **C** (and this leads to difficulties) whereas $\exp(iz)$ is at least bounded on the upper half-plane for if $z = x + iy$ and $y \geqslant 0$, then

$$|\exp(iz)| = \exp(-y) \leqslant 1.$$

Moreover, our choice is adequate, as if $x$ is real,

$$\text{Re}\,[f(x)] = \cos(x)/(1+x^2).$$

Now let $Q$ be the rectangle with vertices $a$, $b$, $b + ih$ and $a + ih$, where $a < -1$, $b > 1$ and $h > 1$ and let $\Gamma$ be the cycle considering of the four boundary segments of $Q$ orientated so that $n(\Gamma, i) = 1$. Note also that $n(\Gamma, -i) = 0$ and so using (9.5.4) and (9.5.5) we find that

$$\int_\Gamma f(z)\,\mathrm{d}z = 2\pi i \,\text{Res}\,(f, i)$$
$$= \pi/e.$$

We now consider $\Gamma$ as four segments and deal with the corresponding integrals separately (see Diagram 9.5.1).

If $\gamma_1$ denotes the 'top' segment of $Q$, then $|1 + z^2| \geqslant h^2 - 1$ on $\gamma_1$ and

$$\left|\int_{\gamma_1} f\right| \leqslant \frac{b-a}{h^2-1}.$$

The integrals over the vertical sides $\gamma_2$ (corresponding to $x = a$) and $\gamma_3$ require a more careful estimate than usual. For $\gamma_2$ we have the following inequality (see Theorem 9.3.1):

$$\left|\int_{\gamma_2} f\right| \leqslant \int_0^h \frac{e^{-y}}{a^2-1}\,\mathrm{d}y$$
$$\leqslant (a^2-1)^{-1}$$

and a similar inequality holds for $\gamma_3$.

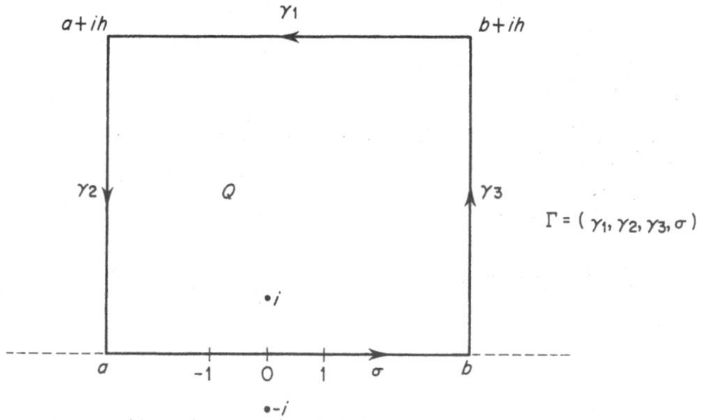


Diagram 9.5.1

We now have the inequality

$$\left| \int_a^b \frac{e^{it}}{1+t^2}\,dt - \frac{\pi}{e} \right| = \left| \sum_{j=1}^{3} \int_{\gamma_j} f \right|$$
$$\leqslant \frac{b-a}{h^2-1} + \frac{1}{a^2-1} + \frac{1}{b^2-1} .$$

First , we let $h \to +\infty$: this inequality therefore remains valid when the term involving $h$ is omitted. It is then clear that

$$\lim_{\substack{a\to-\infty \\ b\to+\infty}} \int_a^b \frac{e^{it}}{1+t^2}\,dt = \frac{\pi}{e}$$

and the stated answer follows by equating the real parts of each side of this equation.

It is worthwhile to examine this and the preceding example in greater detail. If $p$ and $q$ are polynomials with deg $(q) \geqslant 2 +$ deg $(p)$, and if $q$ has no real zeros then the integrals

$$\int_0^\infty \frac{p(t)}{q(t)}\,dt, \quad \int_{-\infty}^0 \frac{p(t)}{q(t)}\,dt \tag{9.5.6}$$

both converge (Exercise 9.5.1) and so

$$\lim_{R\to+\infty} \int_{-R}^{R} \frac{p(t)}{q(t)}\,dt \tag{9.5.7}$$

exists and equals the sum of these two integrals. Note, however, that if deg $(q) < 2 +$ deg $(p)$, then it may happen that the limit (9.5.7) exists without the integrals (9.5.6) existing (Exercise 9.5.2). These comments relate to the technique illustrated by Example 9.5.2: as $|\cos(t)| \leqslant 1$ for all real $t$, they also relate to Example 9.5.3.

It is interesting to note that if $p$ and $q$ are real polynomials with deg $(q) = 1 +$ deg $(p)$, then the integrals

$$\int_0^\infty \frac{p(t)}{q(t)} \cos(t)\,dt, \qquad \int_{-\infty}^0 \frac{p(t)}{q(t)} \cos(t)\,dt$$

(and similar integrals with cos replaced by sin) do converge. Indeed with this hypothesis, $|zp(z)/q(z)| \leqslant M$ for, say, $|z| > \rho$ and using the method illustrated in Example 9.5.3 we readily obtain the estimate

$$\left| \int_a^b \frac{p(t)}{q(t)} [\cos(t) + i\sin(t)]\,dt - 2\pi i \sum_{j=1}^{k} n(\Gamma, z_j) \operatorname{Res}(f, z_j) \right|$$
$$\leqslant \frac{M(b-a)}{h} + \frac{M}{|a|} + \frac{M}{b},$$

where $|a|$, $b$ and $h$ are greater than $\rho$ and the $z_j$ are the zeros of $q$. Again we let $h \to +\infty$ and so neglect the term containing $h$.

If we now take $0 < a < b$ and insist that $a$ is sufficiently large so that no $z_j$ lies inside $Q$, then $n(\Gamma, z_j) = 0$ for each $j$ and so

$$\left| \int_a^b \frac{p(t)}{q(t)} e^{it} \, dt \right| \leqslant \frac{2M}{a}. \tag{9.5.8}$$

This is precisely the criterion (6.3.5) that

$$\lim_{R \to +\infty} \int_0^R \frac{p(t)}{q(t)} e^{it} \, dt$$

should exist.

We have assumed throughout this discussion that $q$ has no real zeros but with care, the same method may be applied when $q$ does have real zeros. Indeed, in this case the inequality (9.5.8) still holds and this shows that the integral

$$\int_a^\infty \frac{p(t)}{q(t)} e^{it} \, dt \tag{9.5.9}$$

exists. We give one example to illustrate this point.

*Example 9.5.4* We show that

$$\int_0^\infty \frac{\sin(x)}{x} \, dx = \pi/2.$$

First, the integrand is continuous on **R** and so for all positive $a$ the integral

$$\int_0^a \frac{\sin(x)}{x} \, dx$$

exists. The comments relating to (9.5.9) are applicable and so the given integral does exist.

We now define $f$ by

$$f(z) = \frac{\exp(iz)}{z} = 1/z + g(z),$$

where $g$ is analytic in **C**. This means, of course, that we cannot integrate $f$ on any cycle which passes through the origin. Because of this we let $\Gamma$, $\Gamma = (\gamma_1, \ldots, \gamma_6)$, be the cycle described in Diagram 9.5.2 $(0 < \delta < a)$ : thus $\Gamma$ differs from the boundary of a rectangle only by a small semi-circular detour (an indentation) around the origin (the zero of $q$).

For brevity, let $I_j$ denote the integral of $f$ around $\gamma_j$. Then

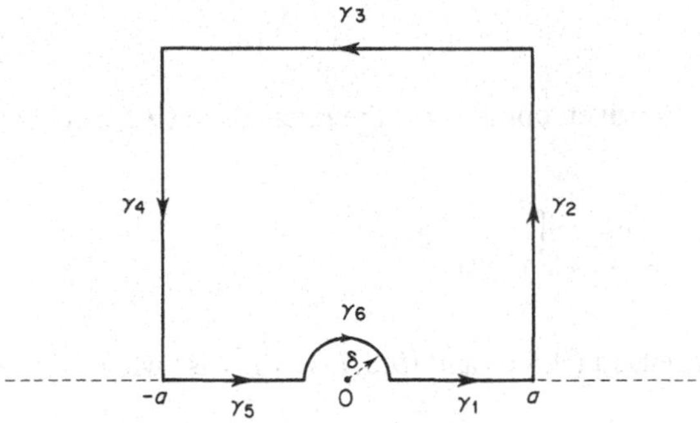


Diagram 9.5.2

$$I_1 + \cdots + I_6 = \int_\Gamma f$$

$$= n(\Gamma, 0) + \int_\Gamma g$$

$$= 0.$$

We now estimate or compute the $I_j$. As before,

$$|I_3| \leqslant 2a/h;$$

$$|I_2| \leqslant 1/a, \qquad |I_4| \leqslant 1/a.$$

Next, a change of variable in $I_5$ gives

$$I_1 + I_5 = 2i \int_\delta^a \frac{\sin(t)}{t}\, \mathrm{d}t.$$

Finally, $g(z) \to i$ as $z \to 0$ and so for $\delta < \delta_0$, say, we have $|g(\delta e^{it})| \leqslant 2$ and therefore

$$I_6 = -\int_0^\pi \left[\frac{1}{\delta e^{it}} + g(\delta e^{it})\right] i\delta e^{it}\, \mathrm{d}t$$

$$= -\pi i + J,$$

where $|J| \leqslant 2\pi\delta$.

We deduce that

$$|(I_1 + I_5) - \pi i| = |-(I_2 + I_3 + I_4 + I_6) - \pi i|$$

$$\leqslant |I_2| + |I_3| + |I_4| + |I_6 + \pi i|$$

$$\leqslant 2/a + 2a/h + 2\pi\delta.$$

Let $h \to +\infty$: then

$$\left| \int_\delta^a \frac{\sin(t)}{t}\, \mathrm{d}t - \frac{\pi}{2} \right| \leqslant \frac{1}{a} + \pi\delta$$

and the result follows by letting $\delta \to 0$ and $a \to +\infty$.

**Exercise 9.5**

1. Prove that under the given conditions, the integrals in (9.5.6) converge: for example

$$\int_0^{\infty} \frac{p(t)}{q(t)}\, dt = \lim_{R \to +\infty} \int_0^{R} \frac{p(t)}{q(t)}\, dt$$

exists.

2. Consider the integrals in (9.5.6) and (9.5.7) in the case when $q(z) = 1$ and $p(z) = z$ in **C**.
3. Show that for each positive integer $m$,

$$\int_0^{\pi/2} [\cos(t)]^{2m}\, dt = \frac{2\pi}{4^{m+1}} \binom{2m}{m}.$$

4. Evaluate (for positive integers $m$ and $n$),

$$\int_0^{2\pi} \sin(nt)\cos(mt)\, dt,$$

$$\int_0^{2\pi} \sin(nt)\sin(mt)\, dt,$$

$$\int_0^{2\pi} \cos(mt)\cos(nt)\, dt.$$

5. Show that

$$\int_0^{\infty} \frac{x^2}{1+x^4}\, dx = \frac{\pi}{2\sqrt{2}}.$$

6. Evaluate

$$\int_0^{\infty} \frac{1}{(x^2+a^2)^2}\, dx.$$

7. Using a cycle $\Gamma$ which is the boundary of a quarter circle, show that

$$\int_0^{\infty} \frac{x}{1+x^4}\, dx = \frac{\pi}{4}.$$

Why cannot you use a semi-circle?

8. Evaluate

$$\int_{-\infty}^{+\infty} \frac{\cos(mx)}{x^2-a^2}\, dx, \qquad (m \in \mathbf{Z}).$$

9. Evaluate

$$\int_0^{\infty} \frac{x \sin(x)}{1+x^2}\, dx.$$

## 9.6 CAUCHY'S THEOREM

Cauchy's Theorem states that under certain conditions

$$\int_\Gamma f = 0. \tag{9.6.1}$$

It is not expected that the reader should see the full significance of this now, but it will soon be apparent that this is one of the central results in complex analysis. There are a variety of conditions (which vary in generality and difficulty) available but in all cases one assumes that $f$ is either differentiable or analytic in a domain $D$. We remarked earlier (Theorem 9.1.2) that these assumptions are equivalent: nevertheless, until this has been proved it is essential to distinguish between these possibilities.

The remaining conditions required are of a topological or geometric character and are concerned solely with the domain $D$ and the cycle $\Gamma$.

This section is concerned with the following two versions of Cauchy's Theorem and their consequences.

*Theorem 9.6.1* *Let $f$ be differentiable in a domain $D$ and let $Q$ be a rectangle contained in $D$. Then*

$$\int_{\partial Q} f = 0.$$

This result is required only in order to prove that a differentiable function is analytic. We therefore assume that $f$ is differentiable (rather than analytic) and it is sufficient to restrict ourselves to the simplest geometrical situation.

*Cauchy's Theorem* *Let $D$ be any domain and $\Gamma$ any cycle in $D$. Then*

(a) $\int_\Gamma f = 0$

*for every $f$ which is analytic in $D$ if and only if*

(b) $n(\Gamma, w) = 0$

*for every $w$ not in $D$.*

This is the most general version of Cauchy's Theorem and we emphasize that it holds for *any domain $D$ without exception.* It is the mutual interaction (b) of the cycle $\Gamma$ and the domain $D$ in the form of the index (and *not* the topological or geometrical qualities of $D$ alone) which provides the criterion for the integral to vanish. Observe that (b) implies that if $z$ is inside $\Gamma$ (that is if $n(\Gamma, z) \neq 0$) then $z$ is inside $D$; in other words (b) is simply the formal statement that the cycle $\Gamma$ and its 'inside' lies in $D$.

The versions of Cauchy's Theorem which involve only topological (or geometrical) assumptions on $D$ (for example, convexity) are obtained by simply imposing sufficient restrictions on $D$ so that the index condition (b) is automatically satisfied (and need not therefore be mentioned). This is the case, for example, if $D$ is a disc, for then (b) is satisfied.

From a more general viewpoint, if the complement $\mathbf{C} - D$ of $D$ is connected and unbounded then $n(\Gamma, w)$ is constant on $\mathbf{C} - D$ (as $\mathbf{C} - D$ is connected) and this constant is zero (as $\mathbf{C} - D$ is unbounded). Thus (a) *holds whenever D has a connected unbounded complement* (in more intuitive language, when $D$ has no 'holes' in it) and this is, perhaps, the most useful form of Cauchy's Theorem.

In general, the domain $D$ may have 'holes' $E_1, E_2, \ldots$ (possibly infinitely many) and the curves $\gamma_j$ which comprise $\Gamma$ may wind about these holes. Nevertheless, provided that (b) holds then so does (a). Indeed, it is precisely this case that is required in order to prove the important Residue Theorem. We illustrate this situation in Diagram 9.6.1: in this case (b) does hold and the vertically shaded region $K$ is the set of points inside $\Gamma$, $\Gamma = (\gamma_1, \gamma_2, \gamma_3)$. The 'holes' $E_j$ have horizontal shading.

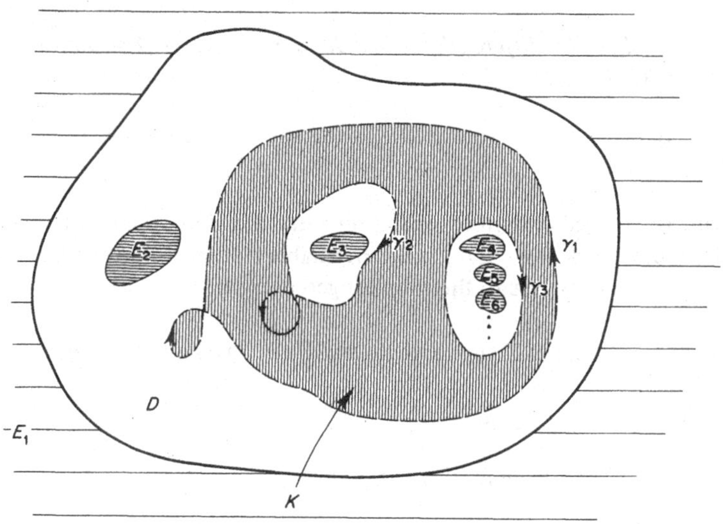


Diagram 9.6.1

No doubt some readers would have preferred to see these results proved earlier and this would have been possible. The results have been deliberately delayed, however, for two reasons. First, the delay can only emphasize the fact that integration plays no essential part in the topological consequences of analyticity and second, we now have a sufficient working knowledge of the index to appreciate the full content of Cauchy's Theorem.

The proof of Cauchy's Theorem given below makes essential and effective use of the index and completely avoids any other topological consideration. Because it is stated for analytic functions its proof does not even require Theorem 9.6.1 as a prerequisite.

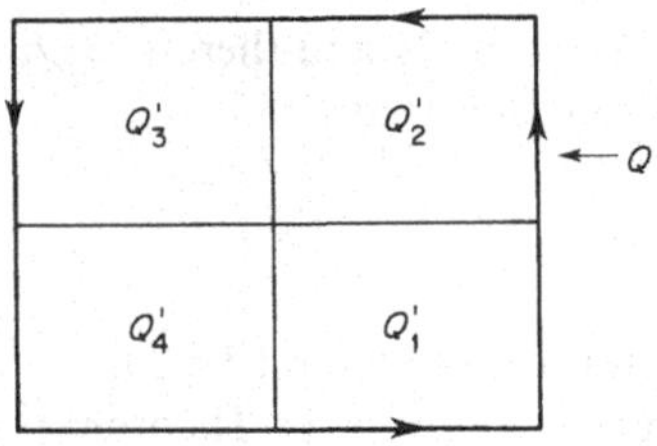


Diagram 9.6.2

*Proof of Theorem* 9.6.1 We recall that $\partial Q$ is the boundary of $Q$ orientated so that if $w$ is inside $\partial Q$, then $n(\partial Q, w) = 1$. The hypothesis '$Q$ is contained in $D$' means that if $w$ is inside $\partial Q$, then $w$ is in $D$. In terms of the index this is equivalent to the implication 'if $w \notin D$, then $n(\partial Q, w) = 0$' and this is condition (b) in Cauchy's Theorem.

We subdivide $Q$ into four conguent rectangles $Q'_j$, $j = 1, \ldots, 4$ as in Diagram 9.6.2 and define $Q_1$ to be that $Q'_j$ which maximizes

$$\left| \int_{\partial Q'_j} f \right|.$$

Then

$$\left| \int_{\partial Q} f \right| = \left| \sum_{j=1}^{4} \int_{\partial Q'_j} f \right|$$

$$\leqslant 4 \left| \int_{\partial Q_1} f \right|.$$

We continue this process of subdivision constructing $Q_{k+1}$ from $Q_k$ in the same way and this yields a sequence of similar rectangles $Q(= Q_0), Q_1, Q_2, \ldots$ with $Q_{k+1} \subset Q_k$ and

$$\left| \int_{\partial Q} f \right| \leqslant 4^k \left| \int_{\partial Q_k} f \right|. \tag{9.6.2}$$

Now let $d_k$ be the diameter (the length of the diagonal) of $Q_k$: so $d_k = 2^{-k} d_0$, $d_k \to 0$ as $k \to \infty$ and there is a unique point $\zeta$ which lies in every $Q_k$ (Theorem 6.4.2). As $f$ is differentiable at $\zeta$ we may write

$$f(z) = f(\zeta) + (z - \zeta) f^{(1)}(\zeta) + \epsilon(z)(z - \zeta):$$

this is valid throughout $D$, $\epsilon$ is continuous in $D$ and $\epsilon(\zeta) = 0$.

The integral of any polynomial over any cycle is zero and with this representation of $f$ we obtain

$$\left| \int_{\partial Q_k} f \right| = \left| \int_{\partial Q_k} \epsilon(z)(z - \zeta)\, dz \right|$$

$$\leqslant L(\partial Q_k) \sup \{ | \epsilon(z)(z - \zeta) | : z \in \partial Q_k \}$$

$$\leqslant L(\partial Q_k) d_k \sup \{ | \epsilon(z) | : z \in \partial Q_k \}$$

$$\leqslant [4^{-k} L(\partial Q_0) d_0] \sup \{ | \epsilon(z) | : | z - \zeta | \leqslant d_k \}.$$

These estimates are valid as $\zeta$ lies in $Q_k$ and therefore $Q_k \subset \mathbf{C}(\zeta, d_k)$. As $d_k \to 0$ as $k \to \infty$ and $\epsilon(z) \to 0$ as $z \to \zeta$, we find that

$$4^k \int_{\partial Q_k} f \to 0$$

as $k \to \infty$ and the desired result follows from (9.6.2).

We are now able to complete the proof of Theorem 9.1.2.

*Proof of Theorem 9.1.2* We know that if $f$ is analytic in $D$, then it is differentiable in $D$. Suppose now that $f$ is differentiable in $D$, select a point $\zeta$ in $D$ and construct a rectangle $Q$ contained in $D$ such that $\zeta$ lies inside $\partial Q$. This may be done so that on $Q$,

$$\left| \frac{f(z) - f(\zeta)}{z - \zeta} - f^{(1)}(\zeta) \right| < 1;$$

hence the first term is bounded near $\zeta$.

Now subdivide the rectangle $Q$ into nine rectangles $Q_1$ (which contains $\zeta$), $Q_2, \ldots, Q_9$ as in Diagram 9.6.3.

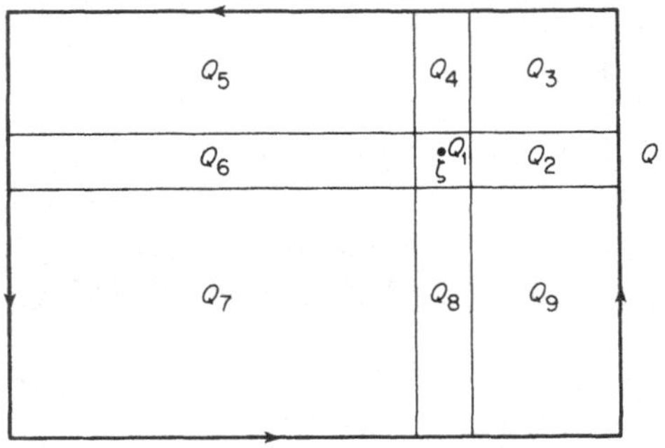


Diagram 9.6.3

The function

$$F(z) = \frac{f(z) - f(\zeta)}{z - \zeta} \tag{9.6.3}$$

is differentiable in $D - \{\zeta\}$ and Theorem 9.6.1 is applicable to $F$ and $Q_j$, $j = 2, \ldots, 9$. Thus

$$\begin{aligned} \left| \int_{\partial Q} F \right| &= \left| \sum_{j=1}^{9} \int_{\partial Q_j} F \right| \\ &= \left| \int_{\partial Q_1} F \right| \\ &\leqslant L(\partial Q_1)[1 + |f^{(1)}(\zeta)|]. \end{aligned}$$

As this subdivision may be constructed with $L(\partial Q_1)$ as small as we please, we must have

$$\int_{\partial Q} F = 0.$$

We now use (9.6.3) and conclude directly from Theorem 9.5.2 that

$$2\pi i n(\partial Q, \zeta) f(\zeta) = \int_{\partial Q} \frac{f(z)}{z - \zeta}\, dz,$$

and hence

$$f(\zeta) = \frac{1}{2\pi i} \int_{\partial Q} \frac{f(z)}{z - \zeta}\, dz. \tag{9.6.4}$$

The right-hand side is an analytic function of $\zeta$ inside $\partial Q$ (Corollary, Theorem 9.3.2) and so $f$ has a power series expansion with centre of convergence $\zeta$. This completes the proof of Theorem 9.1.2.

We remark that the Maximum Modulus Theorem shows that the values of $f$ inside $\partial Q$ are uniquely determined by the values of $f$ on $\partial Q$ and (9.6.4) gives an explicit formula for this determination.

Without doubt, the main use of Theorem 9.1.2 is to provide a simple criterion for analyticity. We already know that finite sums and products of analytic functions are analytic; this is also implied by Theorem 9.1.2. Similarly, Theorem 9.1.2 includes the fact that a power series is analytic. Other straightforward applications of Theorem 9.1.2 (which have not been proved earlier) are

(a) if $f$ is analytic and non-zero in $D$, then $1/f$ is analytic in $D$;
(b) if $g : D_1 \to D_2$ and $f : D_2 \to \mathbf{C}$ are analytic, then so is $f \circ g$;
(c) if $L(z)$ is a branch (a continuous choice) of Log $(z)$, $z \in D$, then $L$ is analytic in $D$ (see Example 9.1.2).

*The proof of Cauchy's Theorem* The proof that (a) implies (b) is trivial for if $w \notin D$, then $f(z) = (z - w)^{-1}$ is analytic in $D$ and so by (a) and Theorem 9.5.2, $n(\Gamma, w) = 0$.

Now suppose that (b) holds. We define $K$ to be the set of points inside or on $\Gamma$: thus

$$K = [\Gamma] \cup \{z \in \mathbf{C} : n(\Gamma, z) \neq 0\}.$$

As (b) holds, $K \subset D$. The complement of $K$ is the set of $z$ for which $n(\Gamma, z) = 0$: this is open and contains $z$ for all sufficiently large $|z|$, thus $K$ is actually a compact subset of $D$. Diagram 9.6.1 again illustrates a typical situation: $K$ is the vertically shaded region, $\Gamma = (\gamma_1, \gamma_2, \gamma_3)$ and the components of $\mathbf{C} - D$ are $E_1, E_2, \ldots$.

We now use Theorem 6.4.3 to infer that

$$\delta = dist\,(K, \mathbf{C} - D) > 0. \tag{9.6.5}$$

Using this $\delta$ we draw the horizontal lines $x = n\delta/2$ ($n \in \mathbf{Z}$) and the vertical lines $y = m\delta/2$ ($m \in \mathbf{Z}$). These lines partition $\mathbf{C}$ into congruent non-overlapping squares and we denote by $Q_j, j \in J$, those squares which meet $K$. As $K$ is compact there are only a finite number of the $Q_j$.

Clearly, if $z_j \in Q_j \cap K$, then from (9.6.5),

$$Q_j \subset \mathbf{C}(z_j, \delta) \subset D.$$

Now select any $\zeta$ *inside*, say $\partial Q_k$. The function $[f(z) - f(\zeta)]/(z - \zeta)$ is analytic in each disc $\mathbf{C}(z_j, \delta), j \in J$, and therefore (Theorem 9.2.1),

$$\int_{\partial Q_j} \frac{f(z) - f(\zeta)}{z - \zeta}\,dz = 0.$$

Exactly as in the proof of Theorem 9.6.1, this gives

$$n(\partial Q_j, \zeta)\, f(\zeta) = \frac{1}{2\pi i} \int_{\partial Q_j} \frac{f(z)}{z - \zeta}\,dz. \tag{9.6.6}$$

We next sum both sides of this equation over $J$ and using

$$n(\partial Q_j, \zeta) = \begin{cases} 1 & \text{if} \quad j = k \\ 0 & \text{if} \quad j \neq k \end{cases}$$

we obtain

$$f(\zeta) = \frac{1}{2\pi i} \sum_{j \in J} \int_{\partial Q_j} \frac{f(z)}{z - \zeta}\,dz. \tag{9.6.7}$$

Each cycle $\partial Q_j$ consists of four segments, namely the sides of $Q_j$. If $\{\sigma_i : i = 1, \ldots, q\}$ denotes the totality of these segments arising from the $Q_j, j \in J$, we may rewrite (9.6.7) as

$$f(\zeta) = \frac{1}{2\pi i} \sum_{i=1}^{q} \int_{\sigma_i} \frac{f(z)}{z - \zeta}\,dz. \tag{9.6.8}$$

If some side $\sigma_i$ meets $K$, then it is a side of two adjacent squares $Q_j$ and so it arises exactly twice in the above sum but with opposite orientations. The contribution from this side to the above sum is therefore zero and so (9.6.8) holds when $\sigma_1, \ldots, \sigma_q$ denote only those sides $\sigma_i$ which do not meet $K$.

This means that both sides of (9.6.8) are continuous functions of $\zeta$ on $K$. As each point of $K$ is the limit of a sequence $\zeta_1, \zeta_2, \ldots$ of points inside some $\partial Q_k$, and as (9.6.8) holds when $\zeta = \zeta_1, \zeta_2, \ldots$, we find that by continuity, (9.6.8) actually holds when $\zeta \in K$. In particular, (9.6.8) *holds when* $\zeta$ *is on* $\Gamma$ and this is the crucial stage of the proof.

Assuming for the moment that we can interchange the order of integration, we have, with $\Gamma = (\gamma_1, \ldots, \gamma_m)$,

$$\int_\Gamma f(\zeta)\,d\zeta = \int_\Gamma \left[\frac{1}{2\pi i}\sum_{i=1}^{q}\int_{\sigma_i}\frac{f(z)}{z-\zeta}\,dz\right]d\zeta$$

$$= \sum_{j=1}^{m}\sum_{i=1}^{q}\int_{\gamma_j}\left(\int_{\sigma_i}\frac{1}{2\pi i}\frac{f(z)}{z-\zeta}\,dz\right)d\zeta$$

$$= \sum_{j=1}^{m}\sum_{i=1}^{q}\int_{\sigma_i}\left(\int_{\gamma_j}\frac{1}{2\pi i}\frac{f(z)}{z-\zeta}\,d\zeta\right)dz$$

$$= \sum_{j=1}^{m}\sum_{i=1}^{q}\int_{\sigma_i}[-n(\gamma_j, z)\,f(z)]\,dz$$

$$= -\sum_{i=1}^{q}\int_{\sigma_i} n(\Gamma, z)f(z)\,dz$$

$$= 0,$$

since if $z$ is on $\sigma_i$, then $z$ is not in $K$ and so $n(\Gamma, z) = 0$.

It only remains to justify the interchange of integrals. We shall use (without proof) the standard result for real integrals, namely that if $g(x, y)$ is real valued and continuous for $a \leqslant x \leqslant b, c \leqslant y \leqslant d$, then

$$\int_a^b\left[\int_c^d g(x,y)\,dy\right]dx = \int_c^d\left[\int_a^b g(x,y)\,dx\right]dy.$$

This remains true for complex-valued $g$ (simply by addition of the real and imaginary parts of $g$). Hence for paths $\sigma$ and $\gamma$,

$$\int_\sigma\left[\int_\gamma F(z, w)\,dz\right]dw = \int_\gamma\left[\int_\sigma F(z,w)\,dw\right]dz$$

provided only that

$$F(\gamma(t), \sigma(s))\gamma^{(1)}(t)\sigma^{(1)}(s)$$

is a continuous function of the variable $(t, s)$. As $\gamma$ and $\sigma$ are paths their derivatives are continuous and it is only necessary to verify that $F(\gamma(t), \sigma(s))$ is continuous. This is easily seen to be true in the application above and the proof is complete.

Cauchy's Theorem is often used in the following invariant form: *if $\Gamma$ and $\Sigma$ are cycles in $D$ such that*

$$n(\Gamma, w) = n(\Sigma, w)$$

*whenever $w \notin D$, then for all $f$ analytic in $D$,*

$$\int_\Gamma f = \int_\Sigma f.$$

This enables us to change from, say, $\Gamma$ to a more convenient cycle $\Sigma$ without changing the value of the integral and for this reason it is most useful. The proof is a direct application of Cauchy's Theorem to the cycle $(\gamma_1, \ldots, \gamma_p, \sigma_1^-, \ldots, \sigma_q^-)$, where $\Gamma = (\gamma_1, \ldots, \gamma_p)$ and $\Sigma = (\sigma_1, \ldots, \sigma_q)$.

**Excercise 9.6**

1. Use the method of the proof of Theorem 9.6.1 to show that if $f$ is differentiable in some domain $D$ and if $T$ is any triangle in $D$, then

$$\int_{\partial T} f = 0.$$

2. Prove Cauchy's Theorem for a convex domain (which we may assume contains the origin) as follows. Let $f$ be differentiable in $D$ and for $z$ in $D$ define

$$F(z) = \int_{[0,z]} f.$$

Use Excercise 9.6.1 to show that

$$\frac{F(z+h) - F(z)}{h} - f(z) = \frac{1}{h} \int_{[z,z+h]} [f(w) - f(z)]\, dw$$

and deduce that $F^{(1)}(z) = f(z)$ in $D$.
Conclude that (9.6.1) holds.

3. Let $D$ be any domain and let $g$ be any function which is continuous on $D$. Choose $\zeta$ in $D$ and let $\Gamma$ be a chain in $D$ which joins $\zeta$ to $z$ in $D$. Write

$$G(z, \Gamma) = \int_{\Gamma} g.$$

Prove that $G(z, \Gamma)$ is independent of $\Gamma$ if and only if

$$\int_{\Sigma} g = 0 \tag{1}$$

for every closed cycle $\Sigma$ in $D$.

Assume that (1) holds and write $G(z)$ for $G(z, \Gamma)$. Prove that $G$ is differentiable in $D$ and that $G^{(1)}(z) = g(z)$ in $D$.

Derive Morera's Theorem: *if $g$ is continuous in $D$ and if* (1) *holds, then $g$ is analytic in $D$.*

## 9.7 APPLICATIONS

Our first application of Cauchy's Theorem is to extend the techniques developed in §9.5 for evaluating real integrals. The method is essentially the same and we merely give two examples.

*Example 9.7.1* If $a > 0$, then

$$\int_0^\infty \frac{\log_e(x)}{x^2 + a^2}\,dx = \frac{\pi \log_e(a)}{2a}\ .$$

We need to specify a domain $D$, a cycle $\Gamma$ in $D$ and a function $f$ analytic in $D$. Let $D$ be the complement of the half-line $L$, $L = \{-it: t \geqslant 0\}$. As $L$ is connected and unbounded,

$$\int_\Gamma F = 0 \tag{9.7.1}$$

for every cycle $\Gamma$ in $D$ and every $F$ analytic in $D$. We shall use the cycle $\Gamma$, $\Gamma = (\gamma_1, \ldots, \gamma_4)$, illustrated in Diagram 9.7.1.

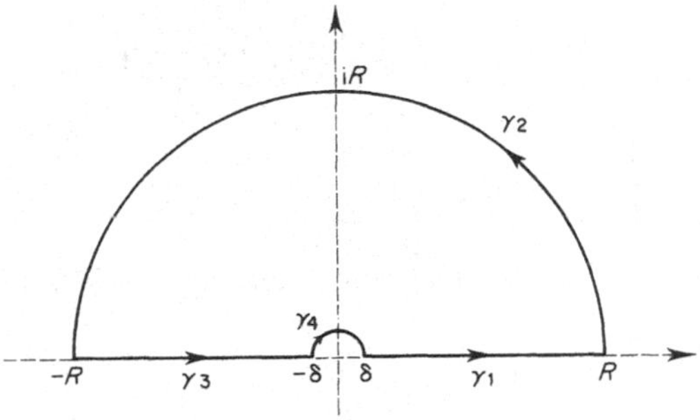


Diagram 9.7.1

The function

$$\log(z) = \log_e(|z|) + i\theta(z) \qquad (z \in D),$$

where $\theta(z)$ is the unique choice of Arg $(z)$ in $(-\pi/2, 3\pi/2)$, is continuous and hence analytic in $D$ (Theorems 5.4.1 and 9.1.2, Example 9.1.3). We take

$$f(z) = \frac{\log(z)}{z^2 + a^2}\ :$$

this is analytic in $D$ except for a pole of order one at $ia$. Using (9.5.4),

$$\text{Res}\,(f, ia) = \frac{\log(ia)}{2ia}$$

and, as in (9.5.2), we may write

$$f(z) = \frac{\log(ia)}{2ia(z - ia)} + F(z)$$

where $F$ is analytic in $D$. We now integrate $f$ around $\Gamma$ and use (9.7.1): this gives

$$\int_\Gamma f = 2\pi i n(\Gamma, ia)\frac{\log(ia)}{2ia}$$

$$= (\pi/a)[\log_e(a) + \tfrac{1}{2}\pi i]\,.$$

As before, we must compute or estimate the integrals over the paths $\gamma_j$. On $\gamma_2$, $|z^2 - a^2| \geqslant |z|^2 - |a|^2$: thus

$$\left|\int_{\gamma_2} f\right| \leqslant \frac{\pi R[\log_e(R) + \pi]}{R^2 - a^2} = \alpha(R)$$

and $\alpha(R) \to 0$ as $R \to +\infty$ (Exercise 9.7.1). Similarly, on $\gamma_4$,

$|z^2 - a^2| \geqslant |a|^2 - |z|^2$: thus

$$\left|\int_{\gamma_4} f\right| \leqslant \frac{\pi\delta[\log_e(\delta) + \pi]}{a^2 - \delta^2} = \beta(\delta)$$

and $\beta(\delta) \to 0$ as $\delta \to 0$ (Exercise 9.7.2).

Finally,

$$\int_{\gamma_1} f + \int_{\gamma_3} f = \int_{\delta}^{R} \frac{\log_e(x)}{x^2 + a^2}\,dx + \int_{-R}^{-\delta} \frac{[\log_e(x) + i\pi]}{x^2 + a^2}\,dx$$

$$= 2\int_{\delta}^{R} \frac{\log_e(x)}{x^2 + a^2}\,dx + i\pi \int_{\delta}^{R} \frac{dx}{x^2 + a^2},$$

and the given integral can now be evaluated by letting $R \to \infty$ and $\delta \to 0$.

*Example 9.7.2* If $-1 < b < 1$, then

$$\int_0^\infty \frac{x^b}{1 + x^2}\,dx = \frac{\pi}{2} \sec\left(\frac{\pi b}{2}\right),$$

where

$$x^b = \exp(b \log_e(x)).$$

We proceed as in the previous example, except that we now consider

$$f(z) = \frac{\exp(b \log(z))}{1 + z^2}.$$

This gives

$$\int_\Gamma f = \pi \exp(\tfrac{1}{2}\pi i b)$$

as $\log(i) = i\pi/2$.

Next, using Theorem 4.1.3(f), if $R > 2$, then

$$\left|\int_{\gamma_2} f\right| \leqslant \frac{\pi R \exp(b \log_e(R))}{R^2 - 1}$$

$$\leqslant \frac{2\pi \exp(b \log_e(R))}{R}$$

$$= 2\pi \exp([b - 1] \log_e(R))$$

and this tends to zero as $R \to \infty$ because $b - 1 < 0$.

Similarly, if $\delta < \frac{1}{2}$, then

$$\left| \int_{\gamma^4} f \right| \leqslant 2\pi \exp([b+1] \log_e(\delta))$$

and this tends to zero as $\delta \to 0$ because $b + 1 > 0$.

As in the previous example we compute the integrals over $\gamma_1$ and $\gamma_3$ and find that

$$\int_{\gamma_1} f + \int_{\gamma_3} f = \int_\delta^R \frac{x^b}{1+x^2}\,dx + \int_\delta^R \frac{x^b \exp(\pi i b)}{1+x^2}\,dx.$$

The result now follows by letting $R \to \infty$ and $\delta \to 0$.

The same technique can be used to evaluate certain infinite series if we create a situation in which the partial sum of the series is essentially the same as the sum of residues. We give one example of this, the idea being developed further in the Exercises.

*Example 9.7.3* We shall show that

$$\sum_{n=1}^{\infty} \frac{1}{n^2} = \frac{\pi^2}{6}.$$

It was shown in Example 8.5.3 that for each positive integer $N$,

$$\sin(\pi z) = z(z^2-1)(z^2-2^2)\cdots(z^2-N^2)g(z),$$

where $g$ is analytic and not zero in $C(0, N+1)$. Thus $g$ and hence $1/g$ is differentiable in this disc and we deduce that $1/g$ is actually analytic there. We may now write

$$\cot(\pi z) = \frac{\cos(\pi z)\,[1/g(z)]}{z(z^2-1)\cdots(z^2-N^2)}$$

and this is analytic in $C(0, N+1)$ except for simple poles at $-N, \ldots, 0, \ldots, N$ (cos is introduced here to provide the required residues).

Now consider the circle given by

$$\gamma(t) = (N+\tfrac{1}{2})e^{it}, \qquad 0 \leqslant t \leqslant 2\pi,$$

and the function

$$f(z) = z^{-2}\cot(\pi z).$$

Using Theorem 4.4.1(b),

$$\left| \int_\gamma f \right| \leqslant 2\pi(N+\tfrac{1}{2})[3/(N+\tfrac{1}{2})^2]$$

and this tends to zero as $N \to \infty$. As $f$ is meromorphic in $C(0, N+1)$ with poles only at $-N, \ldots, 0, \ldots, N$ we find that

$$\lim_{N\to\infty} \sum_{k=-N}^{N} \operatorname{Res}(f, k) = 0. \qquad (9.7.2)$$

It is now only a matter of computing residues. If $k \neq 0$, then $k$ is a pole of order one and we have

$$\begin{aligned}\operatorname{Res}(f,k) &= \lim_{z\to k} \frac{\cos(\pi z)(z-k)}{z^2 \sin(\pi z)}\\ &= \lim_{z\to k} \frac{\cos(\pi z)}{\pi z^2}\left[\frac{\sin(\pi z)-\sin(\pi k)}{\pi(z-k)}\right]^{-1}\\ &= (\pi k^2)^{-1},\end{aligned}$$

as the derivative of sin is cos.

The easiest way to compute Res $(f, 0)$ is to observe that for some $a_n$ and for $0 < |z| < 1$,

$$\frac{\cos(\pi z)}{z^2 \sin(\pi z)} = \frac{a_{-3}}{z^3} + \frac{a_{-2}}{z^2} + \frac{a_{-1}}{z} + a_0 + \cdots).$$

where Res $(f, 0) = a_{-1}$. This shows that for these $z$

$$z\left(1 - \frac{(\pi z)^2}{2!} + \cdots\right) = \left(\pi z - \frac{(\pi z)^3}{3!} + \cdots\right)(a_{-3} + a_{-2}z + a_{-1}z^2 + \cdots).$$

We may multiply the series, the identity necessarily holds throughout **C** and we may equate coefficients. This gives Res $(f, 0) = \pi/3$ and we conclude from (9.7.2) that

$$\frac{-\pi}{3} + \frac{2}{\pi} \lim_{N\to\infty} \sum_{k=1}^{N} \frac{1}{k^2} = 0.$$

Taking a more theoretical viewpoint now, we continue to combine the ideas of §9.5 with Cauchy's Theorem as suggested by (9.5.3). We begin with our final version of the Argument Principle.

*The Argument Principle* *Let $f$ be analytic in a domain $D$ and let $\Gamma$ be a cycle in $D$ such that $n(\Gamma, w) = 0$ whenever $w \notin D$. Then*

$$n(f \circ \Gamma, 0) = \sum_{z, f(z)=0} v_f(z) n(\Gamma, z).$$

This is the same conclusion as we obtained in §8.5: this version applies to *any* domain $D$, however, although there is now the mild restriction that $\Gamma$ must consist of continuously differentiable (and not arbitrary) curves (this arises only because of the use of the integral).

*Proof* The set $K$ of points inside and on $\Gamma$ is a compact subset of $D$ and so we can write

$$f(z) = (z - z_1)^{v_1} \cdots (z - z_s)^{v_s} g(z),$$

where $g$ is analytic in $D$, non-zero in $K$ and where $v_j = v_f(z_j)$. Exactly as in §8.5, this gives

$$n(f \circ \Gamma, 0) = \sum_{j=1}^{s} v_f(z_j) n(\Gamma, z_j) + n(g \circ \Gamma, 0).$$

Observe first that if $\gamma : [a, b] \to D$ is a path occurring in $\Gamma$, then

$$2\pi i n(g \circ \gamma, 0) = \int_a^b \frac{g^{(1)}(\gamma(t))\gamma^{(1)}(t)}{g(\gamma(t))} \, dt$$

$$= \int_\gamma \frac{g^{(1)}}{g} .$$

Thus for the cycle $\Gamma$,

$$2\pi i n(g \circ \Gamma, 0) = \int_\Gamma \frac{g^{(1)}}{g} .$$

Now let $z_{s+1}, z_{s+2}, \ldots$ denote the zeros of $f$ in $D - K$ and let

$$D^* = D - \{z_{s+1}, z_{s+2}, \ldots\}$$

(if $f$ has no zeros in $D - K$ put $D^* = D$). Then $g^{(1)}/g$ is analytic in $D^*$ and $n(\Gamma, w) = 0$ whenever $w \notin D^*$ (for either $w \notin D$ or $w = z_{s+j}$ and $z_{s+j}$ is outside $\Gamma$). Thus by Cauchy's Theorem

$$\int_\Gamma \frac{g^{(1)}}{g} = 0$$

and the proof is complete.

The Argument Principle can be extended without difficulty to functions with poles. If $g$ and $h$ are analytic in $D$ (and $h$ is not identically zero) then $f = g/h$ satisfies $fh = g$ and so

$$n(f \circ \Gamma, 0) + n(h \circ \Gamma, 0) = n(g \circ \Gamma, 0).$$

Further, the zeros of $f$ and $g$ coincide, the zeros of $h$ are the poles of $f$. We may assume that $g$ and $h$ have no common zeros and with the convention that $f(z) = \infty$ if $z$ is a zero of $h$ (of order $v_f(z)$) we have

$$n(f \circ \Gamma, 0) = \sum_{z, f(z)=0} v_f(z) n(\Gamma, z) - \sum_{z, f(z)=\infty} v_f(z) n(\Gamma, z).$$

In almost all applications of the Argument Principle, $n(\Gamma, w)$ is zero or one according to whether $w$ is outside or inside $\Gamma$ and this is usually clear in specific cases. We have stated and proved a general version of the Argument Principle without relying on geometric intuition and the burden now lies with the reader to compute (analytically or intuitively as he pleases) the terms $n(\Gamma, z)$ in any particular application.

Continuing with the same theme we can obtain a general version of (9.6.6) applicable to all derivatives $f^{(k)}$. Let $D$ be any domain, let $\Gamma$ be a cycle in $D$ with $n(\Gamma, w) = 0$ whenever $w \notin D$, let $\zeta$ be in $D$ but not on $\Gamma$ and let $f$ be analytic in $D$.

In some disc $\mathbf{C}(\zeta, t)$, $t > 0$,

$$f(z) = \sum_{n=0}^{\infty} a_n(z - \zeta)^n$$

and after differentiating $k$ times and putting $z = \zeta$ we obtain

$$k!a_k = f^{(k)}(\zeta).$$

This gives the *Taylor expansion* of $f$, namely

$$f(z) = \sum_{n=0}^{\infty} \frac{f^{(n)}(\zeta)}{n!} (z - \zeta)^n$$

and this converges in (at least) the largest disc $\mathbf{C}(\zeta, t)$ which is in $D$.

For each positive integer $k$,

$$\frac{f(z)}{(z - \zeta)^{k+1}} = \frac{f(\zeta)}{(z - \zeta)^{k+1}} + \ldots + \frac{f^{(k)}(\zeta)}{k!(z - \zeta)} + g(z),$$

where $g$ is analytic in $D$ and Cauchy's Theorem together with Theorem 9.5.2 yields

$$\frac{k!}{2\pi i} \int_\Gamma \frac{f(z)}{(z - \zeta)^{k+1}} \mathrm{d}z = n(\Gamma, \zeta) f^{(k)}(\zeta).$$

This is *Cauchy's Integral formula* for $f^{(k)}$: the case $k = 0$ is precisely the general version of (9.6.6). In simplest terms, this formula gives an explicit formula for $f^{(k)}(\zeta)$ solely in terms of the values of $f$ on $\Gamma$. It can also be proved by showing (alternatively, it shows) that one can differentiate both sides of (9.6.6) as functions of $\zeta$, the differentiation being carried out 'under the integral sign'.

An immediate application is to the uniform limit of analytic functions. The partial sums of a power series $f$ do not in general converge uniformly to $f$ on the whole disc of convergence. They do, however, converge uniformly on each compact subset of the disc and this is typical of the general situation. We shall say, then, that $f_n \to f$ *locally uniformly in D* if $f_n \to f$ uniformly on each compact subset of $D$.

*Theorem 9.7.1* *Let $f_1, f_2, \ldots$ be analytic in a domain D and suppose that $f_n \to f$ locally uniformly in D. Then f is analytic in D and for $k = 1, 2, \ldots, f_n^{(k)} \to f^{(k)}$ locally uniformly in D.*

*Proof* Let $\zeta$ be in $D$ and select a positive $r$ such that $\bar{\mathbf{C}}(\zeta, r) \subset D$. Let $\gamma(t) = \zeta + re^{it}$, $0 \leqslant t \leqslant 2\pi$. If $|w - \zeta| < r$, then

$$\begin{aligned} f(w) &= \lim_{n\to\infty} f_n(w) \\ &= \lim_{n\to\infty} \frac{1}{2\pi i} \int_\gamma \frac{f_n(z)}{z - w} \mathrm{d}z \\ &= \frac{1}{2\pi i} \int_\gamma \lim_{n\to\infty} f_n(z)\ (z - w)^{-1}\ \mathrm{d}z \\ &= \frac{1}{2\pi i} \int_\gamma \frac{f(z)}{z - w}\ \mathrm{d}z, \end{aligned}$$

where we have used both Theorem 6.5.1 and Theorem 9.3.2. As this last integral is an analytic function of $w$ in $\mathbf{C}(\zeta, r)$, $f$ is analytic in $\mathbf{C}(\zeta, r)$ and hence in $D$.

Next observe that if $|w - \zeta| < \frac{1}{2}r$, then

$$|f_n^{(k)}(w) - f^{(k)}(w)| = \left| \frac{k!}{2\pi i} \int_\gamma \frac{[f_n(z) - f(z)]}{(z-w)^{k+1}} \, dz \right|$$

$$\leqslant (k!) r^{-k} 2^{k+1} \| f_n - f \|,$$

where

$$\| f_n - f \| = \sup \{ |f_n(z) - f(z)| : |z - \zeta| = r \}$$

and this tends to zero as $n \to \infty$.

This shows that for each $\zeta$ in $D$ there is an open disc on which $f_n^{(k)} \to f^k$ uniformly. The same is then true on any finite union of such discs and hence on any compact subset of $D$.

*Example 9.7.4* The function

$$\sum_{n=1}^{\infty} \frac{1}{z^2 - n^2}$$

is analytic in $\mathbf{C} - \mathbf{Z}$. For each positive integer $N$, if $|z| \leqslant N$ and $n > 2N$, then

$$\frac{1}{|z^2 - n^2|} \leqslant \frac{1}{n^2 - N^2} \leqslant \frac{2}{n^2}.$$

The Weierstrass test (§6.5) shows that the series

$$\sum_{n=N+1}^{\infty} \frac{1}{z^2 - n^2}$$

is uniformly convergent and hence analytic in $\mathbf{C}(0, N)$ and this is sufficient as $N$ is any positive integer.

We end this chapter by reconstructing the earlier ideas and techniques in the general situation and this culminates in the Residue Theorem. The reader should then see, in retrospect, that many of the earler results are special cases of this very general result.

Let $D$ be a domain and let $\zeta_1, \zeta_2, \ldots$ (possibly only finitely many) be points in $D$ which are isolated in the sense that for all $z$ in $D$ there exists a positive $r$ such that $\mathbf{C}(z, r)$ contains only a finite number of the $\zeta_j$. Further, we suppose that $f$ is analytic in $D^*$, where

$$D^* = D - \{\zeta_1, \zeta_2, \ldots\}.$$

The $\zeta_j$ may be the poles of $f$ but *they need not be so*; for example, consider $D = \mathbf{C}$, $f(z) = \exp(1/z)$ and $\zeta_1 = 0$ (Exercise 9.7.3).

Next, let $\Gamma$ be a cycle in $D$ such that $n(\Gamma, w) = 0$ if $w \notin D$ and suppose that the $\zeta_j$, are not on $\Gamma$. As any compact subset of $D$ contains only finitely many $\zeta_j$, we may relabel the $\zeta_j$ so that $\zeta_1, \ldots, \zeta_s$ are inside $\Gamma$ while $\zeta_{s+1}, \ldots$ are outside $\Gamma$. Specifically, we write $n_j = n(\Gamma, \zeta_j)$: then $n_1, \ldots, n_s$ are non-zero while $n_j = 0$ for $j \geqslant s + 1$.

We are now going to use the invariant form of Cauchy's Theorem to replace $\Gamma$ by the cycle $\Sigma$, $\Sigma = (\sigma_1, \ldots, \sigma_s)$, where

$$\sigma_j(t) = \zeta_j + re^{in_j t}, 0 \leqslant t \leqslant 2\pi.$$

We need to show that if $w \notin D^*$, then

$$n(\Gamma, w) = n(\Sigma, w). \tag{9.7.3}$$

If $w \notin D^*$, there are three possibilities, namely (a) $w \notin D$, (b) $w \in \{\zeta_1, \ldots, \zeta_s\}$ and (c) $w \in \{\zeta_{s+1}, \ldots\}$ . In cases (a) and (c) we know that $n(\Gamma, w) = 0$. As the inside of $\sigma_j$ lies entirely in $D$ except for the point $\zeta_j$, we also have $n(\sigma_j, w) = 0, j = 1, \ldots, s$, and so $n(\Sigma, w) = 0$. Thus (9.7.3) holds in cases (a) and (c).

In case (b) $w = \zeta_p$, say, and

$$\begin{aligned} n(\Sigma, w) &= \sum_{j=1}^{s} n(\sigma_j, \zeta_p) \\ &= n(\sigma_p, \zeta_p) \\ &= n_p \\ &= n(\Gamma, \zeta_p) \\ &= n(\Gamma, w). \end{aligned}$$

Thus (9.7.3) holds in all cases and so

$$\int_\Gamma f = \int_\Sigma f = \sum_{j=1}^{s} \int_{\sigma_j} f.$$

It should be clear that if

$$\gamma_j(t) = \zeta_j + re^{it}, \qquad 0 \leqslant t \leqslant 2\pi$$

(the same circle as $\sigma_j$ but traversed only once), then

$$\int_{\sigma_j} f = n_j \int_{\gamma_j} f$$

and so

$$\int_\Gamma f = \sum_{j=1}^{s} n(\Gamma, \zeta_j) \int_{\gamma_j} f.$$

We now *define* the residue of $f$ at $\zeta_j$ by

$$\operatorname{Res}(f, \zeta_j) = \frac{1}{2\pi i} \int_{\gamma_j} f$$

and (by the invariant form of Cauchy's Theorem) this is independent of $r$ (the radius of $\gamma_j$) provided that $r$ is sufficiently small. With this definition we have our final version of the Residue Theorem.

*The Residue Theorem* *With the above hypotheses,*

$$\int_\Gamma f = \sum_j n(\Gamma, \zeta_j) \operatorname{Res}(f, \zeta_j).$$

This is applicable to any domain $D$ without exception: because $n(\Gamma, \zeta_j) = 0$ if $j > s$ we may take this sum over all $j$, and this removes any reference to a specific and irrelevant labelling of the $\zeta_j$. In fact, it is also true (with the natural definition of the residue) if the $\zeta_j$ are isolated components of $\mathbf{C} - D$ (and not necessarily points).

Of course, if $\zeta_j$ is a pole of $f$, that is, if

$$f(z) = \frac{a_{-m}}{(z - \zeta_j)^m} + \cdots + \frac{a_{-1}}{(z - \zeta_j)} + a_0 + a_1(z - \zeta_j) + \cdots, \tag{9.7.4}$$

say, near $\zeta_j$ we see immediately that the new definition of the residue agrees with the earlier definition. We repeat, however, that the $\zeta_j$ need not be poles of $f$. If every $\zeta_j$ is a pole of $f$ we say that $f$ is *meromorphic* in $D$: this is the promised extension of our earlier definition.

It remains only to show that even if $\zeta_j$ is not a pole of $f$ there is still a series expansion of $f$ near $\zeta_j$ similar to (9.7.4). In fact, it is no harder to prove the following more general result.

*Theorem 9.7.2* *Let $f$ be analytic in the annulus*

$$A = \{z \in \mathbf{C} : R_1 < |z - \zeta| < R_2\}.$$

*Then there are numbers $a_n$, $n \in \mathbf{Z}$, depending on $f$, $\zeta$, $R_1$ and $R_2$ such that*

$$f(z) = \sum_{n=-\infty}^{\infty} a_n (z - \zeta)^n, \qquad z \in A.$$

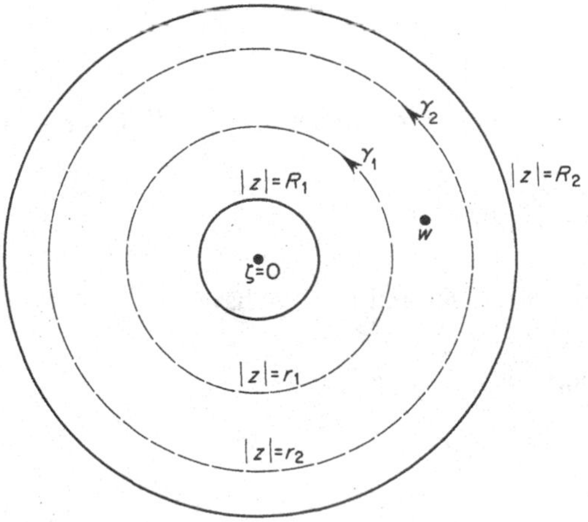


Diagram 9.7.2

*Further, if* $R_1 = 0$, *then* Res $(f, \zeta) = a_{-1}$.

This is called the *Laurent expansion* of $f$ in $A$.

*Proof* We may take $\zeta = 0$. Next, select any numbers $r_1, r_2$ satisfying $R_1 < r_1 < r_2 < R_2$ and let

$$\gamma_j(t) = r_j e^{it}, \qquad 0 \leqslant t \leqslant 2\pi, \quad j = 1, 2$$

(see Diagram 9.7.2).

Now let $\Gamma = (\gamma_1^-, \gamma_2)$ so that if $r_1 < |w| < r_2$, then $n(\gamma_1, w) = 0$, $n(\gamma_2, w) = 1$ and $n(\Gamma, w) = 1$. Cauchy's Integral formula with $k = 0$ gives

$$\begin{aligned} f(w) &= \frac{1}{2\pi i} \int_\Gamma \frac{f(z)}{z - w} \, dz \\ &= \frac{1}{2\pi i} \int_{\gamma_2} \frac{f(z)}{z - w} \, dz - \frac{1}{2\pi i} \int_{\gamma_1} \frac{f(z)}{z - w} \, dz \\ &= \frac{1}{2\pi i} \int_{\gamma_2} \frac{f(z)}{z} \sum_{n=0}^{\infty} \left(\frac{w}{z}\right)^n dz + \frac{1}{2\pi i} \int_{\gamma_1} \frac{f(z)}{z} \sum_{n=1}^{\infty} \left(\frac{z}{w}\right)^n dz. \end{aligned}$$

As these series converge uniformly (as functions of $z$) on $\gamma_2$ and $\gamma_1$ respectively,

$$\begin{aligned} f(w) &= \left( \sum_{n=0}^{\infty} \frac{1}{2\pi i} \int_{\gamma_2} \frac{f(z)}{z^{n+1}} \, dz \right) w^n + \left( \sum_{n=1}^{\infty} \frac{1}{2\pi i} \int_{\gamma_1} \frac{f(z)}{z^{1-n}} \right) w^{-n} \\ &= \sum_{n=0}^{\infty} a_n w^n + \sum_{n=-1}^{-\infty} b_n w^n, \end{aligned}$$

where

$$a_n = \frac{1}{2\pi i} \int_{\gamma_2} \frac{f(z)}{z^{n+1}} \, dz, \qquad b_n = \frac{1}{2\pi i} \int_{\gamma_1} \frac{f(z)}{z^{n+1}} \, dz.$$

The invariant form of Cauchy's Theorem shows that for each integer $m$

$$\int_\gamma \frac{f(z)}{z^m} \, dz$$

is independent of $r$, $R_1 < r < R_2$, where $\gamma(t) = r e^{it}$, $0 \leqslant t \leqslant 2\pi$. Thus if we now write

$$a_n = \frac{1}{2\pi i} \int_\gamma \frac{f(z)}{z^{n+1}} \, dz$$

(this depends only on $f$, $\zeta$, $R_1$ and $R_2$), we have

$$f(z) = \sum_{n=0}^{\infty} a_n z^n + \sum_{n=-1}^{-\infty} a_n z^n.$$

By definition,

$$a_{-1} = \frac{1}{2\pi i} \int_\gamma f(z) \, dz$$

and if $R_1 = 0$ this holds for all sufficiently small $r$. Thus Res $(f, 0) = a_{-1}$.

In fact, if we know that

$$f(z) = \sum_{n=0}^{\infty} A_n z^n + \sum_{n=-1}^{-\infty} A_n z^n,$$

where these series converge uniformly on some circle $|z| = r$, then

$$\int_{|z|=r} \frac{f(z)}{z^{m+1}}\, dz = \sum_{n=-\infty}^{\infty} A_n \int_{|z|=r} z^{n-m-1}\, dz$$

$$= 2\pi i A_m.$$

This shows that the *Laurent coefficients* $a_n$ in Theorem 9.7.2 are uniquely determined by $f$, $\zeta$, $R_1$ and $R_2$: they are *not* uniquely determined by $f$ and $\zeta$ alone (Exercise 9.7.4).

**Exercise 9.7**

1. Prove that $\alpha(R) \to 0$ as $R \to +\infty$ (see Example 9.7.1).
2. Prove that $\beta(\delta) \to 0$ as $\delta \to 0$ (see Example 9.7.1).
3. Prove that $z = 0$ is not a pole of the function $\exp(1/z)$.
4. Let $f(z) = [z(z-1)]^{-1}$. Find the Laurent expansion of $f$ in each of the sets

   $$\{z : 0 < |z| < 1\}, \{z : |z| > 1\}$$

   and hence show that the Laurent coefficients of $f$ do not depend on the centre of the annulus (in this case zero) and $f$ alone.
5. Let $f$ be analytic and satisfy

   $$f(z) = \sum_{n=-\infty}^{+\infty} a_n z^n$$

   in some $\mathbf{C}^*(0, r)$, $r > 0$. Prove that $f$ has a pole at zero if and only if for some negative integer $m$, $a_m \neq 0$ and $a_n = 0$ when $n < m$. Prove also that $f$ is bounded in some $\mathbf{C}^*(0, \epsilon)$ if and only if $a_n = 0$ when $n < 0$, and deduce that in this case $f$ (when $f(0)$ is suitably defined) is analytic in $\mathbf{C}(0, r)$.

   By considering the functions $g$, $g(z) = \delta/[f(z) - w]$, for each complex number $w$ and each positive $\delta$, show that either $\lim f(z)$ exists as $z \to 0$ or that for each complex $w$ there is a sequence $z_n$ with

   $$\lim_{n\to\infty} z_n = 0, \qquad \lim_{n\to\infty} f(z_n) = w.$$

6. Use Cauchy's Integral Formula for $f^{(k)}$ to derive Cauchy's inequalities (8.4.3).
7. Show that if $a > 0$, then

   $$\int_0^{\infty} \frac{[\log_e(x)]^2}{x^2 + a^2}\, dx = \frac{\pi}{a}[\log_e(a)]^2 - \frac{\pi^3}{4a}.$$

8. Show that

   $$\int_0^{\infty} \frac{\log_e(x)}{x^4 + 1}\, dx = -\pi^2/8\sqrt{2}.$$

9. Show that

$$\int_0^\infty \frac{\sqrt{x}}{(1+x^2)^2}\,dx = \frac{\pi}{4\sqrt{2}}.$$

10. Show that if $0 < b < 1$, then

$$\int_0^\infty \frac{x^b}{x(1+x)}\,dx = \frac{\pi}{\sin(\pi b)}.$$

[*Hint:* let $\Gamma$ be the cycle given in Example 9.7.1 with an additional small indentation about the point $-1$ and let $\operatorname{Arg} z$ be in $[0, \pi]$ when $z$ is on $\Gamma$. Let $\Gamma^*$ be the reflection of $\Gamma$ (suitably orientated) in the real axis and let $\operatorname{Arg} z$ be in $[\pi, 2\pi]$ when $z$ is on $\Gamma^*$. Consider the sum of the integrals over $\Gamma$ and $\Gamma^*$.]

11. Let

$$f(z) = \frac{\pi^2}{\sin^2(\pi z)}, \qquad g(z) = \sum_{n=-\infty}^{+\infty} \frac{1}{(z-n)^2}.$$

Show that

(a) for each integer $n$, $f(z) - (z-n)^{-2}$ is analytic in $\mathbf{C}(n, 1)$;
(b) $g$ is analytic in $\mathbf{C} - \mathbf{Z}$;
(c) $\lim_{z\to 0} f(z) - 1/z^2 = \pi^2/3$;
(d) $\lim_{z\to 0} g(z) - 1/z^2 = 2\Sigma_{n=1}^{\infty}(1/n^2)$;
(e) $h$ defined by $h(z) = f(z) - g(z)$ is analytic in $\mathbf{C}$;
(f) $h(0) = 0$ (use Example 9.7.3);
(g) if $N$ is a positive integer and if $|z| = N + \frac{1}{2}$, then

$$|g(z)| \leqslant 4 \sum_{n=0}^{\infty} \frac{1}{(n+\frac{1}{2})^2} < 3\pi^2.$$

Now use Theorem 4.4.1, the Maximum Modulus Theorem and Liouville's Theorem to conclude that $h$ is constant on $\mathbf{C}$. This proves that

$$\frac{\pi^2}{\sin^2(\pi z)} = \sum_{n=-\infty}^{+\infty} \frac{1}{(z-n)^2}.$$

12. Repeat the work in Example 9.7.3 after replacing $z^{-2}$ by any rational function $f$ which satisfies $|z^2 f(z)| \leqslant M$ for $|z| > r$, say.

Deduce that if $f$ has poles $z_1, \ldots, z_s$, then

$$\sum_{k=1}^{s} \operatorname{Res}(F, z_k) + \sum_n{}^* f(n) = 0,$$

where

$$F(z) = f(z)\cot(\pi z)$$

and where $\Sigma^*$ denotes summation over all integers which are distinct from $z_1, \ldots, z_s$. This provides a means of evaluating infinite series.

Show that

$$\sum_{n=1}^{\infty} \frac{1}{n^2+1} = \frac{\pi \cosh(\pi)}{2 \sinh(\pi)} - \frac{1}{2}.$$

13. Use the same technique as in Exercise 12 using cosec instead of cot. Thus prove that, under similar conditions,

$$\sum_{k=1}^{s} \operatorname{Res}(F, z_k) + \sum_{n}{}^{*}(-1)^n f(n) = 0.$$

Show that

$$\sum_{n=1}^{\infty} \frac{(-1)^n}{n^2} = \frac{-\pi^2}{12}.$$

14. Prove that for each even positive integer $k$,

$$\pi^{-k} \sum_{n=1}^{\infty} \frac{1}{n^k}$$

is a rational number.

# Chapter 10

## 10.1 CONFORMAL MAPPING

We recall that a function $f : D \to \mathbf{C}$ is differentiable at $\zeta$ if there exists a number $f^{(1)}(\zeta)$ such that

$$f(z) = f(\zeta) + (z - \zeta)f^{(1)}(\zeta) + (z - \zeta)\epsilon(z) \qquad (10.1.1)$$

where $\epsilon$ is continuous at $\zeta$ and $\epsilon(\zeta) = 0$. According to this definition if a *real-valued* function $u$ is differentiable in $D$, then $u$ is necessarily constant in $D$: this (and much more) follows from Theorem 8.5.2 because if $u(D) \subset \mathbf{R}$ then $u(D)$ cannot be an open subset of $\mathbf{C}$.

It is, however, possible to obtain a real linear approximation of a real function $u$ in a domain $D$ without $u$ necessarily being constant. We say that $u : D \to \mathbf{R}$ is *real-differentiable* at $\zeta$ if and only if there are real numbers $\alpha$ and $\beta$ and a continuous function $\epsilon_1 : D \to \mathbf{R}$ with $\epsilon_1(\zeta) = 0$ so that

$$u(z) = u(\zeta) + (x - \zeta_1)\alpha + (y - \zeta_2)\beta + |z - \zeta|\epsilon_1(z), \qquad z \in D, \qquad (10.1.2)$$

where $\zeta = \zeta_1 + i\zeta_2$ and $z = x + iy$. Obviously, if this holds, then

$$\frac{\partial u}{\partial x}(\zeta) = \lim_{h\to 0,\, h\in\mathbf{R}} \frac{u(\zeta + h) - u(\zeta)}{h} = \alpha, \qquad \frac{\partial u}{\partial y}(\zeta) = \beta$$

and these are the *partial derivatives* of $u$ at $\zeta$.

These three concepts of differentiability are related in the following way.

*Theorem 10.1.1* *Let $\zeta$ be in a domain D, let f be defined in D and let $f = u + iv$. Then f is differentiable at $\zeta$ if and only if both u and v are real-differentiable at $\zeta$ and the Cauchy–Riemann equations*

$$\frac{\partial u}{\partial x} = \frac{\partial v}{\partial y}, \qquad \frac{\partial u}{\partial y} = \frac{-\partial v}{\partial x} \qquad (10.1.3)$$

*hold at $\zeta$.*

*Proof* If (10.1.1) holds we write $f^{(1)}(\zeta) = \alpha + i\beta$ and equate the real and imaginary parts of each side of (10.1.1). The error terms are of the required order of magnitude near $\zeta$; for example, we can write

$$\text{Re }[(z-\zeta)\epsilon(z)] = |z-\zeta|\,\epsilon_1(z),$$

where $|\epsilon_1(z)| < |\epsilon(z)|$, so $u$ and $v$ are real-differentiable at $\zeta$. Moreover, the partial derivatives of $u$ and $v$ can be identified with $\alpha$ and $\beta$ and this gives (10.1.3).

If (10.1.2), a similar approximation for $v$ and (10.1.3) are all true, then we simply compute $u(z) + iv(z)$ and obtain (10.1.1) with

$$f^{(1)}(\zeta) = \frac{\partial u}{\partial x}(\zeta) + i\frac{\partial v}{\partial x}(\zeta).$$

We shall now examine in detail the local behaviour of a differentiable function. To begin with, we suppose only that $f$ is differentiable at a point $\zeta$ in $D$ and that $f^{(1)}(\zeta) \neq 0$. In this case $f$ is approximated near $\zeta$ by $f(\zeta) + (z-\zeta)f^{(1)}(\zeta)$ and this is a composition of a magnification, a rotation and a translation. This suggests that $f$ is *conformal* at $\zeta$, that is that the angle between two smooth curves meeting at $\zeta$ is preserved by $f$, and this is indeed true.

Let $\gamma$ and $\sigma$ be any two curves passing through $\zeta$: we may suppose that $\gamma(0) = \sigma(0) = \zeta$ and we shall assume that $\gamma^{(1)}(0)$ and $\sigma^{(1)}(0)$ are not zero. As $t \to 0$, so

$$\frac{\gamma(t) - \gamma(0)}{t} \to \gamma^{(1)}(0)$$

and, in the obvious sense, Arg $[\gamma(t) - \gamma(0)]$ (which is the angle made by the chord from $\gamma(0)$ to $\gamma(t)$) tends to Arg $\gamma^{(1)}(0)$ (this only requires a continuous choice of Arg near $\gamma^{(1)}(0)$). The angle between the tangents of $\gamma$ and $\sigma$ at $\zeta$ is therefore given by

$$\theta = \text{Arg }[\gamma^{(1)}(0)/\sigma^{(1)}(0)].$$

The image curves are $f \circ \gamma$ and $f \circ \sigma$, and as

$$\frac{(f\circ\gamma)^{(1)}(0)}{(f\circ\sigma)^{(1)}(0)} = \frac{f^{(1)}(\zeta)\gamma^{(1)}(0)}{f^{(1)}(\zeta)\sigma^{(1)}(0)} = \frac{\gamma^{(1)}(0)}{\sigma^{(1)}(0)}$$

we see that $f$ is conformal at $\zeta$: note that we have needed the fact that $f^{(1)}(\zeta) \neq 0$.

*Example 10.1.1* Suppose that $0 < \alpha < 2\pi$ and let $D(\alpha)$ be the 'wedged-shaped' domain given by

$$\{re^{i\theta} : 0 < r, 0 < \theta < \alpha\}$$

as illustrated in Diagram 10.1.1.

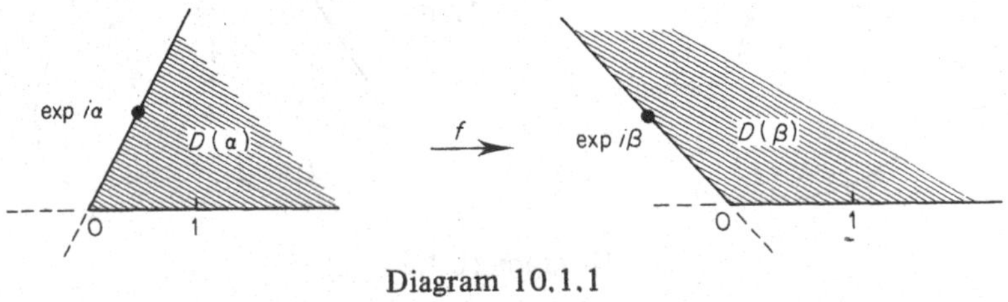


Diagram 10.1.1

Now define

$$f(z) = z^{\beta/\alpha} = \exp\left[\left(\frac{\beta}{\alpha}\right)\log(z)\right],$$

where

$$\log(z) = \log_e(|z|) + i\theta(z)$$

and $\theta(z)$ is the unique choice of Arg $(z)$ in $(0, 2\pi)$. Clearly $f$ is differentiable in $D(\alpha)$ and

$$f^{(1)}(z) = \left(\frac{\beta}{\alpha z}\right)\exp\left[\left(\frac{\beta}{\alpha}\right)\log(z)\right] \neq 0.$$

This shows that $f$ is conformal at each $z$ in $D(\alpha)$: thus $f$ preserves angles at points arbitrarily close to zero. Nevertheless, if we define $f(0) = 0$, $f$ is a continuous map of $\overline{D(\alpha)}$ onto $\overline{D(\beta)}$ and $f$ does not preserve angles at zero.

If $f$ is differentiable throughout $D$ and if $f^{(1)}(\zeta) \neq 0$ we can assert much more than conformality.

*Theorem 10.1.2* *Let $f$ be differentiable in $D$, let $\zeta$ be in $D$ and suppose that $f^{(1)}(\zeta) \neq 0$. Let $\Delta = \mathbf{C}(\zeta, r)$, $r > 0$, be contained in $D$ and let $\gamma(t) = \zeta + re^{it}$, $0 \leqslant t \leqslant 2\pi$. Then for all sufficiently small $r$,*

(a) $f: \Delta \to f(\Delta)$ *is 1–1 and analytic;*
(b) $f^{-1}: f(\Delta) \to \Delta$ *is 1–1 and analytic and*
(c) *if $w \in f(\Delta)$, then*

$$f^{-1}(w) = \frac{1}{2\pi i}\int_\gamma \frac{zf^{(1)}(z)}{f(z) - w}\,dz.$$

Let us assume this result for the moment. As $f$ is differentiable in $D$ it is analytic and by Theorem 8.5.2, $f(\Delta)$ is an open set containing $f(\zeta)$. Each horizontal segment in $f(\Delta)$ is mapped by $f^{-1}$ to a curve in $\Delta$ on which $u(=\text{Re}\,[f])$ is constant and there is exactly one such curve, say $\gamma$, which passes through $\zeta$. Similarly, there is

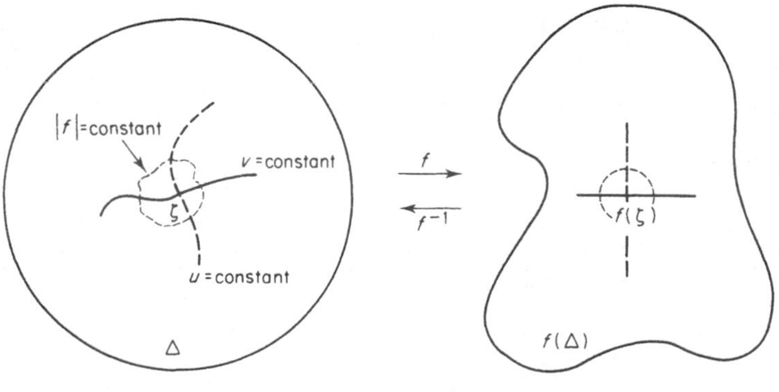


Diagram 10.1.2

exactly one curve $\sigma$ which passes through $\zeta$ and on which $v$ ( = Im $[f]$ is constant and as $f$ is conformal, $\gamma$ and $\sigma$ are orthoganol at $\zeta$. In this way, we can derive information about the so-called *level curves* in $\Delta$ given by $u$ = constant (or $v$ = constant). This is illustrated in Diagram 10.1.2.

*Proof of Theorem 10.1.2* Observe first that we have already proved that for sufficiently small $r, f: \bar{\Delta} \to f(\bar{\Delta})$ is a homeomorphism. This is contained in the second corollary of Theorem 8.5.2 (where we have now written $\Delta$ and $\gamma$ for $\Sigma$ and $\sigma$) because, as $a_1 = f^{(1)}(\zeta)$, $a_1 \neq 0$ if and only if $f^{(1)}(\zeta) \neq 0$.

It remains only to prove that $f^{-1}$ is analytic in $f(\Delta)$ and that (c) holds. Let

$$g(w) = \frac{1}{2\pi i} \int_\gamma \frac{zf^{(1)}(z)}{f(z) - w}\, dz:$$

this is defined when $w \in f(\Delta)$ for, as $f$ is 1–1 on $\bar{\Delta}$, $f(\Delta)$ does not meet $[f \circ \gamma]$. We shall now use the Residue Theorem to evaluate $g(w)$. We know that $f(z) = w$ has precisely one solution inside $\gamma$, say $z_0$, and this shows that the integrand has a simple pole at $z_0$. Thus as $z_0 = f^{-1}(w)$,

$$\begin{aligned} g(w) &= \lim_{z \to z_0} \frac{(z - z_0)zf^{(1)}(z)}{f(z) - w} \\ &= \lim_{z \to z_0} \frac{(z - z_0)zf^{(1)}(z)}{f(z) - f(z_0)} \\ &= z_0 \\ &= f^{-1}(w) \end{aligned}$$

and (c) holds.

There are a variety of ways of showing that $f^{-1}( = g)$ is differentiable and hence analytic in $f(\Delta)$. Using (c) we find (after a little algebra) that

$$\begin{aligned} \frac{g(w_1) - g(w)}{w_1 - w} - \frac{1}{2\pi i} \int_\gamma \frac{zf^{(1)}(z)}{[f(z) - w]^2}\, dz \\ = \frac{(w_1 - w)}{2\pi i} \int_\gamma \frac{zf^{(1)}(z)}{[f(z) - w]^2 [f(z) - w_1]}\, dz. \end{aligned}$$

For some positive $\epsilon$, this integral is bounded when $|w_1 - w| < \epsilon$ and so letting $w_1 \to w$ we find that $g^{(1)}(w)$ exists. The actual form is of no importance: however, as $g(f(z)) = z$,

$$g^{(1)}(f(z))f^{(1)}(z) = 1.$$

We can easily generalize Theorem 10.1.2 to the case where $f^{(1)}(\zeta) = 0$. As $f$ is analytic we may write

$$f(z) = f(\zeta) + (z - \zeta)^m g(z),$$

where $m = v_f(\zeta)$ and $g$ is analytic in $D$ with $g(\zeta) \neq 0$ (Theorem 10.1.2 is the case $m = 1$).

For a suitably small positive $r$, $g$ maps $\mathbf{C}(\zeta, r)$ into some half-plane containing $g(\zeta)$ but not zero, and we can find a continuous branch $G(z)$ of $\log g(z)$ in $\mathbf{C}(\zeta, r)$. The function

$$h(z) = (z - \zeta) \exp\left[\frac{1}{m} G(z)\right]$$

is differentiable (as a composition of differentiable functions) and hence analytic in $\mathbf{C}(\zeta, r)$. Obviously,

$$f(z) = f(\zeta) + [h(z)]^m$$

and by a direct computation

$$h^{(1)}(\zeta) = \exp\left[\frac{1}{m} G(\zeta)\right] \neq 0.$$

Theorem 10.1.2 shows that $h$ is 1–1 on a suitably small disc $\Delta$ containing $\zeta$ and we have obtained the following generalization of Theorem 10.1.2.

*Theorem 10.1.3* *Let $f$ be differentiable in a domain $D$, let $\zeta$ be in $D$ and let $m = v_f(\zeta)$. Then there is an open disc $\Delta$ with centre $\zeta$ and a function $h$, 1–1 and analytic in $\Delta$, such that*

$$f(z) = f(\zeta) + [h(z)]^m. \tag{10.1.4}$$

This result shows that locally, $f$ is a composition of a 1–1 analytic map $h$ (with an analytic inverse), a translation $T$, $T(z) = z + f(\zeta)$, and $p$, $p(z) = z^m$. Thus the multiple-valued character near $\zeta$ of the inverse $f^{-1}$ is essentially the same as that found by taking $m$th roots.

*Example 10.1.2* The representation (10.1.4) of the function $f$, $f(z) = \sin^2 z$, is given by

(a) $h(z) = \sin(z)$, $v_f(\zeta) = 2$ if $\zeta = n\pi$, $n \in \mathbf{Z}$,
(b) $h(z) = i \cos(z)$, $v_f(\zeta) = 2$ if $\zeta = \frac{1}{2}\pi + n\pi$, $n \in \mathbf{Z}$,
(c) $h(z) = \sin^2(z) - \sin^2(\zeta)$, $v_f(\zeta) = 1$ otherwise.

The reader should verify the values of $v_f(\zeta)$ by differentiation ($v_f(\zeta)$ is the smallest $k$ with $f^{(k)}(\zeta) \neq 0$) and then verify (10.1.4) in each case.

**Exercise 10.1**

1. Let $\zeta$ be in a domain $D$ and let $f: D \to \mathbf{C}$ be differentiable at $\zeta$. Define the directional derivative $\mathrm{d}_\alpha f$ ($\alpha$ real) of $f$ at $\zeta$ by

$$\mathrm{d}_\alpha f(\zeta) = \lim_{t \to 0,\, t > 0} \frac{f(\zeta + te^{i\alpha}) - f(\zeta)}{t}.$$

Prove

(a) if $\alpha = 0$, then $d_\alpha f = \dfrac{\partial f}{\partial x}$

(b) if $\alpha = \pi/2$, then $d_\alpha f = \dfrac{\partial f}{\partial y}$

(c) $e^{-i\alpha}\, d_\alpha f$ is independent of $\alpha$.

Derive the Cauchy–Riemann equations from (a), (b) and (c). [Thus (c) is the radially symmetric from of these equations.]

2. Let $f$ be analytic in a domain $D$ and suppose that $\zeta \in D$ and $f(\zeta) \neq 0$. Given a positive integer $m$, construct an analytic $m$th root of $f$ (an analytic function $g$ with $f(z) = [g(z)]^m$) in some $C(\zeta, \epsilon)$.
3. Let $f$ be analytic in a domain $D$ and suppose that $\zeta \in D$. Show that there exists a positive $\epsilon$ such that for all $w$ in $C^*(f(\zeta), \epsilon)$,

(a) $f^{-1}\{w\}$ consists of $v_f(\zeta)$ distinct points $z_j$ and
(b) $v_f(z_j) = 1$.

[This shows that the points where $f$ is not locally 1–1 are isolated in $D$.]

## 10.2 STEREOGRAPHIC PROJECTION

We are now going to adjoin a single point $\infty$ to $\mathbf{C}$ in such a way that we may regard a pole $w$ of $f$ as a point where $f$ is continuous and satisfies $f(w) = \infty$. To achieve this, we select any point not in $\mathbf{C}$, denote it by $\infty$ and form the union $\mathbf{C} \cup \{\infty\}$. This set is called the *extended complex plane* and is denoted by $\mathbf{C}_\infty$.

We do not attempt to extend the algebra of $\mathbf{C}$ to $\mathbf{C}_\infty$; instead, we shall be guided solely by considerations of continuity. The topological structure that we are about to impose on $\mathbf{C}_\infty$ is most easily visualized by specific reference to a model of $\mathbf{C}_\infty$ constructed in three-dimensional Euclidean space $\mathbf{R}^3$. We shall need (and we assume familiarity with) the basic ideas of distance and continuity in $\mathbf{R}^3$.

Let $S$ be the sphere

$$\{(x_1, x_2, x_3) \in \mathbf{R}^3 : x_1^2 + x_2^2 + (x_3 - \tfrac{1}{2})^2 = \tfrac{1}{4}\}:$$

this has centre $(0, 0, \frac{1}{2})$ and the plane $x_3 = 0$ is the tangent plane of $S$ at $(0, 0, 0)$. The plane $x_3 = 0$ is now identified with $\mathbf{C}$ by the natural identification of $x + iy$ or $(x, y)$ with $(x, y, 0)$. The point $(0, 0, 1)$ is in $S$ but not in $\mathbf{C}$ and this is chosen as the point $\infty$.

Any point $\zeta$ in $S$ other than $\infty$ can be projected along the half-line from $\infty$ to $\zeta$ to meet $\mathbf{C}$ in a point $z$ uniquely determined by $\zeta$. This is illustrated in Diagram 10.2.1.

We denote this projection by $\phi$: it is called the *stereographic projection* of $S$ onto $\mathbf{C}_\infty$. Thus if $\zeta = (\zeta_1, \zeta_2, \zeta_3)$, $z = x + iy$ and $\phi(\zeta) = z$, then

$$(x, y, 0) = \phi(\zeta) = (0, 0, 1) + t[(\zeta_1, \zeta_2, \zeta_3) - (0, 0, 1)],$$

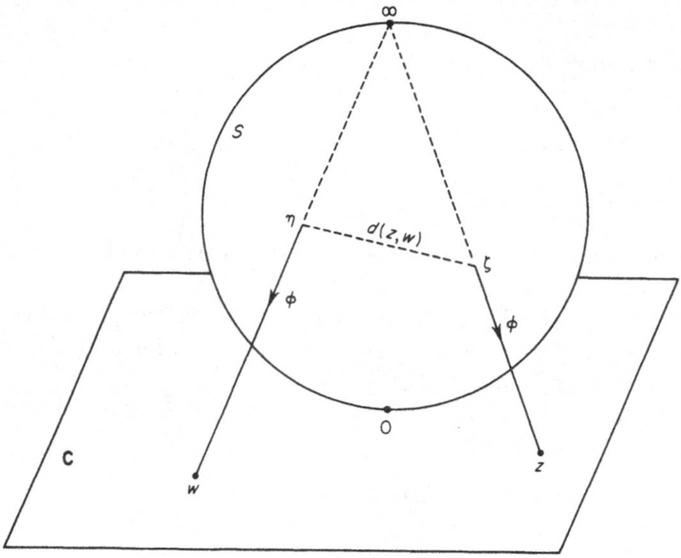


Diagram 10.2.1

where $0 = 1 + t(\zeta_3 - 1)$. This yields an explicit formula for $\phi$, namely

$$\phi(\zeta) = \frac{\zeta_1 + i\zeta_2}{1 - \zeta_3}. \tag{10.2.1}$$

A similar argument (Exercise 10.2.1) gives a formula for $\phi^{-1}$:

$$\phi^{-1}(x + iy) = \left(\frac{x}{1 + |z|^2}, \frac{y}{1 + |z|^2}, \frac{|z|^2}{1 + |z|^2}\right). \tag{10.2.2}$$

It should now be clear that $\phi$ is a 1–1 map of $S - \{\infty\}$ onto **C** and that both $\phi$ and $\phi^{-1}$ are continuous (Exercise 10.2.2). Observe that if $|z|$ is large, then $\phi^{-1}(z)$ is nearly $\infty$. This leads naturally to the definition $\phi(\infty) = \infty$ ($\infty$ is in both $S$ and $\mathbf{C}_\infty$) and $\phi$ is now a 1–1 map of $S$ onto $\mathbf{C}_\infty$.

Next, we define the *chordal distance* $d(z, w)$ between the points $z$ and $w$ in $\mathbf{C}_\infty$ as the length of the chord of $S$ which joins $\phi^{-1}(z)$ and $\phi^{-1}(w)$. Thus

$$\begin{aligned} d(z, w) &= |\phi^{-1}(z) - \phi^{-1}(w)| \\ &= [(\zeta_1 - \eta_1)^2 + (\zeta_2 - \eta_2)^2 + (\zeta_3 - \eta_3)^2]^{1/2} \end{aligned}$$

where $\phi^{-1}(z) = (\zeta_1, \zeta_2, \zeta_3)$ and $\phi^{-1}(w) = (\eta_1, \eta_2, \eta_3)$. Using (10.2.2), we obtain the expression

$$d(z, w) = \begin{cases} \dfrac{|z - w|}{[(1 + |z|^2)(1 + |w|^2)]^{1/2}} & \text{if } z, w \neq \infty; \\[2ex] \dfrac{1}{(1 + |z|^2)^{1/2}} & \text{if } z \neq \infty, w = \infty \end{cases} \tag{10.2.3}$$

and, of course, $d(\infty, \infty) = 0$ (Exercise 10.2.3).

The question of continuity in relation to $\mathbf{C}_\infty$ is now quite straightforward. If $E \subset \mathbf{C}_\infty$, $w \in E$ and $f: E \to \mathbf{C}_\infty$ is any function we say that *f is continuous at w* (*with respect to the chordal distance*) if and only if for every positive $\epsilon$ there is a positive $\delta$ such that

$$d(f(z), f(w)) < \epsilon$$

whenever $z \in E$ and $d(z, w) < \delta$. Of course, $w$ or $f(w)$ (or both) may be $\infty$.

We can transfer the situation from $\mathbf{C}_\infty$ to $S$ and express this continuity in terms of functions on $S$. Given $f$ as above, we define $f^* = \phi^{-1} \circ f \circ \phi$: this maps $\phi^{-1}(E)$ (a subset of $S$) to $S$. If $z = \phi(\zeta)$ and $w = \phi(\eta)$, then

$$d(z, w) = |\zeta - \eta|$$

and

$$d(f(z), f(w)) = |f^*(\zeta) - f^*(\eta)|,$$

and we see at once that $f$ is continuous at $w$ with respect to the chordal distance if and only if $f^*$ is continuous at $\eta$.

Now suppose that $w$ and $f(w)$ are distinct from $\infty$. As $\phi$ and $\phi^{-1}$ are continuous, we also see that $f^*$ is continuous at $\eta$ if and only if $f$ is continuous (in the earlier sense) at $w$. We conclude that $f$ is continuous at $w$ with respect to the chordal distance if and only if it is continous at $w$ in the earlier sense. It is no longer necessary, then, to distinguish between the two definitions of continuity.

*Example 10.2.1* Let $J: \mathbf{C}_\infty \to \mathbf{C}_\infty$ be defined by $J(0) = \infty$, $J(\infty) = 0$ and $J(z) = 1/z$ when $z \neq 0, \infty$. A direct substitution in (10.2.3) shows that

$$d(J(z), J(w)) = d(z, w) \tag{10.2.4}$$

and so $J$ is continuous on $\mathbf{C}_\infty$. We shall reserve the symbol $J$ for this function throughout this Chapter. Observe that $J^{-1} = J$.

Consider now a function $f$ which is meromorphic in some domain $D$ with a pole at $w$. Then $1/f(z) \to 0$ as $z \to w$ and so if we define $f(w) = \infty$ we see that $J(f(z)) \to J(f(w))$ as $z \to w$. This shows that $J \circ f$ and hence $f$ are continuous at $w$.

The same idea can be used to clarify the behaviour of a function $f$ at and near $\infty$. If $f$ is defined on some set

$$D = \{z \in \mathbf{C} : |z| > r\} \cup \{\infty\},$$

then $f \circ J$ is defined on $J(D)$, $(f \circ J)(z) = f(1/z)$ and $J(D) = \mathbf{C}(0, r^{-1})$. We say that

(a) $f$ is *analytic at* $\infty$ if $f \circ J$ is analytic in $J(D)$ and
(b) $f$ has a *pole at* $\infty$ if $f \circ J$ has a pole at zero ($= J(\infty)$).

The valency of $f$ at $\infty$ is defined to be the corresponding valency of $f \circ J$ at zero. We say that $f$ is *analytic in* $\mathbf{C}_\infty$ if it is analytic in $\mathbf{C}$ and at $\infty$: it is *meromorphic in* $\mathbf{C}_\infty$ if it is meromorphic in $\mathbf{C}$ and if (a) or (b) holds.

*Example 10.2.2* Let

$$p(z) = a_0 + a_1 z + \ldots + a_d z^d,$$

where $a_d \neq 0$ and $d \geqslant 1$. As $|p(z)| \to +\infty$ as $|z| \to +\infty$ we define $p(\infty) = \infty$. Then for $z \neq 0$,

$$p(J(z)) = a_0 + \frac{a_1}{z} + \ldots + \frac{a_d}{z^d}$$

and so $p$ has a pole of order $d$ at $\infty$. This shows that $p$ is meromorphic in $\mathbf{C}_\infty$ and for each $w$ in $\mathbf{C}_\infty$, $p(z) = w$ has exactly $d$ solutions in $\mathbf{C}_\infty$. The $d$ solutions of $p(z) = \infty$ all occur at $\infty$.

*Example 10.2.3* Each rational function $R$ is meromorphic in $\mathbf{C}_\infty$. Indeed, it is only necessary to show that $R \circ J$ is analytic or has a pole at zero and this is clear as (with a suitable definition of $R(\infty)$) $R \circ J$ is itself rational.

Example 10.2.3 has a converse.

*Theorem 10.2.1* (a) *If $f$ is analytic in $\mathbf{C}_\infty$, then $f$ is constant*; (b) *if $f$ is meromorphic in $\mathbf{C}_\infty$, then $f$ is a rational function.*

*Proof* First suppose that $f$ is analytic in $\mathbf{C}_\infty$. Then $f(1/z)$ $(= f \circ J(z))$ is analytic and hence bounded on some $\mathbf{C}(0, \delta)$, $\delta > 0$. Thus $f$ is bounded on $\{z \in \mathbf{C} : |z| > \delta^{-1}\}$ and as $f$ is bounded on the compact set $\bar{\mathbf{C}}(0, \delta^{-1})$ we see that $f$ is actually bounded on $\mathbf{C}$. Thus by Liouville's Theorem, $f$ is constant on $\mathbf{C}$ and hence (by continuity) on $\mathbf{C}_\infty$.

To prove (b) we first show that $f$ has only a finite number of poles in $\mathbf{C}_\infty$. Indeed as $f$ is meromorphic in, say, $\{z \in \mathbf{C} : |z| > r\} \cup \{\infty\}$, $f \circ J$ is meromorphic in $\mathbf{C}(0, r^{-1})$. We deduce that for some positive $\delta$, $f \circ J$ has no poles in $\mathbf{C}^*(0, \delta)$ and this means that $f$ itself has no poles in $\{z : |z| > \delta^{-1}\}$. As the poles of $f$ in $\mathbf{C}$ are isolated, $f$ can only have a finite number of poles, say $z_1, \ldots, z_s$, in $\mathbf{C}$ and possibly a pole at $\infty$.

We may now write

$$f(z) = \sum_{j=1}^{s} P_j\left(\frac{1}{z - z_j}\right) + g(z),$$

where the $P_j$ are polynomials and $g$ is analytic in $\mathbf{C}$. This shows that $g \circ J$ differs from $f \circ J$ by a rational function and as $f \circ J$ is meromorphic in $\mathbf{C}(0, r^{-1})$ so is $g \circ J$.

As $g$ is analytic in $\mathbf{C}$ we may write

$$g(z) = \sum_{0}^{\infty} a_n z^n$$

and so in $\mathbf{C}^*(0, r^{-1})$,

$$g \circ J(z) = g(1/z) = \sum_{n=0}^{\infty} a_n z^{-n}.$$

As $g \circ J$ is meromorphic in $\mathbf{C}(0, r^{-1})$, $a_n = 0$ for $n > k$, say; thus $g$ is a polynomial and $f$ is rational.

We can say more than is given in Theorem 10.2.1. The total number of zeros of the rational function

$$R(z) = \frac{a_0 + a_1 z + \cdots + a_n z^n}{b_0 + b_1 z + \cdots + b_m z^m}, \qquad a_n b_m \neq 0, \tag{10.2.5}$$

is $n$ (in $\mathbf{C}$) plus $\max\{0, m-n\}$ (at $\infty$) (we assume that the numerator and denominator have no common zeros). Similarly, $R$ has exactly $m + \max\{0, n-m\}$ poles in $\mathbf{C}_\infty$. We now define the degree $d$ of $R$ to be $\max\{m, n\}$ and $R$ has exactly $d$ zeros and $d$ poles in $\mathbf{C}_\infty$. For each $w$ in $\mathbf{C}$, the rational function $R(z) - w$ is also of degree $d$ (Exercise 10.2.4) and so this too has exactly $d$ zeros. Thus for each $w$ in $\mathbf{C}_\infty$, $R(z) = w$ has exactly $d$ solutions in $\mathbf{C}_\infty$ and $R$ maps $\mathbf{C}_\infty$ onto itself in a $d - 1$ manner.

In particular, the only 1–1 meromorphic maps of $\mathbf{C}_\infty$ into itself are the rational functions of degree one. These are of the form

$$r(z) = \frac{az + b}{cz + d};$$

they actually map $\mathbf{C}_\infty$ *onto* itself and are studied in the next section.

**Exercise 10.2**

1. Prove (10.2.2).
2. Prove that $\phi$ and $\phi^{-1}$ (as defined on p. 186) are continuous
3. Use (10.2.2) to prove (10.2.3). Prove that (in $\mathbf{R}^3$)

$$|\zeta - \infty||z - \infty| = 1, \qquad \infty = (0, 0, 1),$$

   (see Diagram 10.2.1) and hence give an easier proof (using similar triangles) of (10.2.3).
4. Prove that if $R$ is a rational function of degree $d$, then so is $R - w$ for every complex number $w$. [It is obvious that $R - w$ is rational and has degree at most $d$: why is the degree equal to $d$?]
5. A subset $E$ of $\mathbf{C}_\infty$ is *open* if and only if it is some union of chordal discs

$$S(a, r) = \{z \in \mathbf{C}_\infty : d(z, a) < r\}.$$

   A subset $E$ is *closed* if $\mathbf{C}_\infty - E$ is open.

   Prove

   (a) if $E_1$ and $E_2$ are closed and disjoint, then

$$\inf\{d(z_1, z_2) : z_1 \in E_1, z_2 \in E_2\} > 0;$$

   (b) if $E$ is an open subset of $\mathbf{C}$ (§6.1), it is an open subset of $\mathbf{C}_\infty$ (§10.2);
   (c) if $E$ is a compact subset of $\mathbf{C}$ (§6.4), it is a closed subset of $\mathbf{C}_\infty$ (§10.2);
   (d) $\mathbf{C}$ is a closed subset of $\mathbf{C}$ but is not a closed subset of $\mathbf{C}_\infty$.

6. A subset $E$ of $\mathbf{C}_\infty$ is connected if and only if each continuous function $f: E \to \mathbf{Z}$ is constant. Prove
   (a) $\mathbf{C}_\infty$ is connected;
   (b) if $E \subset \mathbf{C}$, then $E$ is connected in this sense if and only if it is connected in the sense defined in Chapter 6;
   (c) if $E$ is closed and not connected then $E = E_1 \cup E_2$ where $E_1$ and $E_2$ are disjoint closed sets;
   (d) as for (c) with 'closed' replaced by 'open'.

## 10.3 MÖBIUS TRANSFORMATIONS

We have seen that the only 1–1 meromorphic maps of $\mathbf{C}_\infty$ onto itself are the non-constant rational functions of degree one. These are of the form

$$f(z) = \frac{az + b}{cz + d} \tag{10.3.1}$$

and are known variously as Möbius, linear fractional and bilinear transformations. In order that $f$ should be non-constant and of degree one the coefficients must satisfy (a) if $a = 0$, then $bc \neq 0$; (b) if $c = 0$, then $ad \neq 0$ and (c) $ac \neq 0$, then $b/a \neq d/c$. These three conditions are equivalent to the single algebraic condition $ad - bc \neq 0$. We recall that if $c \neq 0$ then $f(-d/c) = \infty$ and $f(\infty) = a/c$: if $c = 0$ then $f(\infty) = \infty$.

As $f$ maps $\mathbf{C}_\infty$ 1–1 onto itself there is an inverse mapping $f^{-1}$. This inverse is itself a Möbius transformation and is given by

$$f^{-1}(z) = \frac{dz - b}{-cz + a}.$$

The composition of any functions is associative, the composition of Möbius transformations is again a Möbius transformation and the identity $I$, $I(z) = z$, is a Möbius transformation ($a = d = 1$, $b = c = 0$). We conclude that the set of Möbius transformations forms an algebraic group with respect to the usual composition of functions.

There are four basic and particularly simple types of Möbius transformations. These are the *translations* $t(z) = z + b$, the *rotations* $r(z) = az$, $|a| = 1$, the *magnifications* $m(z) = az$, $a > 0$ and $J$, $J(z) = z^{-1}$. In fact, these four types actually generate the group of Möbius transformations, for if $f$ is given by (10.3.1) and if $c \neq 0$, then

$$f(z) = \frac{a}{c} - \left(\frac{ad - bc}{c^2}\right)\left(z + \frac{d}{c}\right)^{-1}$$

and this is a composition of these four basic transformations. If $c = 0$, then $ad \neq 0$ and

$$f(z) = \frac{a}{d}\left(z + \frac{b}{a}\right).$$

The function $f$ in (10.3.1) has four coefficients; however, one must only expect 'three degrees of freedom' as the right-hand side of (10.3.1) is a homogeneous expression of the coefficients. This idea is made quite precise by the next result.

*Theorem 10.3.1* *Let $z_1, z_2$ and $z_3$ be distinct elements of $\mathbf{C}_\infty$ and let $w_1, w_2$ and $w_3$ be distinct elements of $\mathbf{C}_\infty$. Then there is a unique Möbius transformation $f$ such that $f(z_j) = w_j, j = 1, 2, 3$.*

*Corollary* *If a Möbius transformation fixes three distinct points then it is the identity.*

*Proof of Theorem 10.3.1* If no $z_j$ is $\infty$ then the function

$$f(z : z_1, z_2, z_3) = \frac{(z - z_1)(z_2 - z_3)}{(z - z_3)(z_2 - z_1)}$$

is a Möbius transformation which maps $z_1, z_2, z_3$ to $0, 1, \infty$ respectively. If $z_1, z_2$ or $z_3$ is $\infty$ we define $f$ by

$$\frac{z_2 - z_3}{z - z_3}, \quad \frac{z - z_1}{z - z_3}, \quad \frac{z - z_1}{z_2 - z_1}$$

respectively and in these cases too it is a Möbius transformation which maps $z_1, z_2, z_3$ to $0, 1, \infty$ respectively. The existence of an $f$ with $f(z_j) = w_j$ is now clear; explicitly it is $f(z : z_1, z_2, z_3)$ followed by the inverse of $f(z : w_1, w_2, w_3)$.

To prove the uniqueness we consider first the general $f$ defined by (10.3.1). If this fixes 0, 1 and $\infty$ then necessarily $b = c = 0$ and $a = d$, so $f$ is the identity. Now let $f$ be any Möbius transformation which maps $z_j$ to $w_j$, and write

$$f_1(z) = f(z : z_1, z_2, z_3), \qquad f_2(z) = f(z : w_1, w_2, w_3).$$

Then $f_2 \circ f \circ f_1^{-1}$ fixes each of the points 0, 1 and $\infty$ and so is the identity. Thus $f = f_2^{-1} \circ f_1$ and the uniqueness of $f$ is established.

Any three distinct points $z_1, z_2$ and $z_3$ in $\mathbf{C}_\infty$ determine a circle $Q(z_1, z_2, z_3)$ which passes through these points provided that we adopt the following convention. If $z_1, z_2$ and $z_3$ are in $\mathbf{C}$ and are not collinear, then $Q(z_1, z_2, z_3)$ is the Euclidean circle passing through these points: if $z_1, z_2$ and $z_3$ are collinear or if one $z_j$ is $\infty$ then $Q(z_1, z_2, z_3)$ is the union of $\{\infty\}$ and the Euclidean straight line which contains the finite $z_j$. In the notation of Theorem 10.3.1, not only does $f$ map each $z_j$ to $w_j$, but (as we shall see) it also maps $Q(z_1, z_2, z_3)$ onto $Q(w_1, w_2, w_3)$. In this sense, each Möbius transformation maps each circle onto another circle.

*Theorem 10.3.2* *Let $z_1, z_2$ and $z_3$ be distinct points in $\mathbf{C}_\infty$ and let $f$ be any Möbius transformation. Then $f$ maps $Q(z_1, z_2, z_3)$ onto $Q(w_1, w_2, w_3)$ where $w_j = f(z_j)$.*

Before we illustrate this with several examples we shall make one observation. If $f$ is any Möbius transformation and if $z \neq f^{-1}(\infty)$ or $\infty$ then $f^{(1)}(z)$ exists and is non-zero and so $f$ is conformal at $z$. As we shall see, this is most useful when used in conjunction with Theorem 10.3.2.

*Example 10.3.1* Consider the image $J(L)$ of the line $L$ given by $y = 1$. Theorem 10.3.2 implies that we can find $J(L)$ by selecting any three points on $L$ and computing their $J$-images. For example, as $i$ and $\infty$ are on $L$, $J(L)$ contains $-i$ and 0. The image of any other point on $L$ will now determine $J(L)$. We can, however, avoid any further computation. The imaginary axis $A$ and the line $L$ are orthogonal, so $J(L)$ is orthogonal to $J(A) = A$, so

$$J(L) = \{z : |z + \tfrac{1}{2}i| = \tfrac{1}{2}\}.$$

*Example 10.3.2* Let us construct a Möbius transformation which maps $Q$, $Q = \{z : |z| = 1\}$ onto $\mathbf{R} \cup \{\infty\}$. We select three points on $Q$, say $1, i$ and $-1$, and three points on $\mathbf{R} \cup \{\infty\}$, say 0, 1 and $\infty$, and construct (as in the proof of Theorem 10.3.1) a Möbius transformation $f$ which maps $1, i, -1$ to $0, 1, \infty$ respectively. This is

$$f(z) = -i\left(\frac{z-1}{z+1}\right)$$

and according to Theorem 10.3.2, $f(Q) = \mathbf{R} \cup \{\infty\}$.

*Example 10.3.3* Let $Q_1$ and $Q_2$ be two non-intersecting circles: we shall show that there exists a Möbius transformation $f$ such that $f(Q_1)$ and $f(Q_2)$ are concentric circles. As in the previous example we can find some Möbius transformation $f_1$ such that $f_1(Q_1)$ is the real axis. Next, by a suitable real translation $f_2$ (and, if necessary, a rotation by $\pi$) we may assume that $Q_1'(= f_2 f_1(Q_1))$ is the real axis and $Q_2'(= f_2 f_1(Q_2))$ is the circle $\{z : |z - ik| = r\}$, $k > 0$. As $Q_1$ and $Q_2$ do not meet, $k > r$.

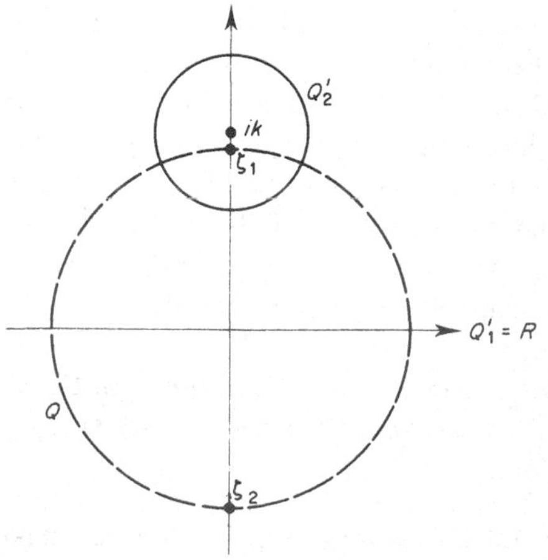


Diagram 10.3.1

The circle $Q$ given by $|z| = (k^2 - r^2)^{1/2}$ is orthogonal to both $Q_1'$ and $Q_2'$ as is the imaginary axis $A$. Now let $\zeta_1$ and $\zeta_2$ be the points where $Q$ meets $A$ (see Diagram 10.3.1) and define $f_3$ by $f_3(z) = (z - \zeta_1)/(z - \zeta_2)$. As $f_3(\zeta_1) = 0$ and $f_3(\zeta_2) = \infty$, $f_3$ maps $Q$ and $A$ to distinct straight lines $L_1$ and $L_2$ which meet at the origin. Moreover, $f_3$ maps $Q_1'$ and $Q_2'$ to circles which are both orthogonal to both $L_1$ and $L_2$; thus $f_3(Q_1')$ and $f_3(Q_2')$ both have their centre at the origin and so are concentric images of $Q_1$ and $Q_2$.

*Proof of Theorem 10.3.2* We know that any Möbius transformation is a composition of magnifications, translations, rotations and $J$. Elementary considerations show that (10.3.2) is satisfied if $f$ is a magnification, a rotation or a translation (see Exercise 10.3.3); thus we need only prove (10.3.2) in the case $f = J$.

We may express any circle or straight line $Q$ as the set of $x + iy$ satisfying some equation

$$a(x^2 + y^2) + b_1 x + b_2 y + c = 0$$

where the coefficients are real. If $a = 0$, then $b_1$ and $b_2$ are not both zero and $Q$ is a line which, by convention, contains $\infty$. This equation may be rewritten as

$$az\bar{z} + bz + \bar{b}\bar{z} + c = 0, \qquad z = x + iy,$$

and hence as

$$a + \bar{b}J(z) + b\overline{J(z)} + cJ(z)\overline{J(z)} = 0.$$

This shows that if $z \in Q$, then $J(z)$ is on the circle (or line) $Q'$ given by the equation

$$cw\bar{w} + \bar{b}w + b\bar{w} + a = 0.$$

We have shown that $J(Q) \subset Q'$: we have not yet proved that $J(Q) = Q'$. The same argument shows that $J(Q') \subset Q$ and so

$$Q' = J(J(Q')) \subset J(Q) \subset Q':$$

thus $J(Q) = Q'$.

The preceding discussion can be viewed entirely in terms of stereographic projection from $S$ and this provides a more compelling reason for our definition of $Q(z_1, z_2, z_3)$ in the case when the $z_j$ are collinear. By virtue of (10.2.2), $z$ satisfies the equation

$$a|z|^2 + b_1 x + b_2 y + c = 0$$

if and only if the inverse projection of $z$, namely $\zeta$, satisfies

$$b_1\zeta_1 + b_2\zeta_2 + (a - c)\zeta_3 + c = 0.$$

This latter condition is exactly the condition that $\zeta$ lies on some plane $\Pi$ in $\mathbf{R}^3$. It follows that if $Q$ is any circle in $\mathbf{C}$, then $\phi^{-1}(Q)$ is the intersection of $S$ with a plane in $\mathbf{R}^3$ and so $\phi^{-1}(Q)$ is a Euclidean circle on $S$. Conversely, if $Q'$ is a circle on $S$, then $Q' = S \cap \Pi$ for some plane $\Pi$ in $\mathbf{R}^3$ and $\phi(Q')$ is a circle in $\mathbf{C}$. This is illustrated in Diagram 10.3.2.

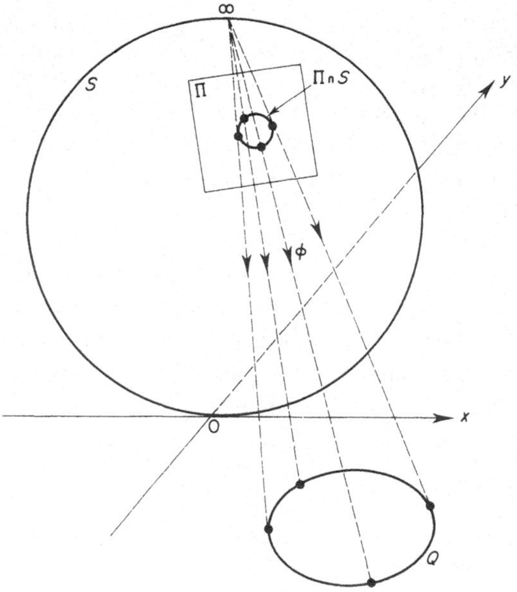


Diagram 10.3.2

With this argument there is no need for the convention described earlier for it is automatically implied by the geometry of the projection. Specifically, if $Q'$ is a circle on $S$ and if $\infty \in Q'$, then $Q' = S \cap \Pi$, say, for some plane $\Pi$ and $\infty \in \Pi$. Thus $\phi(Q') = L \cup \{\infty\}$ where $L = \Pi \cap \mathbf{C}$ (the reader should draw a diagram). Conversely, if $L$ is a straight line in $\mathbf{C}$, let $\Pi$ be the plane in $\mathbf{R}^3$ through $L$ and $\infty$ (=(0, 0, 1)). Then $\phi^{-1}(L \cup \{\infty\}) = \Pi \cap S$ and this is a circle on $S$ which contains $\infty$.

This provides another proof that Theorem 10.3.2 holds when $f = J$. As $J$ preserves the chordal distance (10.2.3), $\phi^{-1} \circ J \circ \phi$ preserves the Euclidean distance on $S$, for if $\zeta$ and $\eta$ are on $S$, then using (10.2.4),

$$\begin{aligned} |\phi^{-1} \circ J \circ \phi(\zeta) - \phi^{-1} \circ J \circ \phi(\eta)| &= d(J \circ \phi(\zeta), J \circ \phi(\eta)) \\ &= d(\phi(\zeta), \phi(\eta)) \\ &= |\phi^{-1}(\phi(\zeta)) - \phi^{-1}(\phi(\eta))| \\ &= |\zeta - \eta|. \end{aligned}$$

Thus $\phi^{-1} \circ J \circ \phi$ maps circles on $S$ to other circles on $S$. If $Q$ is now any circle in $\mathbf{C}_\infty$, then $\phi^{-1}(Q)$ is a circle on $S$, $\phi^{-1} \circ J \circ \phi(\phi^{-1}(Q))$ is another circle on $S$ and this projects to the circle $J(Q)$.

These results imply that in relation to Möbius transformations, it is the projections of circles from $S$ and not the Euclidean circles that are important. For the rest of this section a circle shall mean the projection of a circle on $S$: a Euclidean circle will be referred to as an e-circle.

Before proceeding to other examples, let us resolve one simple but important issue. Let $Q_1$ and $Q_2$ be e-circles and let $f$ be a Möbius transformation such that $f(Q_1) = Q_2$. Can we guarantee that the inside of $Q_1$ is mapped onto the inside or onto the outside of $Q_2$? If so, how can we establish which of these is true?

The answer to the first question depends on connectedness. Let $\Delta_j$ be the inside of $Q_j$ and $D_j$ the outside. Then $\Delta_1$ and $D_1$ are disjoint connected sets and (as $f$ is 1–1 and continuous) so are $f(\Delta_1)$ and $f(D_1)$. Thus $f(\Delta_1)$ and $f(D_1)$ are disjoint connected open sets whose union is $\mathbf{C} - Q_2$ and so $f(\Delta_1)$, for example, must be either $\Delta_2$ or $D_2$. The second question is most easily answered by checking the action of $f$ at a single point. If, for example, $\zeta \in \Delta_1$ and $f(\zeta) \in D_2$, then necessarily $f(\Delta_1) = D_2$. The same ideas apply to configurations of circles and half-planes and we shall no longer concern ourselves with details of this nature: the point needed to be made and that has been done.

*Example 10.3.4* Let us construct a 1–1 analytic map $f$ of the semi-circle $D$,

$$D = \{z : |z| < 1, \operatorname{Im}[z] > 0\}$$

onto the unit disc $\mathbf{C}(0, 1)$. Observe that such an $f$ cannot be conformal at 1 or $-1$ and so cannot be a Möbius transformation. As $f$ is 1–1 in $D$ it will, however, be conformal in $D$.

We first construct the Möbius transformation $f_1$ which maps $-1, 0, 1$ to $0, 1, \infty$ respectively. Thus $f_1$ preserves the real axis, the circle $\{z : |z| = 1\}$ is mapped to the imaginary axis (the circle orthogonal to the real axis at $f(-1)$) and, as the reader should show,

$$D_1 = f_1(D) = \{z : \operatorname{Re}[z] > 0, \operatorname{Im}[z] > 0\}.$$

Now let $g(z) = z^2$ : so

$$D_2 = g(D_1) = \{z : \operatorname{Im}[z] > 0\}.$$

Finally, let $f_2$ be the map constructed in Example 10.3.2. As $f_2(0) \in D_2$, we find that $f_2$, maps $\mathbf{C}(0, 1)$ onto $D_2$ and the mapping we require is $f_2^{-1} \circ g \circ f_1$ (see Diagram 10.3.3). The reader should perform these constructions explicitly.

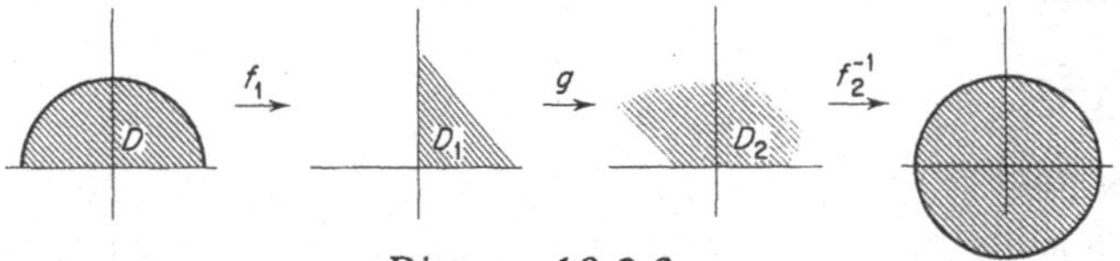


Diagram 10.3.3

Our last result characterizes *all* one-to-one analytic maps of the unit disc $\mathbf{C}(0, 1)$ onto itself. These maps are not at first assumed to be Möbius transformations: the fact that they necessarily are is an important part of the Theorem.

*Theorem 10.3.3* *Let $f$ be any 1–1 analytic map of* $\mathbf{C}(0, 1)$ *onto itself. Then for some $a$, $|a| < 1$, and some real $\theta$,*

$$f(z) = e^{i\theta}\left(\frac{z - a}{\bar{a}z - 1}\right). \tag{10.3.2}$$

*Proof* For each $a$, $|a| < 1$, define

$$f_a(z) = \frac{z - a}{\bar{a}z - 1}.$$

If $|z| = 1$, then

$$|\bar{a}z - 1| = \bar{z}(\bar{a}z - 1)| = |\bar{a} - \bar{z}| = |a - z|$$

and so $|f_a(z)| = 1$. As $f_a$ is a Möbius transformation and as $f_a(a) = 0$, we see that $f_a$ is a 1–1 map of $\mathbf{C}(0, 1)$ onto itself.

Now let $f$ be any 1–1 map of $\mathbf{C}(0, 1)$ onto itself and let $g = f \circ f_a^{-1}$ where $a = f^{-1}(0)$. Then $g$ and $g^{-1}$ are both 1–1 analytic maps of $\mathbf{C}(0, 1)$ onto itself and $g(0) = g^{-1}(0) = 0$. Thus by Schwarz's Lemma (§8.4) for all $z$ in $\mathbf{C}(0, 1)$, $|g(z)| \leqslant |z|$ and $|g^{-1}(z)| \leqslant |z|$. The second inequality gives

$$|z| = |g^{-1}(g(z))| \leqslant |g(z)|$$

and so $|z| = |g(z)|$. Again, by Schwarz's Lemma, $g(z) = e^{i\theta}z$ and this proves Theorem 10.3.3. Note the important part played by Schwarz's Lemma in this proof.

Almost all of the material in this section can be efficiently described in terms of inverse points and we end with a brief discussion of this topic. First, we describe the conjugation map $g(z) = \bar{z}$ geometrically, but not in the usual trivial way. Given any $z$ (not real or $\infty$), $\bar{z}$ is the unique point with the property that all circles through $z$ and $\bar{z}$ are orthogonal to the real axis: $\bar{z}$ is the *reflection* of $z$ in $\mathbf{R}$.

This property generalizes to all circles. Given any circle $Q$ and any $z$ not on $Q$, we call $z^*$ a reflection of $z$ in $Q$ if all circles through $z$ and $z^*$ are orthogonal to $Q$. As any Möbius transformation $f$ preserves circles and orthogonality we immediately have an invariance principle: *$z^*$ is a reflection of $z$ in $Q$ if and only if $f(z^*)$ is a reflection of $f(z)$ in $f(Q)$.*

In particular we can choose $f$ so that $f(Q)$ is the real axis: then necessarily, $f(z^*) = \overline{f(z)}$ and so

$$z^* = f^{-1}(\overline{f(z)}). \qquad (10.3.3)$$

This shows that there is exactly one reflection $z^*$ of $z$ in $Q$: it also shows that $(z^*)^* = z$, which is obvious anyway. If $z$ is real, then $z = \bar{z}$. It is natural to define $z^* = z$ in the case when $z \in Q$ and then (10.3.3) holds for all $z$ in $\mathbf{C}_\infty$. We now say that $z$ and $z^*$ are *inverse points* with respect to $Q$.

If $Q$ is the circle $\{z: |z - z_0| = r\}$, then directly from the definition of $z^*$, $(z_0)^* = \infty$: thus the centre $z_0$ and $\infty$ are inverse points with respect to $Q$.

The inverse point $z^*$ of $z$ with respect to the unit circle $\{z : |z| = 1\}$ can easily be found by using (10.3.3) and Example 10.3.2. A simple computation shows that $z^* = 1/\bar{z}$: this is the same point as $J(\bar{z})$ and $\overline{J(z)}$. If $f$ is any Möbius transformation which maps the unit disc $\mathbf{C}(0, 1)$ onto itself, then for some $a$ in $\mathbf{C}(0, 1)$, $f(a) = 0$. The invariance principle shows that if $a$ and $a^*$ are reflections of each other in the unit circle, then so are $0(=f(a))$ and $f(a^*)$. Thus $f(a^*) = \infty$ and for some constant $k$,

$$f(z) = k\left(\frac{z-a}{z-a^*}\right) = k\bar{a}\left(\frac{z-a}{\bar{a}z-1}\right).$$

We conclude (as in the proof of Theorem 10.3.3) that $|k\bar{a}| = 1$ and $f$ is then of the form (10.3.2).

Next, if $a$ and $a^*$ are inverse points with respect to $Q$, then $g$, $g(z) = (z-a)/(z-a^*)$, maps $Q$ to some circle $g(Q)$ such that $0\ (=g(a))$ and $\infty\ (=g(a^*))$ are inverse points with respect to $g(Q)$. Thus $g(Q)$ is of the form $\{w : |w| = p\}$ and if $z \in Q$, then $|g(z)| = p$. In other words, $Q$ has equation

$$\left(\frac{z-a}{z-a^*}\right) = p \qquad (10.3.4)$$

(This is a straight line if and only if $p = 1$).

Conversely, if $Q$ is the set of $z$ satisfying (10.3.4), then $g(Q)$ is the circle $\{w : |w| = p\}$ and so $Q$ itself is a circle. Moreover, $g^{-1}(0)$ and $g^{-1}(\infty)$ are reflections of each other in $Q$; that is $a$ and $a^*$ are inverse points with respect to $Q$.

Finally, we return to Theorem 10.3.2. If $Q$ is any circle, we can find finite points $a$ and $a^*$ which are reflections of each other in $Q$. Thus $Q$ has an equation of the form (10.3.4). A trivial computation using (10.3.1) shows that if $z \in Q$, then

$$\left|\frac{f(z)-f(a)}{f(z)-f(a^*)}\right| = p'$$

for some constant $p'$. This shows that $f(Q)$ is itself a circle and also that $f(a)$ and $f(a^*)$ are reflections of each other in $f(Q)$.

### Exercise 10.3

1. Show directly that each of the transformations

$$f_1(z) = z + a, \qquad f_2(z) = kz, \qquad k \neq 0,$$

maps $Q(z_1, z_2, z_3)$ onto some $Q(w_1, w_2, w_3)$.

2. Verify that $f$ given by (10.3.1) satisfies

$$f^{(1)}(z) = \frac{ad-bc}{(cz+d)^2}.$$

This shows that $f$ is conformal except at $-d/c$ and that $f$ is constant if and only if $ad = bc$.

3. Let $f$ be any Möbius transformation which maps $\mathbf{R} \cup \{\infty\}$ onto itself and suppose that $f(\infty) \neq \infty$. Show that $f$ maps $Q(-1+i, i, 1+i)$ onto a circle tangent to the real axis. [Consider the intersection of this $Q$ with $\mathbf{R} \cup \{\infty\}$.]
4. Let $f$ be given by (10.3.1). Show that if $f$ is not the identity map, then $f$ has either (a) exactly one or (b) exactly two fixed points.

Prove that in Case (a)

$$g \circ f \circ g^{-1}(z) = z + k,$$

where $f(w) = w$, $g(z) = (z - w)^{-1}$ and $k$ is some constant.

If $f^n$ denotes the $n$th iterate of $f$, show that

$$f^n(z) = g^{-1}(nk + g(z)).$$

Compute $f^3(z)$ when

$$f(z) = \frac{2z - 1}{z}.$$

What can be said in Case ($b$)? [If $f(w_j) = w_j$, $j = 1, 2$, let

$$g(z) = \frac{z - w_1}{z - w_2}.]$$

5. Find a Möbius transformation which maps $\{z : |z| > 1\} \cup \{\infty\}$ onto $\{x + iy : x > 0\}$.
6. Find a Möbius transformation which maps the domain

$$\{z : |z - 1| < \sqrt{2}\} \cap \{z : |z + 1| < \sqrt{2}\}$$

onto the quadrant $\{x + iy : x > 0\}$.

7. Given a sequence of four distinct points $(z_1, \ldots, z_4)$ define their cross-ratio as

$$[z_1, z_2, z_3, z_4] = \frac{(z_1 - z_3)(z_2 - z_4)}{(z_1 - z_4)(z_2 - z_3)}.$$

If $z_j = \infty$, use this formula with $z_j \neq \infty$ and then let $z_j \to \infty$.

(a) Prove that for any Möbius transformation $f$

$$[z_1, z_2, z_3, z_4] = [f(z_1), f(z_2), f(z_3), f(z_4)].$$

(b) Deduce that the $z_j$ are concyclic if and only if their cross-ratio is real.
(c) Which permutations of the $z_j$ leave the cross-ratio unchanged?
(d) How many distinct values can the cross-ratio take as the $z_j$ are permuted among themselves?
(e) Place 0, 1, $\infty$ and $z$ in a suitable order so that their cross-ratio is equal to $z$.

# PART III

*Interactions with Plane Topology*

# Chapter 11

## 11.1 SIMPLY CONNECTED DOMAINS

A domain $D$ is simply connected if every closed curve in $D$ can be continuously deformed (in a sense to be made precise later) to a point in $D$ in such a way that throughout the deformation the curve remains closed and in $D$. This suggests, for example, that an open disc is simply connected but that an annulus is not: a simply connected domain cannot have any 'holes' in it.

In order to describe the idea of a continuous deformation of a closed curve $\gamma : [a, b] \to D$ to a point $\zeta$ we can imagine the deformation to take place during the 'time' interval $[0, 1]$. Thus at each time $\tau$ in $[0, 1]$ the position of the deformed curve is described by another closed curve $\sigma_\tau : [a, b] \to D$. The deformation starts when $\tau = 0$ with the curve $\gamma$ $(= \sigma_0)$ and ends when $\tau = 1$ at the point curve $\zeta$ $(= \sigma_1)$. For all $\tau$, $\sigma_\tau$ is a closed curve in $D$ and so $\sigma_\tau(a) = \sigma_\tau(b)$.

We have still to introduce continuity. So far, for each $t$ in $[a, b]$ and each $\tau$ in $[0, 1]$ we have a point $\sigma_\tau(t)$ in $D$. For a fixed $\tau$ we obtain (as a function of $t$) the position $\sigma_\tau(t)$ of the deformed curve at time $\tau$: for a fixed $t$ we obtain (as a function of $\tau$) the path described during the process by the point labelled by $t$ and starting at $\gamma(t)$. It is more symmetric to write $\sigma(t, \tau)$ for $\sigma_\tau(t)$ (we use both notations interchangeably) and we now insist that $\sigma(t, \tau)$ is a continuous function of $(t, \tau)$ in the rectangle

$$\{(t, \tau) : a \leqslant t \leqslant b, 0 \leqslant \tau \leqslant 1\}.$$

We are now ready for a formal definition. The function $\sigma$ is called a homotopy and we use 'null' to describe the degenerate nature of the resulting point curve. Note that we do not insist that $D$ is a domain.

*Definition 11.1.1* *Let $D$ be any subset of* $\mathbf{C}$ *and let $\gamma : [a, b] \to D$ be a closed curve in $D$. Then $\gamma$ is null-homotopic in $D$ if and only if there exists a continuous function $\sigma : [a, b] \times [0, 1] \to D$ and a point $\zeta$ in $D$ such that*

(a) $\sigma(a, \tau) = \sigma(b, \tau)$ if $\tau \in [0, 1]$;

(b) $\sigma(t, 0) = \gamma(t)$ if $t \in [a, b]$;

(c) $\sigma(t, 1) = \zeta$ if $t \in [a, b]$.

*Definition 11.1.2* *A subset D of* **C** *is simply connected if and only if every closed curve in D is null-homotopic in D.*

*Example 11.1.1* Each open disc $\mathbf{C}(w, r), r > 0$, is simply connected (we include the case $r = +\infty$, $\mathbf{C}(w, r) = \mathbf{C}$ here). For each closed curve $\gamma$ in $\mathbf{C}(w, r)$, define

$$\sigma(t, \tau) = w + (1 - \tau)(\gamma(t) - w).$$

It is easy to see that $\sigma$ provides a deformation of $\gamma$ to $w$; moreover, this deformation occurs in $\mathbf{C}(w, r)$ as for all $t$ and $\tau$,

$$| \sigma(t, \tau) - w | \leqslant | \gamma(t) - w | < r.$$

As $\tau$ increases from 0 to 1 each point $\gamma(t_0)$ on the curve $\gamma$ moves to $w$ along the straight-line segment joining $\gamma(t_0)$ to $w$.

The same argument shows that $\bar{\mathbf{C}}(w, r)$ is simply connected.

The previous example shows that each open disc is simply connected: it is not so easy, however, to give a direct and rigorous proof that an annulus is not simply connected. The next result gives equivalent definitions of simple connectivity and with these available it is then clear that, for example, an annulus is not simply connected. Yet more equivalent definitions occur in §11.3.

*Theorem 11.1.1* *Let D be a subdomain of* **C**. *The following are equivalent*:

(a) *D is homeomorphic to* $\mathbf{C}(0, 1)$;

(b) *D is simply connected*;

(c) *for every cycle* $\Gamma$ *in D and each w in* $\mathbf{C} - D$, $n(\Gamma, w) = 0$;

(d) $\mathbf{C}_\infty - D$ *is connected*;

(e) *if f is analytic and non-zero in D, then there exists a branch of* Log *f in D.*

This theorem gives five different characterizations of simple connectivity and the reader should, as far as possible, provide for himself intuitive reasons for their equivalence. The equivalence depends, however, in a crucial way on $D$ being a subdomain of **C** and fails in a more general context. Definition 11.1.2 does not require $D$ to be a domain and is capable of immediate generalization to other spaces: (c) and (d) are not because they refer explicitly to the complement of $D$, (e) is not because it requires analyticity and (a) fails, for example, for a closed disc (see Example 11.1.1).

*Proof of Theorem 11.1.1* If (a) holds there is a homeomorphism $\phi$ of $D$ onto $\mathbf{C}(0, 1)$ with $\phi$ and $\phi^{-1}$ both continuous. If $\gamma$ is any closed curve in $D$, then $\phi \circ \gamma$ is a closed curve in $\mathbf{C}(0, 1)$ and, as in Example 11.1.1,

$$\sigma(t, \tau) = (1 - \tau)\phi(\gamma(t))$$

continuously deforms $\phi \circ \gamma$ to zero. It is easily checked that $\phi^{-1} \circ \sigma$ provides a deformation of $\gamma$ to $\phi^{-1}(0)$ in $D$ and this shows that (a) implies (b). This is illustrated in Diagram 11.1.1.

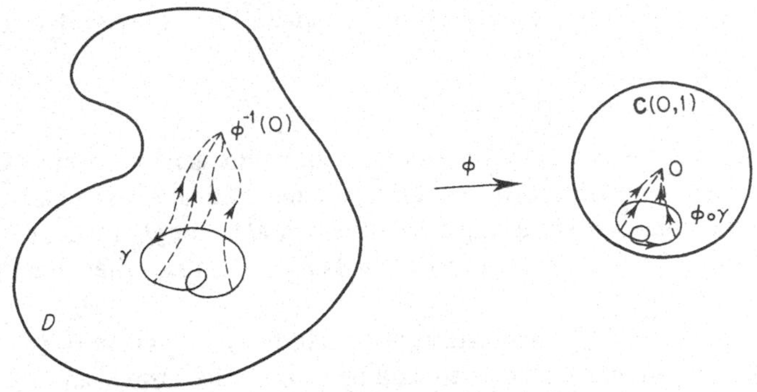


Diagram 11.1.1

The proof that (b) implies (c) is based on the following idea. If (b) holds there is a homotopy $\sigma$ as in Definition 11.1.1 of each closed curve $\gamma$ into a point $\zeta$. At each time $\tau$ in $[0, 1]$ we can compute the index $n(\sigma_\tau, w)$ and as the deformation is continuous, this should be a continuous function of $\tau$. If so, it is a continuous integer-valued function on $[0, 1]$ and so is constant. Thus as $\gamma = \sigma_0$ and $\zeta = \sigma_1$,

$$\begin{aligned} n(\gamma, w) &= n(\sigma_0, w) \\ &= n(\sigma_1, w) \\ &= n(\zeta, w) \\ &= 0 \end{aligned}$$

as $\zeta$ is a point curve. As $n(\gamma, w) = 0$ for each closed curve in $D$, we see that $n(\Gamma, w) = 0$ for each cycle $\Gamma$ in $D$.

It remains only to prove that $n(\sigma_\tau, w)$ is a continuous function of $\tau$. First, let $Q = [a, b] \times [0, 1]$. Then $Q$ is compact and $\sigma(Q)$ is a compact subset of $D$. By Theorem 6.4.3, if $w \notin D$, then

$$\epsilon = \text{dist}\,(w, \sigma(Q)) > 0.$$

The uniform continuity of $\sigma$ on $Q$ implies that there is a positive $\delta$ such that

$$|\,\sigma(t, \tau) - \sigma(t', \tau')\,| < \epsilon$$

whenever $|\,(t, \tau) - (t', \tau')\,| < \delta$. Taking $t = t'$, we find that if $|\,\tau - \tau'\,| < \delta$, then

$$|\,\sigma(t, \tau) - \sigma(t, \tau')\,| < \epsilon \leqslant |\,\sigma(t, \tau) - w\,|\,.$$

This with I6 shows that if $|\,\tau - \tau'\,| < \delta$, then

$$n(\sigma_\tau, w) = n(\sigma'_\tau, w)$$

and this certainly implies that $n(\sigma_\tau, w)$ is a continuous function of $\tau$.

We now prove that if (d) fails, then so does (c); equivalently, (c) implies (d). At this point the reader should recall Exercise 10.2.5. If (d) fails, then $\mathbf{C}_\infty - D$ is not connected and so it is the disjoint union of non-empty closed subsets $E_1$ and $E_2$ of $\mathbf{C}_\infty$. One of these sets, say $E_2$, contains $\infty$. The other set $E_1$ is then a closed subset of $\mathbf{C}_\infty$ which does not contain $\infty$ and so is necessarily a bounded subset of $\mathbf{C}$.

We deduce that $E_1$ is a compact subset of $\mathbf{C}$, that $E_2 - \{\infty\}$ is a closed subset of $\mathbf{C}$ and so

$$\epsilon = \text{dist}\,(E_1, E_2 - \{\infty\}) > 0.$$

We now proceed much as we did in the proof of the general form of Cauchy's Theorem. First, we construct (using horizontal and vertical lines) a partition of $\mathbf{C}$ into congruent squares each having sides of length $\epsilon/4$. Let $Q_i, i \in A$, be those squares meeting $E_1$: these cannot meet $E_2$. As $E_1$ is compact there are only a finite number of the $Q_i$.

We now denote by $\partial Q_i$ the positively oriented boundary curve of $Q_i$. The partition may clearly be constructed so that some $w$ in $E_1$ lies inside, say, $\partial Q_k$ and then

$$\sum_i n(\partial Q_i, w) = n(\partial Q_k, w) = 1.$$

Each $\partial Q_i$ may be replaced by four linear segments $\gamma_{i,j}$ which bound $Q_i$. If one such segment meets $E_1$, then it occurs with the opposite orientation on the boundary of some $Q_l$. The cycle consisting of all $\partial Q_i$ may therefore be replaced (by repeated cancellation) by the cycle $\Gamma$ consisting of all those $\gamma_{i,j}$ which do not meet $E_1$. These segments cannot meet $E_2$, for no $Q_i$ does, and so $\Gamma$ is actually a cycle in $D$ ($= \mathbf{C}_\infty - (E_1 \cup E_2)$). This yields

$$n(\Gamma, w) = \sum_i n(\partial Q_i, w) = 1$$

and this contradicts (c).

Next we prove that (d) implies (e). If $\Gamma$ is a cycle in $D$, then $n(\Gamma, w)$ is a continuous integer-valued function of $w$ on $\mathbf{C} - [\Gamma]$ and hence on $\mathbf{C} - D$. Moreover, $n(\Gamma, w) \to 0$ as $w \to \infty$, and so if we define $n(\Gamma, \infty) = 0$ we obtain $n(\Gamma, w)$ as a continuous integer-valued function on $\mathbf{C}_\infty - D$. As $\mathbf{C}_\infty - D$ is connected this function is constant, hence zero, and we have verified (b) of Cauchy's Theorem. We deduce that

$$\int_\Gamma g = 0$$

for every cycle $\Gamma$ in $D$ and every $g$ which is analytic in $D$.

If $f$ is analytic and non-zero in $D$, we write $g = f^{(1)}/f$ and so $g$ is analytic in $D$. This means that if we select a point $z_0$ in $D$, then

$$F(z) = \int_{z_0}^{z} f^{(1)}/f$$

defines a continuous function of $z$ in $D$, the integral being taken over and independent of any chain joining $z_0$ to $z$ in $D$. It is easy to see that in $D$,

$$F^{(1)}(z) = f^{(1)}(z)/f(z)$$

(Exercise 11.1.2). Thus in $D$,

$$\frac{\mathrm{d}}{\mathrm{d}z}[f(z) \exp(-F(z))] = 0,$$

and therefore

$$f(z) = f(z_0) \exp(F(z)).$$

If $L$ is now any particular choice of $\operatorname{Log} f(z_0)$, then $L + F(z)$ is a branch of $\operatorname{Log} f(z)$, $z \in D$.

It remains to prove that (e) implies (a), and different proofs are required depending on whether $D = \mathbf{C}$ or not.

First, let $D = \mathbf{C}$. Then (a) holds as

$$\phi(z) = \frac{z}{1 + |z|}$$

is a homeomorphism of $\mathbf{C}$ onto $C(0, 1)$. We have shown that (b) holds and (c) is vacuously true as there is no $w$ in $\mathbf{C} - D$ to violate (c). As $\mathbf{C}_\infty - D = \{\infty\}$, (d) holds and finally, (d) implies (e) as above. Thus when $D = \mathbf{C}$ all five statements are true.

The proof that (e) implies (a) when $D$ is a proper subdomain of $\mathbf{C}$ (that is when $D \subset \mathbf{C}, D \neq \mathbf{C}$) is much harder and is derived as a corollary of a much stronger result in the next section.

We end this section with an application of Theorem 11.1.1 (we assume the validity of this) which will be needed later. First, however, we must provide a definition of simply connected subsets of $\mathbf{C}_\infty$ (rather than $\mathbf{C}$).

The easiest definition for our purposes is to say that *a subset $D$ of $\mathbf{C}_\infty$ is simply connected* if and only if $D = \mathbf{C}_\infty$ or $D$ is homeomorphic to some simply connected subset of $\mathbf{C}$. An alternative (and equivalent) approach is to define homotopy in terms of the chordal metric.

*Theorem 11.1.2* *Let $K$ be a connected compact subset of $\mathbf{C}_\infty$. Then each component of $\mathbf{C}_\infty - K$ is a simply connected domain.*

*Proof* Let $D$ be a component of $\mathbf{C}_\infty - K$. As $D$ is a component it is connected and as $K$ is closed, $D$ is open. Thus $D$ is a domain in $\mathbf{C}_\infty$. We may assume (by applying a homeomorphism of $\mathbf{C}_\infty$ onto itself) that $\infty \in K$: the case $K = \phi$ is trivial.

If $D = \mathbf{C}_\infty - K$, then by Theorem 11.1.1 (d), $D$ is simply connected.

If not, let the components of $\mathbf{C}_\infty - K$ be $D$ and $D_\alpha$, $\alpha \in A$, and consider any continuous function $f : K \cup (\bigcup_\alpha D_\alpha) \to \mathbf{Z}$. As $K$ is connected, $f$ is constant on $K$, say $f = k$ there. Similarly, $f$ is constant, say $f = k_\alpha$, on $D_\alpha$, $\alpha \in A$. Obviously $\bar{D}_\alpha \cap K$ contains a point $\zeta$ (actually, $\partial D_\alpha \subset K$) and so as $f$ is continuous,

$$k_\alpha = \lim_{z \to \zeta,\, z \in D_\alpha} f(z) = f(\zeta) = k.$$

Thus $f$ is constant on $K \cup (\bigcup_\alpha D_\alpha)$ and this set is connected. This set is $\mathbf{C}_\infty - D$ and again by Theorem 11.1.1 (d), $D$ is simply connected.

**Exercise 11.1**

1. Show that no two of the sets

$$\mathbf{C}(0,1), \quad \bar{\mathbf{C}}(0,1), \quad \mathbf{C}(0,1) \cup \{1\}, \quad \mathbf{C}^*(0,1)$$

are homeomorphic.
2. Let $f$ and $F$ be as in the proof of Theorem 11.1.1, (d) implies (e). By considering $[F(z+h)-F(z)]/h$, prove that $F^{(1)} = f^{(1)}/f$.
3. Prove that if $K$ is a compact subset of $\mathbf{C}_\infty$ and if each component of $\mathbf{C}_\infty - K$ is simply connected, then $K$ is connected.

## 11.2 THE RIEMANN MAPPING THEOREM

First, we conform to common usage and say that an analytic function is *univalent* if it is 1–1 on its domain of definition. Next, we say that two domains $D_1$ and $D_2$ are *conformally equivalent* if and only if there is a univalent analytic function $f$ which maps $D_1$ onto $D_2$ ($f^{-1}$ is then a univalent analytic map of $D_2$ onto $D_1$). In particular, if $D_1$ and $D_2$ are conformally equivalent then they are homeomorphic. The converse is false for $\mathbf{C}$ is homeomorphic to but not conformally equivalent to $\mathbf{C}(0, 1)$ (Exercise 11.2.1).

*The Riemann Mapping Theorem* *Let $D$ be a simply connected subdomain of* $\mathbf{C}$. *Then either* (a) $D = \mathbf{C}$ *or* (b) *$D$ is conformally equivalent to* $\mathbf{C}(0, 1)$.

There are several aspects of this important result which are worthy of special mention. First, the preceding remarks show that (a) and (b) cannot both be true. Nevertheless, in both cases $D$ is homeomorphic to $\mathbf{C}(0, 1)$: thus *any two simply connected subdomains of* $\mathbf{C}$ *are homeomorphic.*

The Theorem implies that all simply connected *proper* subdomains of $\mathbf{C}$ can be mapped in a 1–1 angle-preserving way onto $\mathbf{C}(0, 1)$ and this is indeed remarkable when one considers the possible complexity of the boundary of a simply connected domain. Note that there are no assumptions made and no conclusions drawn about any relation between the boundaries of $D$ and of $\mathbf{C}(0, 1)$. In fact, the topological nature of the boundary of $D$ can be quite different from that of $\mathbf{C}(0, 1)$: examples of this occur in Exercise 11.2.2.

Finally, in our proof of Riemann's Theorem we shall assume only that $D$ satisfies (e) of Theorem 11.1.1 (we have proved that this is implied by the simple connectivity of $D$). The reason we do this is that the proof will then contain a proof that (e) implies (a) in Theorem 11.1.1 when $D \neq \mathbf{C}$. Thus at the same time, we shall both prove Riemann's Theorem and complete the proof of Theorem 11.1.1 With this in mind, let us observe now that if Theorem 11.1.1 (e) holds and if $f$ is analytic and non-zero in $D$, then there is a branch $g(z)$ of $\operatorname{Log} f(z)$ in $D$. Now $g$ is analytic in $D$ (Exercise 11.2.3) and so

$$s(z) = \exp\left(\tfrac{1}{2} g(z)\right)$$

is analytic and satisfies $s^2(z) = f(z)$ in $D$. We say, then, that $s$ is an *analytic square root* of $f$ in $D$. Observe that if $s(z_1) = s(z_2)$, then $f(z_1) = f(z_2)$: thus if $f$ is univalent, so is $s$.

*Proof* The proof is long and involves several ideas which are important in themselves. We begin by selecting any point $\zeta$ in $D$. Now let $\mathscr{C}$ be the family of all functions $f$ which are univalent analytic maps of $D$ into $\mathbf{C}(0, 1)$ with $f(\zeta) = 0$ (the condition $f(\zeta) = 0$ is simply a convenient normalization). It is sufficient, of course, to show that there is some $f$ in $\mathscr{C}$ which maps $D$ *onto* $\mathbf{C}(0, 1)$. The proof can best be summarized by the next three results. First, we consider the local magnification of elements in $\mathscr{C}$ at $\zeta$.

*Lemma 1* $\mathscr{C} \neq \phi$ *and*

$$\mu = \sup \{ |g^{(1)}(\zeta)| : g \in \mathscr{C} \}$$

*is finite.*

Next, the image $f(D)$, $f \in \mathscr{C}$, is maximal if $f$ has a maximal magnification at $\zeta$.

*Lemma 2* *Let $f$ be in $\mathscr{C}$. Then $f$ maps $D$ onto* $\mathbf{C}(0, 1)$ *if and only if* $|f^{(1)}(\zeta)| = \mu$.

Finally, this maximal magnification is attained.

*Lemma 3* *There is an $f$ in $\mathscr{C}$ with* $|f^{(1)}(\zeta)| = \mu$.

These results, when considered together, clearly constitute a proof of Reimann's Theorem. The fact that $\mu$ is finite and the 'only if' part of Lemma 2 are not needed for the proof; nevertheless, the extra detail can only provide greater motivation and understanding.

*Proof of Lemma 1* As $D$ is a proper subdomain of $\mathbf{C}$ there is some $w$ in $\mathbf{C}$ but not in $D$ and so there is an analytic square root $s(z)$ of $z - w$, $z \in D$. This function $s$ is analytic and non-zero in $D$ and satisfies

$$[s(z) - s(z')]\,[s(z) + s(z')] = z - z', \quad z, z' \in D.$$

We can draw two conclusions from this. First, if $s(z) = s(z')$ then $z = z'$ and so $s$ is univalent in $D$. Next, if $z$ and $z'$ (not necessarily distinct) are in $D$, then $s(z) \neq -s(z')$ as otherwise $z = z'$, $s(z) = s(z')$ and so $s(z) = 0$. This shows that the images of $D$ by the analytic functions $s$ and $-s$ are disjoint: hence $s(D)$ does not meet the open set $(-s)(D)$. The open set $(-s)(D)$ contains a closed disc $Q$ which can be mapped by a Möbius transformation $m$ onto the complement of $\mathbf{C}(0, 1)$. Thus $m(s(D)) \subset \mathbf{C}(0, 1)$ and by choosing $m$ correctly (or by applying another Möbius transformation) we can ensure that $\zeta$ maps to zero. Thus $m \circ s \in \mathscr{C}$ and $\mathscr{C}$ is not empty.

It is easy to see that $\mu$ is finite. For some positive $r$, $\bar{\mathbf{C}}(\zeta, r) \subset D$ and we put

$\gamma(t) = \zeta + re^{it}, 0 \leqslant t \leqslant 2\pi$. Then for all $g$ in $\mathscr{C}$,

$$|g^{(1)}(\zeta)| = \left| \frac{1}{2\pi i} \int_\gamma \frac{g(z)}{(z-\zeta)^2} \, dz \right| \leqslant \frac{(2\pi r) \cdot 1}{2\pi(r^2)}$$

(as $|g| \leqslant 1$ in $D$) and so $\mu \leqslant r^{-1}$.

*Proof of Lemma 2* We suppose first that $f$ is in $\mathscr{C}$ and maps $D$ onto $\mathbf{C}(0, 1)$. For $g$ in $\mathscr{C}$, we define $h = g \circ f^{-1}$ and this is a univalent analytic map of $\mathbf{C}(0, 1)$ into itself. As $h(0) = 0$, Schwarz's Lemma is applicable and we conclude that $|h^{(1)}(0)| \leqslant 1$. Writing $g = h \circ f$, we find that

$$|g^{(1)}(\zeta)| = |h^{(1)}(0)| \cdot |f^{(1)}(\zeta)| \leqslant |f^{(1)}(\zeta)|$$

and this proves that $|f^{(1)}(\zeta)| = \mu$.

We now have to prove that if $f$ is in $\mathscr{C}$ and if $|f^{(1)}(\zeta)| = \mu$, then $f$ maps $D$ onto $\mathbf{C}(0, 1)$. We achieve this by showing that if $g$(in $\mathscr{C}$) does not map $D$ onto $\mathbf{C}(0, 1)$, then $|g^{(1)}(\zeta)| < \mu$. We assume, then, that $g - w$ is non-zero in $D$, where $w$ is some point in $\mathbf{C}(0, 1)$.

Let

$$g_1(z) = \frac{g(z) - w}{1 - \bar{w}g(z)}.$$

As $g_1$ is non-zero in $D$, we can find an analytic square root $g_2$ in $D$; thus

$$[g_2(z)]^2 = g_1(z).$$

Finally, define $G$ in $D$ by

$$G(z) = \frac{g_2(z) - g_2(\zeta)}{1 - \overline{g_2(\zeta)}g_2(z)}.$$

Observe now that as $g$ is univalent in $D$, so is $g_1$. As $g_1$ is univalent, so is its analytic square root $g_2$ and so also is $G$. Next, as $g_1 = m \circ g$, where $m$ is a Möbius transformation that preserves $\mathbf{C}(0, 1)$, we find that $g_1$ maps $D$ into $\mathbf{C}(0, 1)$. Thus in $D$,

$$|g_2(z)| \leqslant |g_1(z)| < 1$$

and so $g_2$ and hence $G$ (as for $g_1$) maps $D$ into $\mathbf{C}(0, 1)$. Finally, it is clear that $G(\zeta) = 0$ and so $G \in \mathscr{C}$.

The construction of $G$ from $g$ involves square roots. By reversing the construction we obtain

$$\left[\frac{G(z) + g_2(\zeta)}{1 + G(z)\overline{g_2(\zeta)}}\right]^2 = \frac{g(z) - w}{1 - \bar{w}g(z)},$$

and so $g$ can be expressed in terms of $G$, namely

$$g(z) = G(z)\left[\frac{G(z) + \lambda}{1 + \bar{\lambda}G(z)}\right],$$

where

$$|\lambda| = \frac{2|g_2(\zeta)|}{1+|g_2(\zeta)|^2} < 1$$

(this inequality is the arithmetic–geometric mean inequality and does not depend in any way on the form of $g_2(\zeta)$).

We deduce by direct computation that

$$\begin{aligned} |g^{(1)}(\zeta)| &= |\lambda||G^{(1)}(\zeta)| \\ &< |G^{(1)}(\zeta)| \\ &\leqslant \mu \end{aligned}$$

and this completes the proof of Lemma 2.

*Proof of Lemma 3* First, there exists a sequence $f_1, f_2, \ldots$ of elements in $\mathscr{C}$ such that as $k \to \infty$,

$$|f_k^{(1)}(\zeta)| \to \mu.$$

We shall show that there is some subsequence $f_{k_q}$ which converges locally uniformly in $D$ to some analytic function $f$. Assuming this for the moment and (for brevity) relabelling the sequence $f_{k_q}$ as $f_1, f_2, \ldots$ we have

$$f(\zeta) = \lim_{k\to\infty} f_k(\zeta) = 0$$

and (using Theorem 9.7.1) that

$$|f^{(1)}(\zeta)| = \lim_{k\to\infty} |f_k^{(1)}(\zeta)| = \mu.$$

Next it is easy to see that $f$ must be univalent in $D$. Indeed, if $z_1$ and $z_2$ are distinct points in $D$ with $f(z_1) = f(z_2) = w$, say, we can construct (in the usual way) suitably small circles $\gamma_1$ and $\gamma_2$ with the properties

(a) $\gamma_j$ has centre $z_j$,
(b) $z_i$ lies outside $\gamma_j$ $(i \neq j)$ and
(c) $f(z) \neq w$ on $\gamma_j$.

The uniform convergence of $f_k$ to $f$ implies that for all sufficiently large $k$,

$$|f_k(z) - f(z)| < |f(z) - w|$$

on $\gamma_1$ and so by I6,

$$n(f_k \circ \gamma_1, w) = n(f \circ \gamma_1, w).$$

As $f(z) - w$ has at least one zero, namely $z_1$, inside $\gamma_1$ we conclude from the Argument Principle that $f_k(z) - w$ also has at least one zero, say $z_1'$, inside $\gamma_1$. Similarly, there is a $z_2'$ inside $\gamma_2$ with $f_k(z_2') = w$. Thus $z_1' \neq z_2'$,

$$f_k(z_1') = w = f_k(z_2')$$

and this contradicts the fact that $f_k$ is univalent in $D$. This shows that $f$ is univalent in $D$.

We now know that $f$ is analytic and univalent in $D$, that $f(\zeta) = 0$ and $|f^{(1)}(\zeta)| = \mu$. As $|f_k(z)| < 1$ for each $k$ and each $z$ in $D$ we see that $|f(z)| \leqslant 1$ in $D$. As $f(D)$ is actually an open subset of $\mathbf{C}$ we can conclude that $f(D) \subset \mathbf{C}(0, 1)$. Thus $|f(z)| < 1$ in $D$ and so $f \in \mathscr{C}$.

It remains only to establish the existence of $f$ and this is a direct consequence of the next result (in which $D$ need not be simply connected and the $f_k$ need not be univalent).

*Theorem 11.2.1* *Let $D$ be any domain and let $f_k$, $k = 1, 2, \ldots$, be analytic and satisfy $|f_k(z)| \leqslant 1$ in $D$. Then there is some subsequence of $f_1, f_2, \ldots$ which converges locally uniformly in $D$ to some function $f$.*

*Proof* Our first task is to construct the desired subsequence. To do this we first consider the countable set consisting of all points $z_n$, $z_n = x_n + iy_n$, $n = 1, 2, \ldots$ in $D$ with $x_n$ and $y_n$ rational. As $f_1(z_1), f_2(z_1), \ldots$ is a bounded sequence there is a subsequence, say, $(f_n(z_1) : n \in Z_1)$ which converges at $z_1$. As the sequence $(f_n(z_2) : n \in Z_1)$ is also bounded, it has a subsequence, say $(f_n(z_2) : n \in Z_2)$, converging at $z_2$. Observe that $Z_1$ and $Z_2$ are infinite sets with $Z_2 \subset Z_1$.

This argument may be continued and in this way we construct infinite sets $Z_1, Z_2, \ldots$ of positive integers with

$$Z_1 \supset Z_2 \supset \cdots$$

and such that for each $j$, the sequence $(f_n(z_j) : n \in Z_j)$ converges.

Now select $n_1 \in Z_1$, $n_2 \in Z_2$ with $n_2 > n_1$ and so on so that $n_j \in Z_j$ and

$$n_1 < n_2 < \cdots$$

and define $Z^* = \{n_1, n_2, \ldots\}$. For each positive integer $j$,

$$\{n_j, n_{j+1}, n_{j+2}, \ldots\} \subset Z_j$$

and so the sequence $(f_n : n \in Z^*)$ converges at $z_j$. The sequence $(f_n : n \in Z^*)$ therefore *converges at each point* $z_j$, $j = 1, 2, \ldots$, and this is the subsequence we are seeking. The above construction is known as the 'diagonal process'.

Consider now any positive $\epsilon$ and any compact subset $K$ of $D$. We define the positive numbers $\eta$ and $t$ by

$$3\eta = \text{dist}\,(K, \mathbf{C} - D)$$

and

$$t = \min\,\{\epsilon\eta/8, \eta\}.$$

As $t > 0$, each point of $D$ lies within a distance $t$ of some $z_j$. Thus the family $\{\mathbf{C}(z_j, t) : j = 1, \ldots\}$ is an open cover of $D$ and hence of $K$. As $K$ is compact, there is an integer, say $s$, with

$$K \subset \bigcup_{j=1}^{s} \mathbf{C}(z_j, t).$$

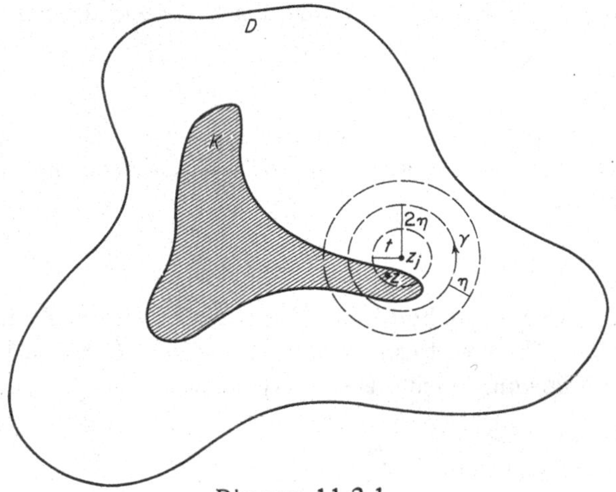


Diagram 11.2.1

Thus if $z \in K$, then for some $j$, $1 \leqslant j \leqslant s$, $z \in \mathbf{C}(z_j, t)$. The disc $\bar{\mathbf{C}}(z_j, 2\eta)$ is in $D$ and if $\gamma$ is its boundary circle, then as $t \leqslant \eta$ (see Diagram 11.2.1),

$$
\begin{aligned}
|f_n(z) - f_n(z_j)| &= \left| \frac{1}{2\pi i} \int_\gamma \frac{f_n(w)}{w - z} - \frac{f_n(w)}{w - z_j} \, dw \right| \\
&= \left| \frac{1}{2\pi i} \int_\gamma \frac{f_n(w)(z - z_j)}{(w - z)(w - z_j)} \, dw \right| \\
&\leqslant \frac{|z - z_j| \, 2\pi(2\eta)}{2\pi(\eta^2)} \\
&\leqslant \frac{2t}{\eta} \\
&\leqslant \frac{\epsilon}{4}.
\end{aligned}
$$

We conclude that for each $n$ and $m$,

$$
\begin{aligned}
|f_n(z) - f_m(z)| &\leqslant |f_n(z) - f_n(z_j)| + |f_n(z_j) - f_m(z_j)| + |f_m(z_j) - f_m(z)| \\
&< \frac{\epsilon}{4} + \sum_{i=1}^{s} |f_n(z_i) - f_m(z_i)| + \frac{\epsilon}{4}.
\end{aligned}
$$

The integer $s$ depends only on $t$ and $K$ and hence ultimately only on $\epsilon$ and $K$. If $m$ and $n$ are in $Z^*$ and sufficiently large, say $m \geqslant n \geqslant q$, then

$$|f_n(z_i) - f_m(z_i)| < \frac{\epsilon}{2s}, \qquad i = 1, \ldots, s,$$

and so

$$|f_n(z_i) - f_m(z)| < \epsilon.$$

As this $z$ was any point in $K$, we deduce that if $m \geqslant n \geqslant q$ ($q$ depends only on $K$ and $\epsilon$), then

$$\| f_n - f_m \|_K \leqslant \epsilon.$$

The General Principle of Uniform Convergence now implies that the sequence $(f_n : n \in Z^*)$ is uniformly convergent on $K$, and as $K$ was any compact subset of $D$ the proof is complete.

**Exercise 11.2**

1. Prove that $\mathbf{C}$ and $\mathbf{C}(0, 1)$ are homeomorphic but not conformally equivalent. [Apply Liouville's Theorem to any analytic function $f : \mathbf{C} \to \mathbf{C}(0, 1)$.]
2. Show that the following domains are simply connected.

   (a) $\mathbf{C}(0, 1) - \{x : x \geqslant 0\}$,

   (b) $\mathbf{C} - (E_1 \cup E_2)$, where $E_1 = [-i, i]$ and $E_2 = \{x + iy : x > 0, y = \sin(1/x)\}$

   (c) $\{x + iy : x > 0, 0 < y < 1\} - \bigcup_{n=2}^{\infty} L_n$,

   where

$$L_n = \begin{cases} \left[\frac{1}{n}, \frac{1}{n} + \frac{3i}{4}\right] & \text{if } n \text{ is even,} \\ \left[\frac{1}{n} + \frac{i}{4}, \frac{1}{n} + i\right] & \text{if } n \text{ is odd.} \end{cases}$$

3. Show that if $f$ is analytic and not zero in a domain $D$ and if $g$ is a branch of $\text{Log} f$ in $D$, then $g$ is also analytic in $D$.

## 11.3 BRANCHES OF THE ARGUMENT

We have already discussed four situations in which branches of the argument do or do not exist, namely

(a) there does not exist a branch of $\text{Arg}\, z$ on $\{z : |z| = 1\}$ (Example 5.4.1);
(b) there exists a branch of $\text{Arg}\, z$ on each $\mathbf{C} - L_\alpha$ (Theorem 5.4.1);
(c) if $\gamma$ is non-zero on $[a, b]$, there exists a branch of $\text{Arg}\, \gamma$ on $[a, b]$ (Theorem 7.2.1);
(d) if $f$ is non-zero and analytic in a simply connected domain $D$, then there exists a branch of $\text{Arg} f$ in $D$ (Theorem 11.1(e)).

These results have been the basic material with which we have developed the subject. In this section we shall pursue these ideas still further and obtain refinements of (a)–(d) which lead to the deeper topological results which are of interest in complex analysis.

Let $E$ be any subset of $\mathbf{C}_\infty$ and let $f : E \to \mathbf{C}_\infty$ be any continuous function. We shall be interested in conditions which imply the existence of a branch (a continuous choice) of Arg $f$ on $E$ and obviously we must insist that $f$ does not take the values zero or $\infty$ in $E$. We define

$$\mathbf{C}^* = \mathbf{C} - \{0\} = \mathbf{C}_\infty - \{0, \infty\}$$

and so we can write (in a concise way) that '$f : E \to \mathbf{C}^*$ is continuous'. In the case $f(z) = z$, this condition reduces to $0, \infty \notin E$ and, as usual, we shall refer to Arg $z$ (the purists can use Arg $I$, $I(z) = z$).

*Example 11.3.1* Let $[a, b]$ be a compact real interval and let $\gamma : [a, b] \to \mathbf{C}^*$ be continuous, then there exists a branch of Arg $\gamma$ on $[a, b]$. This is precisely Theorem 7.2.1 (see (c) above).

*Example 11.3.2* Let $\gamma : \mathbf{R} \to \mathbf{C}^*$ be continuous, then there exists a branch of Arg $\gamma$ on $\mathbf{R}$. To prove this we select any value, say $\alpha$, of Arg $\gamma(0)$. For each $m$, $m = 1, 2, \ldots$, there is a branch $\theta_m$ of Arg $\gamma$ on $[-m, m]$ and by adding a suitable integral multiple of $2\pi$ we may assume that $\theta_m(0) = \alpha$.

Now define $\theta$ on $\mathbf{R}$ by

$$\theta = \theta_m \text{ on } [-m, m], \qquad m = 1, \ldots .$$

This is properly defined, for if $m > q$, then $\theta_m = \theta_q$ on $[-q, q]$ because $\phi$, $\phi = (2\pi)^{-1} [\theta_m - \theta_q]$, is continuous and integer valued on $[-q, q]$ and $\phi(0) = 0$. As each $\theta_m$ is continuous, so is $\theta$ and $\theta$ is the desired branch.

The proof of Theorem 7.2.1 and its extension to Example 11.3.2 embody the same basic idea and it is worthwhile to consider this in the general situation.

*Proposition 11.3.1* *Let $A$ and $B$ be subsets of* $\mathbf{C}$ *which are both open or both closed and suppose that $A \cap B$ is connected. If $f : A \cup B \to \mathbf{C}^*$ is continuous and if there exists a branch $\theta_A$ of* Arg $f$ *on $A$ and a branch $\theta_B$ of* Arg $f$ *on $B$, then there exists a branch of* Arg $f$ *on $A \cup B$.*

*Proof* If $A \cap B = \emptyset$ we define $\theta = \theta_A$ on $A$ and $\theta = \theta_B$ on $B$. If not, we select any $\zeta$ in $A \cap B$ and (as in Example 11.3.2) we may assume that $\theta_A(\zeta) = \theta_B(\zeta)$. We deduce that $\theta_A = \theta_B$ on $A \cap B$ (as in Example 11.3.2 and using the fact that $A \cap B$ is connected) and we again define $\theta$ by $\theta = \theta_A$ on $A$ and $\theta = \theta_B$ on $B$.

It remains to show that $\theta$ is continuous on $A \cup B$. Select any $w$ in $A \cup B$; we may assume that $w \in A$. If $A$ and $B$ are open, then $\theta = \theta_A$ near $w$ and so $\theta$ is continuous at $w$. If $A$ and $B$ are closed observe that either $w \notin B$ or $w \in A \cap B$. In the first case $\theta = \theta_A$ near $w$, while in the second case both $\theta_A$ and $\theta_B$ are continuous at $w$. In all cases, then, $\theta$ is continuous at $w$ (the reader is asked to provide the formal details).

*Example 11.3.3* Let $Q$ be any square $[a, b] \times [a, b]$ and let $f : Q \to \mathbf{C}^*$ be continuous, then there exists a branch of Arg $f$ on $Q$. The proof of this uses the same idea as in the proof of Theorem 7.2.1.

As $f(Q)$ is compact and does not contain zero we have

$$\epsilon = \text{dist}\,(f(Q), 0) > 0.$$

The uniform continuity of $f$ on $Q$ provides a positive $\delta$ such that if $z$ and $z'$ are in $Q$ with $|z - z'| < \delta$, then $|f(z) - f(z')| < \epsilon$.

We now subdivide $Q$ in the usual manner into congruent non-overlapping squares $Q_{ij}$, $1 \leqslant i, j \leqslant s$, of diameter less than $\delta$ and it follows that each $f(Q_{ij})$ lies in some disc $\mathbf{C}(\zeta, \epsilon)$ which does not contain zero. If $\theta$ is a branch of Arg $w$, $w \in \mathbf{C}(\zeta, \epsilon)$, then $\theta \circ f$ is a branch of Arg $f$ on $Q_{ij}$. Thus we have constructed a branch of Arg $f$ on each $Q_{ij}$.

We may suppose that the $Q_{ij}$ are labelled so that $Q_{ij}$ is the square in the $i$th row and $j$th colum in the subdivision of $Q$. Using Proposition 11.3.1 (and exactly as in the proof of Theorem 7.2.1) we can construct, in succession, a branch of Arg $f$ on each of the sets

$$Q_{i1}, Q_{i1} \cup Q_{i2}, \ldots, Q_{i1} \cup \cdots \cup Q_{is}, \qquad i = 1, \ldots, s.$$

Let $Q_i = Q_{i1} \cup \cdots \cup Q_{is}$. The same argument enables us to construct branches of Arg $f$ on the sets

$$Q_1, Q_1 \cup Q_2, \ldots, Q_1 \cup \cdots \cup Q_s$$

and $Q = Q_1 \cup \cdots \cup Q_s$.

*Example 11.3.4* Let $f : \mathbf{C} \to \mathbf{C}^*$ be continuous, then there exists a branch of Arg $f$ on $\mathbf{C}$. This follows from Example 11.3.3. exactly as Example 11.3.2 followed from Example 11.3.1. Select any value $\alpha$ of Arg $f(0)$ and let $\theta_m$ be the branch of Arg $f$ on $[-m, m] \times [-m, m]$ with $\theta_m(0) = \alpha$. Then $\theta$ defined on $\mathbf{C}$ by $\theta = \theta_m$ on $[-m, m] \times [-m, m]$ $(m = 1, 2, \ldots)$ is the desired branch.

The four examples given above become more interesting once one has the next result available (see also Theorem 6.5.1).

*Proposition 11.3.2* *Let $E_1$ be any subset of $\mathbf{C}_\infty$ and let $\phi : E_1 \to E_2$ be a homeomorphism of $E_1$ onto $E_2$. If $E_1$ has the property that for every continuous function $f : E_1 \to \mathbf{C}^*$ there is a branch of* Arg $f$ *on $E_1$, then $E_2$ also has this property.*

*Proof* Let $g : E_2 \to \mathbf{C}^*$ be continuous. Then $g \circ \phi : E_1 \to \mathbf{C}^*$ is continuous and there exists a branch, say $\theta$, of Arg $(g \circ \phi)$ on $E_1$. We conclude that $\theta \circ \phi^{-1}$ is a branch of Arg $g$ on $E_2$.

As illustrations of the use of Proposition 11.3.2, we prove two more interesting results.

*The Fixed Point Theorem* *Let $f$ be a continuous map of $\bar{\mathbf{C}}(0, 1)$ into itself. Then $f$ has a fixed point, that is there is some $z$ with $f(z) = z$.*

*Proof* As $\bar{\mathbf{C}}(0, 1)$ is homeomorphic to $[-1, 1] \times [-1, 1]$ (Exercise 11.3.1) we see that if $g : \bar{\mathbf{C}}(0, 1) \to \mathbf{C}^*$ is continuous then there exists a branch of Arg $g$ on $\bar{\mathbf{C}}(0, 1)$.

Now suppose that $f$ has no fixed points and put $g(z) = f(z) - z$. Thus there exists a branch $\theta$ of Arg $g$ on $\bar{C}(0, 1)$. If $\gamma(t) = e^{it}$, $0 \leqslant t \leqslant 2\pi$, then $\theta \circ \gamma$ is a branch of Arg $g \circ \gamma$ on $[0, 2\pi]$ and so

$$n(g \circ \gamma, 0) = (2\pi)^{-1} [\theta(\gamma(2\pi)) - \theta(\gamma(0))] = 0.$$

We now put $h(z) = [f(z)/z] - 1$ so $g(z) = zh(z)$ and then

$$\begin{aligned} 0 &= n(g \circ \gamma, 0) \\ &= n(\gamma, 0) + n(h \circ \gamma, 0) \\ &= 1 + n(h \circ \gamma, 0). \end{aligned}$$

Finally, observe that $h$ maps $[\gamma]$ into the closed disc $\bar{C}(-1, 1)$ (this is just $|f(z)| \leqslant |z| = 1$ on $\gamma$) and $h \neq 0$ on $\gamma$. Thus $h$ maps $[\gamma]$ into the left half-plane and so there exists a branch of $h \circ \gamma$ on $[0, 2\pi]$. This gives $n(h \circ \gamma, 0) = 0$ and this is a contradiction.

The same technique gives *a topological version of the Argument Principle.* If $f: \bar{C}(0, 1) \to \mathbf{C}^*$ is continuous then, with $\gamma$ as above, $n(f \circ \gamma, 0) = 0$. Equivalently, if $f$ is continuous on $\bar{C}(0, 1)$ and if $n(f \circ \gamma, 0) \neq 0$ then $f$ is zero at some point of the disc. Of course, we cannot always count the zeros of $f$ for this depends in an essential way on $f$ being analytic.

*Theorem 11.3.1* *Let $D$ be a subdomain of* $\mathbf{C}$ *(possibly $D = \mathbf{C}$). Then $D$ is simply connected if and only if for every continuous function $f: D \to \mathbf{C}^*$ there exists a branch of* Arg *$f$ on $D$.*

*Proof* If $D = \mathbf{C}$, then $D$ is simply connected and from Example 11.3.4, there exists a branch of Arg $f$ on $D$.

Now let $D$ be a proper subdomain of $\mathbf{C}$. If $D$ is simply connected then $D$ is homeomorphic to $\mathbf{C}$ and the existence of a branch of Arg $f$ follows from Proposition 11.3.2 and Example 11.3.4. If there always exists a branch of Arg $f$, then there does so whenever $f$ is analytic and so by Theorem 11.1.1, $D$ is simply connected.

The next result is much deeper.

*Theorem 11.3.2* (Eilenberg) (a) *Let $E$ be a compact subset of* $\mathbf{C}$ *and suppose that $z_0 \in \mathbf{C} - E$. Then there exists a branch of* Arg $(z - z_0)$ *on $E$ if and only if $z_0$ and $\infty$ lie in the same component of $\mathbf{C}_\infty - E$.*

(b) *If $E$ is a compact subset of $\mathbf{C}_\infty$ and if $z_1$ and $z_2$ are distinct points in $\mathbf{C} - E$, then there exists a branch of*

$$\operatorname{Arg}\left(\frac{z - z_1}{z - z_2}\right), \qquad z \in E,$$

*on $E$ if and only if $z_1$ and $z_2$ lie in the same component of $\mathbf{C}_\infty - E$.*

*Proof* Without loss of generality we may take $z_0 = 0$. We suppose first that 0 and $\infty$ lie in the same component $D$ of $\mathbf{C}_\infty - E$. As $E$ is compact, $D$ is open and as $D$ is also connected there is a curve $\gamma : [a, b] \to D$ with $\gamma(a) = 0$ and $\gamma(b) = \infty$.

As $[\gamma]$ is compact and connected in $\mathbf{C}_\infty$ we may apply Theorem 11.1.2 and write

$$\mathbf{C}_\infty - [\gamma] = \bigcup_{\alpha \in A} D_\alpha,$$

where the $D_\alpha$, $\alpha \in A$, are mutually disjoint simply connected domains. As $D_\alpha$ is simply connected and as 0 and $\infty$ are not in $D_\alpha$, we deduce from Theorem 11.1.1(e) that there exists a branch $\theta_\alpha$ of Arg $z$ on $D_\alpha$.

The function $\theta$ defined on $\bigcup_\alpha D_\alpha$ by $\theta = \theta_\alpha$ on $D_\alpha$, $\alpha \in A$, is a branch of Arg $z$ on $\bigcup_\alpha D_\alpha$. As $E \subset \bigcup_\alpha D_\alpha$, there exists a branch of Arg $z$ on $E$.

We now assume that there is a branch $\theta$ of Arg $z$ on $E$ and that 0 and $\infty$ lie in different components of $\mathbf{C}_\infty - E$. Our aim is to reach a contradiction and we do this by reducing the situation to that discussed in Example 5.4.1.

Let $D$ be the component of $\mathbf{C}_\infty - E$ that contains zero but not $\infty$. As $E$ is a bounded subset of $\mathbf{C}$, $D$ is also bounded, since otherwise $\infty \in D$ (Exercise 11.3.2). We now extend the domain of definition of $\theta$ from $E$ to $E \cup D$, so that $\theta$ becomes continuous on $E \cup D$. This is achieved by the Tietze Extension Theorem (see Exercise 11.3.3). Note that the values of $\theta(z)$, $z \in D$, are not necessarily values of Arg $z$: we simply require that $\theta$ is unchanged on $E$ and that $\theta : E \cup D \to \mathbf{R}$ is continuous.

Next, define $f : \mathbf{C} \to \mathbf{C}^*$ by

$$f(z) = \begin{cases} z/|z| & \text{if } z \in \mathbf{C} - (E \cup D), \\ z/|z| \; [= \exp(i\theta(z))] & \text{if } z \in E, \\ \exp(i\theta(z)) & \text{if } z \in D. \end{cases}$$

It is easy to see that $f$ is continuous on $\mathbf{C}$ (Exercise 11.3.4) and as $f$ is never zero, there exists a branch $\theta_0$ of Arg $f$ on $\mathbf{C}$ (Example 11.3.4). Let $K = \{z : |z| = r\}$, where $r$ is chosen so that $E \cup D$ lies in $\mathbf{C}(0, r)$. As $z$ and $f(z)$ have the same arguments when $z \in K$, $\theta_0$ is actually a branch of Arg $z$ on $K$ and (as in Example 5.4.1) this cannot be so. We have now proved (a).

The situation in (b) easily reduces to (a). We let

$$m(z) = \frac{z - z_1}{z - z_2}$$

and this is a homeomorphism of $\mathbf{C}_\infty$ onto itself. Thus $z_1$ and $z_2$ lie in the same component of $\mathbf{C}_\infty - E$ if and only if 0 and $\infty$ lie in the same component of $\mathbf{C}_\infty - m(E)$. This is so if and only if there exists a branch $\theta$ of Arg $w$, $w \in m(E)$; equivalently, if and only if there exists a branch $\phi$ $(= \theta \circ m)$ of Arg $m$ on $E$.

We can now profitably study the notion of a simple curve. A curve $\gamma : [a, b] \to \mathbf{C}_\infty$ is a *simple curve* if $\gamma$ is a homeomorphism on $[a, b]$. Geometrically speaking a simple curve cannot 'meet itself' for if it does there are distinct points $t$ and $t'$ in $[a, b]$ with $\gamma(t) = \gamma(t')$. In fact, this is a necessary and sufficient condition for a curve $\gamma$ to be a simple curve (Exercise 11.3.5).

*Theorem 11.3.3* *Let $\gamma : [a, b] \to \mathbf{C}_\infty$ be a simple curve, then $\mathbf{C}_\infty - [\gamma]$ is a simply connected domain.*

*Proof* We first prove that $\mathbf{C}_\infty - [\gamma]$ is connected and we may assume that $\infty \in [\gamma]$. As $[\gamma]$ is homeomorphic to $[a, b]$ (by $\gamma$) we can use Example 11.3.1 and Proposition 11.3.2 to deduce that if $f : [\gamma] \to \mathbf{C}^*$ is continuous, then there exists a branch of $\operatorname{Arg} f$ on $[\gamma]$. In particular if $z_1$ and $z_2$ are any two distinct points in $\mathbf{C}_\infty - [\gamma]$, there exists a branch of

$$\operatorname{Arg}\left(\frac{z - z_1}{z - z_2}\right)$$

on $[\gamma]$. We conclude from Theorem 11.3.2(b) that $z_1$ and $z_2$ are in the same component of $\mathbf{C}_\infty - [\gamma]$ and so this set is connected.

As $[\gamma]$ is compact and contains $\infty$, the set $D$, $D = \mathbf{C}_\infty - [\gamma]$, is open and hence is a subdomain of $\mathbf{C}$. As $\mathbf{C}_\infty - D = [\gamma]$, which is connected, $D$ is simply connected (Theorem 11.1.1). An alternative proof that $D$ is simply connected is to apply Theorem 11.1.2.

Finally, we combine Proposition 11.3.1 and Theorem 11.3.2.

*Theorem 11.3.4* (Janiszewski) *Let $A$ and $B$ be compact subsets of $\mathbf{C}_\infty$ such that $A \cap B$ is connected. If $z_1$ and $z_2$ lie in the same component of $\mathbf{C}_\infty - A$ and in the same component of $\mathbf{C}_\infty - B$, then they lie in the same component of $\mathbf{C}_\infty - (A \cup B)$.*

*Proof* By applying a Möbius transformation we may assume that $z_1 = 0$ and $z_2 = \infty$. By Theorem 11.3.2, there exists a branch $\theta_A$ of $\operatorname{Arg} z$ on $A$ and a branch $\theta_B$ of $\operatorname{Arg} z$ on $B$. By Proposition 11.3.1, there is a branch of $\operatorname{Arg} z$ on $A \cup B$ and using Theorem 11.3.2 again, $z_1$ and $z_2$ lie in the same component of $\mathbf{C}_\infty - (A \cup B)$.

**Exercise 11.3**

1. Prove that $\bar{\mathbf{C}}(0, 1)$ and $[-1, 1] \times [-1, 1]$ are homeomorphic. [Use Theorem 6.5.1.]
2. Let $E$ be a non-empty compact subset of $\mathbf{C}$ and suppose that $D$ is a component of $\mathbf{C}_\infty - E$ that does not contain $\infty$. Prove that $E \cup D$ is a compact subset of $\mathbf{C}$.
3. Prove Tietze's Extension Theorem in the following form. Suppose that $E$ and $D$ are disjoint subsets of $\mathbf{C}$, that $E$ is compact and that $f : E \to \mathbf{R}$ is continuous. Then there exists a continuous function $F : E \cup D \to \mathbf{R}$ with $f = F$ on $E$. The proof divides into four parts.

   (a) For some $m$, $|f(z)| \leqslant m$ on $E$. Define

   $$A = \{z \in E : f(z) \leqslant -m/3\}, \qquad B = \{z \in E : f(z) \geqslant m/3\}$$

   and suppose that these sets are not empty. Define $f_1$ on $E \cup D$ by

   $$f_1(z) = \frac{m[\operatorname{dist}(z, A) - \operatorname{dist}(z, B)]}{3[\operatorname{dist}(z, A) + \operatorname{dist}(z, B)]}.$$

Prove that $f_1$ is defined and continuous on $E \cup D$ and that

$$|f_1(z)| \leqslant m/3 \qquad \text{on } E \cup D,$$

$$|f(z) - f_1(z)| \leqslant 2m/3 \qquad \text{on } E.$$

(b) Modify the definition of $f_1$ (a constant function will suffice) in the case when $A$ or $B$ (or both) are empty so that the conclusions in (a) remain true.

(c) Prove (by induction on $n$) that there are functions $f_1, \ldots, f_n$ which are continuous on $E \cup D$ and which satisfy

$$|f_k(z)| \leqslant (2/3)^k m \qquad \text{on } E \cup D,$$

$$|f(z) - [f_1(z) + \cdots + f_n(z)]| \leqslant (2/3)^n m \qquad \text{on } E.$$

(d) Deduce that

$$F = \sum_{n=1}^{\infty} f_n$$

has the desired properties.

4. Prove that $f$ (as defined in the proof of Theorem 11.3.2) is continuous on $\mathbf{C}$. [It is obviously continuous on the open set $\mathbf{C} - E$, so consider $\lim f(z)$ as $z \to \zeta$, $\zeta \in E$.]
5. Prove that a curve $\gamma : [a, b] \to \mathbf{C}_\infty$ is simple if and only if $\gamma$ is 1–1 on $[a, b]$. [Use Theorem 6.5.1.]

## 11.4 THE JORDAN CURVE THEOREM

A Jordan curve is a closed curve which is topologically equivalent to a circle. More precisely, $\gamma : [a, b] \to \mathbf{C}$ is a *Jordan curve* if there is a homeomorphism $\phi$ of $\{z : |z| = 1\}$ onto $[\gamma]$ such that if $t \in [a, b]$, then

$$\gamma(t) = \phi\left(\exp\left(2\pi i\left[\frac{t-a}{b-a}\right]\right)\right) \qquad (11.4.1)$$

This is illustrated in Diagram 11.4.1.

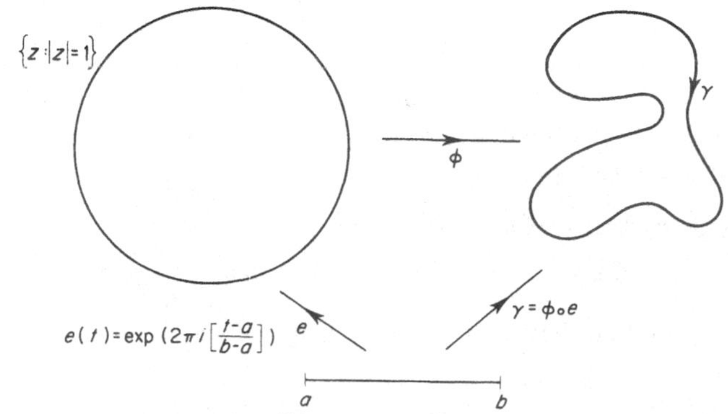


Diagram 11.4.1

If $\gamma$ is a Jordan curve, then $\gamma(a) = \gamma(b)$ and so $\gamma$ is necessarily closed. Further, it is easy to see that if $\gamma(t) = \gamma(t')$, then either $t = t'$ or $\{t, t'\} = \{a, b\}$. Thus $\gamma$ is closed and does not 'cross itself' except in the sense that its end-points coincide. This actually characterizes Jordan Curves (Exercise 11.4.1).

There is no loss of generality in assuming that $a = 0$ and $b = 1$ and we shall do this. It is also convenient to extend the definition of $\gamma$ to $\mathbf{R}$ by periodicity so that for all $t$ in $\mathbf{R}$ and all integers $n$,

$$\gamma(t) = \gamma(t + n). \tag{11.4.2}$$

We may now regard $\gamma$ as being defined on any interval $[s, s + 1]$. In the strictest sense, this does change the curve. However, it does not change $n(\gamma, w)$ or $[\gamma]$ and, for the moment, these are our only concern.

Although the following result seems self-evident, any proof of it must refer specifically to the formal definitions of the terms involved and herein lies part of the difficulty in the proof. Observe that we suppose that $\gamma$ lies in $\mathbf{C}$.

*The Jordan Curve Theorem* *Let* $\gamma : [0, 1] \to \mathbf{C}$ *be a Jordan curve. Then*

(a) $\mathbf{C}_\infty - [\gamma]$ *has exactly two components*;
(b) *each component is a simply connected domain*;
(c) $[\gamma]$ *is the boundary of each component*;
(d) *if the components are* $D_0$ *and* $D_1$ *with, say,* $\infty \in D_0$, *then*

$$n(\gamma, w) = \begin{cases} \pm 1 & \text{if } w \in D_1 \\ 0 & \text{if } w \in D_0, w \neq \infty. \end{cases}$$

In terms of the complex plane, $D_1$ is a bounded simply connected domain and is called the *interior* of $\gamma$. The plane domain $D_0 - \{\infty\}$ is unbounded and is called the *exterior* of $\gamma$. Observe that (d) implies (in our earlier terminology) that *the interior of* $\gamma$ *is precisely the set of points inside* $\gamma$.

Note that even (c) is not trivial. It is (or should be) clear that if $D$ is any component of $\mathbf{C}_\infty - [\gamma]$, then $\partial D \subset [\gamma]$: it is not so clear that $\partial D = [\gamma]$ (see Exercise 11.4.2).

*Proof* As $[\gamma]$ is compact and connected, (b) follows directly from Theorem 11.1.2. Next, as $[\gamma]$ is a compact subset of $\mathbf{C}$, one component of $\mathbf{C}_\infty - [\gamma]$, say $D_0$, contains $\infty$. In this case $D_0 - \{\infty\}$ is an unbounded subdomain of $\mathbf{C}$ and $n(\gamma, w) = 0$ if $w \in D_0 - \{\infty\}$.

We next show that $\mathbf{C}_\infty - [\gamma]$ has at least two components. The properties (a), (b) and (c) are invariant under a homeomorphism of $\mathbf{C}_\infty$ onto itself and so by applying a Möbius transformation we may assume that 0 and $\infty$ lie on $\gamma$. Moreover, by using (11.4.2) we may assume that $\gamma(0) = 0$ and $\gamma(c) = \infty$, where $0 < c < 1$.

Now let $\gamma_1$ and $\gamma_2$ be the restrictions of $\gamma$ to $[0, c]$ and $[c, 1]$ respectively. Then $\gamma_1$ and $\gamma_2$ are two simple curves each joining 0 to $\infty$ and not intersecting except at 0 and $\infty$. Theorem 11.3.3 implies that $\mathbf{C}_\infty - [\gamma_1]$ and $\mathbf{C}_\infty - [\gamma_2]$ are both simply connected subdomains of $\mathbf{C}$ and so by Theorem 11.1.1 there exist branches of

Arg $z$ on each of these domains. As

$$\mathbf{C}_\infty - [\gamma] = (\mathbf{C}_\infty - [\gamma_1]) \cap (\mathbf{C}_\infty - [\gamma_2]),$$

the assumption that $\mathbf{C}_\infty - [\gamma]$ is connected leads (using Proposition 11.3.1) to the existence of a branch of Arg $z$ on the union of $\mathbf{C}_\infty - [\gamma_1]$ and $\mathbf{C}_\infty - [\gamma_2]$. But this union is $\mathbf{C}^*$ and so this branch cannot exist. We conclude that $\mathbf{C}_\infty - [\gamma]$ *has at least two components*. This is now an established topological fact and it is no longer necessary to continue to assume that, for example, $\gamma(0) = 0$.

For the moment, we leave the proof of (a) and turn our attention to (c). Let $D$ and $D'$ be any two components of $\mathbf{C}_\infty - [\gamma]$. Clearly $\partial D \subset [\gamma]$ and we shall now examine the consequences of the assumption that $\partial D$ is not equal to $[\gamma]$. First, $\partial D$ is a compact subset of $[\gamma]$ and, using (11.4.2), we may assume that $\partial D$ lies in the image of $\gamma$ of $[0, c]$, where $0 < c < 1$. The restriction $\gamma_3$ of $\gamma$ to $[0, c]$ is a simple curve and so by Theorem 11.3.3, $\mathbf{C}_\infty - [\gamma_3]$ is a (simply connected) domain. It follows that any point in $D$ can be joined to any point in $D'$ by a curve $\sigma$ in $\mathbf{C}_\infty - [\gamma_3]$. As this curve starts in $D$ and does not meet $\partial D$ (for it does not meet $[\gamma_3]$) it must be entirely in $D$ and so $D = D'$. This contradicts the fact that $\mathbf{C}_\infty - [\gamma]$ has at least two components and so we are led to the conclusion that for each component $D$ of $\mathbf{C}_\infty - [\gamma]$, $\partial D = [\gamma]$. This proves (c).

We can now complete the proof of (a). As $\mathbf{C}_\infty - [\gamma]$ has at least two components there is one component, say $D_1$, that is bounded (Exercise 11.3.2). Let $D_\alpha$, $\alpha \in A$, be the remaining components (one of which is $D_0$).

Select any point $z^*$ in $D_1$ and let $[z_1, z_2]$ be the largest horizontal segment containing $z^*$ which lies, apart from its end-points, in $D_1$ (we assume that $\mathrm{Re}\,[z_1] < \mathrm{Re}\,[z_2]$: see Diagram 11.4.2 and Exercise 11.4.4). We define $\sigma$ by

$$\sigma(t) = z^* + t, \qquad -t_1 \leqslant t \leqslant t_2,$$

where $z_1 = z^* - t_1$ and $z_2 = z^* + t_2$. Observe that $z_1$ and $z_2$ are on $\gamma$ and $[\sigma] = [z_1, z_2]$. Using (11.4.2), we may assume that $z_1 = \gamma(0)$ and that $z_2 = \gamma(c)$, where $0 < c < 1$.

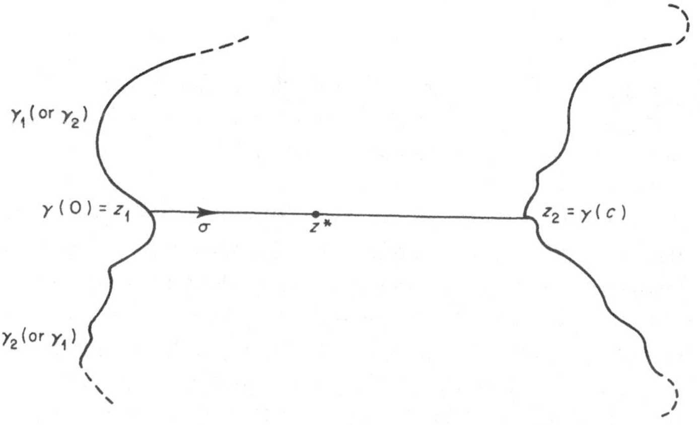


Diagram 11.4.2

Now let $\gamma_1$ and $\gamma_2$ be the restriction of $\gamma$ to $[0, c]$ and $[c, 1]$ respectively; then $\gamma_1$ is a simple curve with initial point $z_1$ and final point $z_2$ and $\gamma_2$ is a simple curve with initial point $z_2$ and final point $z_1$. We must not make any assumptions about any 'direction' induced on $[\gamma]$ by $\gamma$.

Consider now any $w$ in $\cup_\alpha D_\alpha$. The set

$$\bigcup_\alpha D_\alpha \cup ([\gamma_2] - \{z_1, z_2\})$$

is connected (because of (c)), it contains $w$ and $\infty$ and it is disjoint from $[\sigma] \cup [\gamma_1]$. Thus $w$ and $\infty$ lie in the same component of $\mathbf{C}_\infty - ([\sigma] \cup [\gamma_1])$ and, for similar reasons, in the same component of $\mathbf{C}_\infty - ([\sigma] \cup [\gamma_2])$. According to Theorem 11.3.4, $w$ and $\infty$ lie in the same component of $\mathbf{C}_\infty - ([\sigma] \cup [\gamma])$ and hence in the same component of $\mathbf{C}_\infty - [\gamma]$ (which is larger). This shows that $w \in D_0$ and so $D_0 = \cup_\alpha D_\alpha$. Thus $\mathbf{C}_\infty - [\gamma]$ has exactly two components, namely $D_0$ and $D_1$.

We have now established (a), (b), (c) and the part of (d) relating to $D_0$. It remains to prove that if $w \in D_1$, then

$$|n(\gamma, w)| = 1.$$

Let $\gamma_1, \gamma_2, z^*$ and $\sigma$ be as above and select a positive $r$ so that $\bar{\mathbf{C}}(z^*, r) \subset D_1$. Now let $\sigma_1, \sigma_2$ and $\sigma_3$ be the restrictions of $\sigma$ to $[-t_1, -r]$, $[-r, r]$ and $[r, t_2]$ respectively and let

$$\tau(t) = z^* + r\,e^{it}, \qquad 0 \leqslant t \leqslant \pi$$

(thus $\tau$ is a semi-circle with diameter $[\sigma_2]$). Finally, let

$$\zeta_1 = z^* + \tfrac{1}{2}ir \qquad \text{and} \qquad \zeta_2 = z^* - \tfrac{1}{2}ir.$$

These are all illustrated in Diagram 11.4.3.

We emphasize that the following argument is based on computations and not on Diagram 11.4.3. For each $w$ not on any of the cycles

$$\Sigma_1 = (\gamma_2, \sigma),$$
$$\Sigma_2 = (\gamma_2, \sigma_1, \tau^-, \sigma_3),$$
$$\Sigma_3 = (\sigma_2, \tau)$$

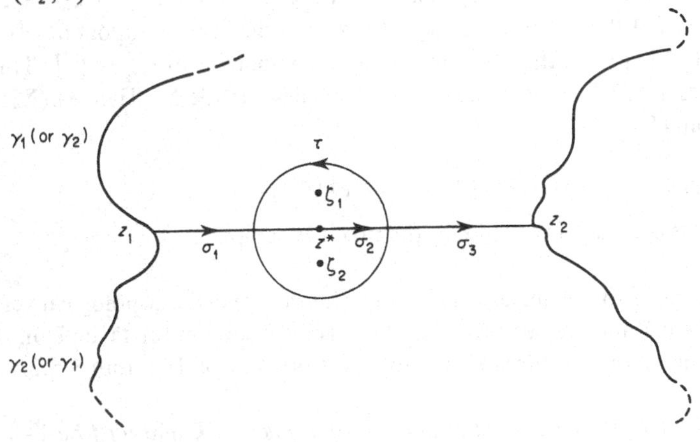


Diagram 11.4.3

we have, by cancellation,

$$n(\Sigma_1, w) = n(\Sigma_2, w) + n(\Sigma_3, w).$$

We know that

$$n(\Sigma_3, \zeta_1) = 1, \qquad n(\Sigma_3, \zeta_2) = 0$$

and as the segment $[\zeta_1, \zeta_2]$ does not meet $\Sigma_2$, we have

$$n(\Sigma_2, \zeta_1) = n(\Sigma_2, \zeta_2).$$

We deduce that

$$n(\Sigma_1, \zeta_1) - n(\Sigma_1, \zeta_2) = 1. \tag{11.4.3}$$

If $\zeta_1$ and $\zeta_2$ lay in the same component of the Jordan curve described by $\Sigma_1$, the left-hand side of (11.4.3) would be zero. Thus exactly one of the points $\zeta_1$ and $\zeta_2$ lies inside $\Sigma_1$, we call this $w_1$, and the other, we call this $w_2$, lies outside $\Sigma_1$. Thus $n(\Sigma_1, w_2) = 0$ and so from (11.4.3)

$$|n(\Sigma_1, w_1)| = 1. \tag{11.4.4}$$

We do not know whether $w_1 = \zeta_1$ or $w_1 = \zeta_2$, for this depends on the orientation that $\gamma$ induces on $[\gamma]$.

We now define the cycle $\Sigma_4$ by

$$\Sigma_4 = (\gamma_1, \sigma^-)$$

and a similar argument yields the analogue of (11.4.3), namely

$$n(\Sigma_4, \zeta_1) - n(\Sigma_4, \zeta_2) = -1. \tag{11.4.5}$$

As $\{\zeta_1, \zeta_2\} = \{w_1, w_2\}$, exactly one of these points lies inside $\Sigma_4$ and exactly one lies outside $\Sigma_4$.

It is intuitively clear that $w_2$ lies inside $\Sigma_4$, but we must prove that this is so. Indeed if $w_2$ lies outside $\Sigma_4$, then it lies outside $\Sigma_1$ and $\Sigma_4$; that is $w_2$ and $\infty$ lie in the same component of $\mathbf{C}_\infty - [\Sigma_1]$ and in the same component of $\mathbf{C}_\infty - [\Sigma_4]$. Theorem 11.3.4 then shows that $w_2$ and $\infty$ lie in the same component of $\mathbf{C}_\infty - ([\gamma] \cup [\sigma])$ and therefore in the same component of $\mathbf{C}_\infty - [\gamma]$. This is false; thus $w_2$ lies inside $\Sigma_4$ and, using (11.4.5), $w_1$ lies outside $\Sigma_4$. Hence $n(\Sigma_4, w_1) = 0$ and so from (11.4.4),

$$|n(\gamma, w_1)| = |n(\Sigma_1, w_1) + n(\Sigma_4, w_1)| = 1.$$

As $n(\gamma, w)$ is constant for $w$ in $D_1$, the proof is complete.

As an application of the Jordan Curve Theorem and the topological version of the Argument Principle, we prove the Invariance Theorem for Plane Domains. The reader should compare this with the second Corollary of Theorem 8.5.2.

*Theorem 11.4.1* (Brouwer) *Let $D$ be a subdomain of $\mathbf{C}$ and let $f$ be 1–1 and continuous on $D$. Then $f(D)$ is a domain and $f: D \to f(D)$ is a homeomorphism.*

*Proof* Obviously $f(D)$ is connected. Now take any $w$ in $D$, let $r$ be positive and such that $\bar{C}(w, r) \subset D$ and let $\gamma(t) = w + r e^{it}, 0 \leqslant t \leqslant 2\pi$.

The restriction of $f$ to $\bar{C}(w, r)$ is a homeomorphism (Theorem 6.5.1) and so $f \circ \gamma$ is a Jordan curve with interior, say, $\Delta$. If $\zeta \in \Delta$, then

$$n(f \circ \gamma - \zeta, 0) = n(f \circ \gamma, \zeta) = \pm 1$$

and so (see the comment preceding Theorem 11.3.1), $f(z) = \zeta$ at some point $z$ of $C(w, r)$. This shows that

$$\Delta \subset f(C(w, r)).$$

As $f(C(w, r))$ is connected and does not meet $[f \circ \gamma]$, it lies in one of the components of $C_\infty - [f \circ \gamma]$. Thus

$$\Delta = f(C(w, r))$$

and so for some positive $\epsilon$,

$$C(f(w), \epsilon) \subset \Delta \subset f(D).$$

This shows that $f(D)$ is open and hence is a domain.

Because $f$ restricted to $C(w, r)$ is a homeomorphism, so is $f^{-1} : \Delta \to C(w, r)$, and so $f^{-1}$ is continuous at $f(w)$. Thus $f^{-1}$ is continuous on $f(D)$.

We shall need two other results concerning Jordan curves.

*Theorem 11.4.2* *Let $\gamma : [0, 1] \to C$ be a Jordan curve with interior $D$ and suppose that $\zeta \in [\gamma]$. Then for all positive $\epsilon$ there is a positive $\delta$ such that if $z$ and $w$ are in $D \cap C(\zeta, \delta)$, there exists a curve $\sigma$ joining $z$ to $w$ in $D \cap C(\zeta, \epsilon)$.*

Roughly speaking, if $z$ and $w$ are in $D$ and sufficiently close to $\zeta$, then they can be joined by a curve $\sigma$ in $D$ which is also close to $\zeta$. This is not true for all closed curves (see Exercise 11.4.3).

*Proof* We may assume that $\epsilon$ is sufficiently small so that $C(\zeta, \epsilon)$ does not contain $[\gamma]$. Now let $\gamma^*$ be the restriction of $\gamma$ to some compact subinterval of $[0, 1]$ so that $\gamma^*$ lies in $C(\zeta, \epsilon)$ and contains $\zeta$ but not as an endpoint (so $\gamma^*$ is a sub-arc of $\gamma$).

Next, define

$$E = [\gamma] - [\gamma^*],$$
$$\delta = \text{dist}(\zeta, E),$$
$$Q = \{z : |z - \zeta| = \epsilon\}.$$

A typical situation is illustrated in Diagram 11.4.4.

These definitions imply that $0 < \delta < \epsilon$ (because $E$ contains $[\gamma] - C(\zeta, \epsilon)$),

$$E \cap [\gamma] = E, \quad E \cup [\gamma] = [\gamma], \quad Q \cap [\gamma] \subset E$$

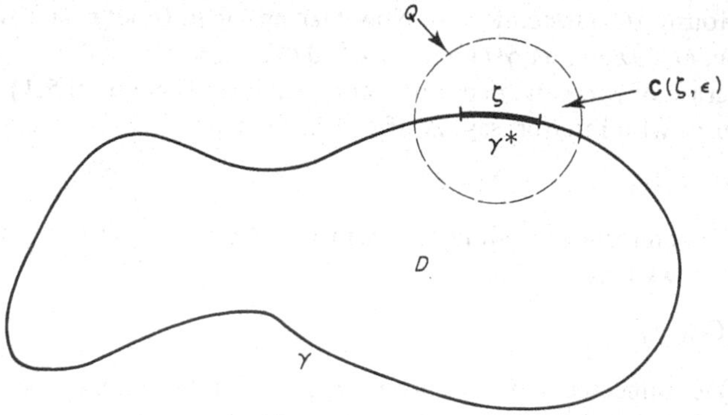


Diagram 11.4.4

and finally,

$$\begin{aligned} E &= E \cap [\gamma] \\ &\subset (E \cup Q) \cap [\gamma] \\ &= (E \cap [\gamma]) \cup (Q \cap [\gamma]) \\ &= E \cup (Q \cap [\gamma]) \\ &\subset E \cup E \\ &= E. \end{aligned}$$

If $z$ and $w$ are in $\mathbf{C}(\zeta, \delta)$ and also in $D$ we can join $z$ to $w$ by a straight segment $\sigma$ in $\mathbf{C}(\zeta, \delta)$. Thus $[\sigma]$ does not meet $E$ or $Q$ and so $z$ and $w$ lie in the same component of $\mathbf{C}_\infty - (E \cup Q)$. But $z$ and $w$ also lie in the same component, namely $D$, of $\mathbf{C}_\infty - [\gamma]$. As $(E \cup Q) \cap [\gamma] = E$, it is connected and Theorem 11.3.4 shows that $z$ and $w$ lie in the same component, say $D^*$, of $\mathbf{C}_\infty - (E \cup Q \cup [\gamma])$, that is of $\mathbf{C}_\infty - ([\gamma] \cup Q)$. Thus $D^*$ does not meet $[\gamma]$ (so $D^* \subset D$) or $Q$ (so $D^* \subset \mathbf{C}(\zeta, \epsilon)$) : hence $D^* \subset D \cap \mathbf{C}(\zeta, \epsilon)$. As $D^*$ is a domain we can certainly join $z$ to $w$ by a curve $\sigma$ in $D^*$, and the proof is complete.

*Theorem 11.4.3* *Let $\gamma : [0, 1] \to \mathbf{C}$ be a Jordan curve with interior $D$. Suppose that $\zeta_1$ and $\zeta_2$ are distinct points on $\gamma$ and that $\sigma$ is a simple curve joining $\zeta_1$ to $\zeta_2$ in $D \cup \{\zeta_1, \zeta_2\}$. Then $D - [\sigma]$ is the disjoint union of two non-empty domains.*

*Proof* We may assume that $\zeta_1 = \gamma(0)$ and that $\zeta_2 = \gamma(c)$ where $0 < c < 1$. We denote by $\gamma_1$ the restriction of $\gamma$ to $[0, c]$ and by $\gamma_2$ the restriction of $\gamma$ to $[c, 1]$ and we define the cycles $\Gamma_1$ and $\Gamma_2$ by

$$\Gamma_1 = (\gamma_1, \sigma^-), \qquad \Gamma_2 = (\gamma_2, \sigma).$$

Clearly, we may regard the $\Gamma_j$ as Jordan curves and

$$n(\Gamma, z) = n(\Gamma_1, z) + n(\Gamma_2, z). \tag{11.4.6}$$

As $[\Gamma_1] \subset [\Gamma] \cup D$, the exterior of $\Gamma$ does not meet $[\Gamma_1]$. However, the exteriors of $\Gamma$ and $\Gamma_1$ certainly meet (near $\infty$) and so (being connected) the exterior of $\Gamma$ is contained in the exterior of $\Gamma_1$. By symmetry, the same is true of $\Gamma_2$. Thus for all $z$

in $\mathbf{C} - ([\Gamma] \cup [\sigma])$,

$$n(\Gamma, z) = 0 \qquad \text{implies} \qquad n(\Gamma_1, z) = n(\Gamma_2, z) = 0$$

which, by virtue of (11.4.6), can be strengthened to

$$n(\Gamma, z) = 0 \qquad \textit{if and only if} \qquad n(\Gamma_1, z) = n(\Gamma_2, z) = 0. \tag{11.4.7}$$

Now let $D_j$ be the interior of $\Gamma_j$. If $z \in D - [\sigma]$, then $n(\Gamma, z) \neq 0$ and by (11.4.7), $z \in D_1$ or $z \in D_2$. Conversely, if $z$ is in $D_1$ or $D_2$, then again by (11.4.7) $n(\Gamma, z) \neq 0$ and so $z$ is in $D$. Thus

$$D - [\sigma] = D_1 \cup D_2.$$

With this available we see that if $z$ is in $D_1 \cap D_2$, then all three terms in (11.4.6) are each +1 or −1, and this is impossible. Hence $D_1$ and $D_2$ are disjoint and the proof is complete.

We mention one other important result which can be proved by similar methods. It is, however, an immediate corollary of the Riemann Mapping Theorem and the theorem in the next section.

*Theorem 11.4.4* *Let $\gamma$ be a Jordan curve with interior $D$ and suppose that $z \in D$ and $\zeta \in [\gamma]$. Then there exists a simple curve joining $z$ to $\zeta$ in $D \cup \{\zeta\}$.*

It is worth noting now that we shall use a much weaker version of this which is easy to prove. A point $\zeta$ on $\gamma$ is said to be *linearly accessible* if there exists a segment $[z, \zeta]$, $z \in D$, which lies in $D \cup \{\zeta\}$. Given any $\zeta$ on $\gamma$ and any positive $\epsilon$, we can find a $z$ in $D \cap C(\zeta, \frac{1}{2}\epsilon)$. Now let $\zeta'$ be a point on $\gamma$ which is as near to $z$ as any other point on $\gamma$. Then $[z, \zeta']$ is in $D \cup \{\zeta'\}$ and so $\zeta'$ is linearly accessible. Moreover,

$$\begin{aligned} |\zeta - \zeta'| &\leqslant |\zeta - z| + |z - \zeta'| \\ &\leqslant 2|\zeta - z| \\ &< \epsilon: \end{aligned}$$

thus there are linearly accessible points arbitrarily close to any $\zeta$.

**Exercise 11.4**

1. Let $\gamma : [a, b] \to \mathbf{C}$ be any closed curve. Prove that $\gamma$ is a Jordan curve if and only if for all $t$ and $t'$ in $[a, b]$, $\gamma(t) = \gamma(t')$ implies that $t = t'$ or $\{t, t'\} = \{a, b\}$.
2. Let $\gamma : [0, 2 + 2\pi] \to \mathbf{C}$ be defined by

$$\gamma(t) = \begin{cases} t & \text{if } 0 \leqslant t \leqslant 1, \\ e^{i(t-1)} & \text{if } 1 \leqslant t \leqslant 1 + 2\pi, \\ 2 + 2\pi - t & \text{if } 1 + 2\pi \leqslant t \leqslant 2 + 2\pi. \end{cases}$$

Prove that $\gamma$ is a closed curve and that $\mathbf{C} - [\gamma]$ has exactly two components. Let $D_1$ be the unbounded component : show that $\partial D_1 \neq [\gamma]$.

3. Let $\gamma$ be as in Exercise 2 and let $\zeta = \frac{1}{2}$. Prove (rigorously) that

$$z_n = \zeta + i/n, \qquad w_n = \zeta - i/n, \qquad n = 3, 4, \ldots,$$

are in the same component $D$ of $\mathbf{C} - [\gamma]$ and converge to $\zeta$ on $\gamma$ but that they cannot be joined by a curve in $D \cap \mathbf{C}(\zeta, \frac{1}{2})$.
4. Prove that $E$ (as defined in the proof of Theorem 11.4.2) is connected.
5. Let $D$ be the simply connected domain given in Exercise 11.2.2(c). Prove that $z$ (in $D$) cannot be joined to $i/2$ (in $\partial D$) by a curve in $D \cup \{i/2\}$. [Compare this with Theorem 11.4.4.]
6. Let $\gamma$ be a Jordan curve with interior $D$ and let $\zeta$ be on $\gamma$. Let $\delta_n$ correspond to the choice $\epsilon = 1/n$ in Theorem 11.4.2 (we may assume that $\delta_n \geqslant \delta_{n+1}$) and select $z_n$ in $D$ with $|z_n - \zeta| < \delta_n$. Use Theorem 11.4.2 to show that there is a curve joining $z_1$ (through $z_2, z_3, \ldots$) to $\zeta$ in $D \cup \{\zeta\}$. [Compare this with Theorem 11.4.4]

## 11.5 CONFORMAL MAPPING OF A JORDAN DOMAIN

We end this text with a result which plays a central part in the theory of conformal mapping and which depends in an essential way on the topology of the plane. In view of our earlier concern for plane topology we can give a more explicit proof of this result than is usually offered.

*Theorem 11.5.1* *Let $\gamma$ and $\Gamma$ be Jordan curves in $\mathbf{C}$ with interiors $D$ and $\Delta$ respectively and let $f$ be a univalent analytic function that maps $D$ onto $\Delta$. Then $f$ may be extended to a homeomorphism of $D \cup [\gamma]$ onto $\Delta \cup [\Gamma]$.*

Observe that the Jordan Curve Theorem implies that $\bar{D} = D \cup [\gamma]$ and $\bar{\Delta} = \Delta \cup [\Gamma]$. The proof can easily be reduced to a proof of the following (apparently much simpler) result.

*Proposition 11.5.1* *Let the hypotheses of Theorem 11.5.1 hold. Then for all $\zeta$ on $\gamma$,*

$$\lim_{z \to \zeta, z \in D} f(z)$$

*exists.*

*Proof of Theorem 11.5.1* We assume the validity of Proposition 11.5.1 and as $[\gamma] = \partial D$ we may define $f$ on $[\gamma]$ by

$$f(\zeta) = \lim_{z \to \zeta, z \in D} f(z).$$

Now suppose that $\zeta$ is on $\gamma$ and let $z_n$, $n = 1, 2, \ldots$, be any sequence in $\bar{D}$ that converges to $\zeta$. If $z_n \in D$ we let $z_n' = z_n$: if $z_n \notin D$ we find a point $z_n'$ in $D$ with

$$|z_n - z_n'| < \frac{1}{n}, \qquad |f(z_n) - f(z_n')| < \frac{1}{n}.$$

Then $z'_n$ converges to $\zeta$ from within $D$ and so

$$\lim_{n\to\infty} f(z_n) = \lim_{n\to\infty} f(z'_n) = f(\zeta).$$

This shows that $f$ is continuous in $\bar{D}$.

Next, $f(\bar{D})$ is compact and contains $\Delta$ so $\bar{\Delta} \subset f(\bar{D})$. On the other hand, $f(D) \subset \bar{\Delta}$ and so from the definition of $f$,

$$f(\bar{D}) \subset \overline{f(D)} = \bar{\Delta}.$$

Thus $f(\bar{D}) = \bar{\Delta}$, that is $f$ maps $\bar{D}$ onto $\bar{\Delta}$.

We have shown that $f$ extends to a continuous map of $\bar{D}$ onto $\bar{\Delta}$. We write $g$ for $f^{-1}$ in $\Delta$ and likewise, $g$ extends to a continuous map of $\bar{\Delta}$ onto $\bar{D}$.

It remains only to prove that $g : \bar{\Delta} \to \bar{D}$ is the inverse of $f : \bar{D} \to \bar{\Delta}$. If $z \in \bar{D}$ we can select $z_n$ in $D$ with $z_n \to z$. Then $f(z_n) \in \Delta$, $g(f(z_n)) = z_n$, $f(z_n) \to f(z)$ and hence

$$\begin{aligned} g(f(z)) &= \lim_{w\to f(z), w\in\Delta} g(w) \\ &= \lim_{n\to\infty} g(f(z_n)) \\ &= \lim_{n\to\infty} z_n \\ &= z. \end{aligned}$$

Thus $g = f^{-1}$ on $\bar{\Delta}$ and so $f : \bar{D} \to \bar{\Delta}$ is a homeomorphism.

If we had wished to do so, we could have reduced the problem in Proposition 11.5.1 to the case when either $D$ or $\Delta$ is $\mathbf{C}(0, 1)$. The symmetry of $f$ and $f^{-1}$ exploited above would then no longer have been valid and an alternative argument for this part of the proof would have had to be found.

*Proof of Proposition 11.5.1* If this is false (which we assume) there is a point $\zeta$ on $\gamma$ and two sequences, say $z_n$ and $w_n$ ($n = 1, 2, \ldots$) each in $D$ and converging to $\zeta$, with

$$f(z_n) \to \alpha, \qquad f(w_n) \to \beta$$

as $n \to \infty$ and $\alpha \neq \beta$. Obviously, $\alpha$ and $\beta$ are in $\bar{\Delta}$. In fact, $\alpha$ and $\beta$ are in $[\Gamma]$ for if, say, $\alpha \in \Delta$, then $f^{-1}$ is continuous at $\alpha$, $f^{-1}(\alpha) \in D$ and

$$\begin{aligned} \zeta &= \lim_{n\to\infty} z_n \\ &= \lim_{n\to\infty} f^{-1}(f(z_n)) \\ &= f^{-1}(\alpha). \end{aligned}$$

Thus $\zeta \in D$ which is false.

For the moment we may assume that $\Gamma$ is defined on $[0, 1]$ and that $\Gamma(0) = \alpha$, $\Gamma(c) = \beta$, where $0 < c < 1$. Now choose values $c_1$ and $c_2$ satisfying

$$0 < c_1 < c < c_2 < 1$$

and such that $\xi_1^*(=\Gamma(c_1))$ and $\xi_2^*(=\Gamma(c_2))$ are linearly accessible. This means that we may join $\xi_1^*$ to $\xi_2^*$ in $D \cup \{\xi_1^*, \xi_2^*\}$ by a polygonal curve $\tau$ and (if necessary by the deletion of segments) we may assume that $\tau$ is a simple polygonal curve (Exercise 11.5.1). Finally, let $\Gamma_1$ and $\Gamma_2$ be the restrictions of $\Gamma$ to $[0, c]$ and $[c, 1]$ respectively: these are the arcs of $\Gamma$ from $\alpha$ to $\beta$ and from $\beta$ to $\alpha$. Note that $\xi_j^* \in [\Gamma_j]$.

We must now select a positive number $r$ with certain properties related to the geometry of the situation described above. Precisely, we select any positive $r$ with the properties

(a) the discs $\mathbf{C}(\alpha, 2r)$, $\mathbf{C}(\beta, 2r)$, $\mathbf{C}(\xi_1^*, 2r)$, $\mathbf{C}(\xi_2^*, 2r)$ are pairwise disjoint;
(b) $\mathbf{C}(\alpha, 2r)$ and $\mathbf{C}(\beta, 2r)$ do not meet $[\tau]$;
(c) $\mathbf{C}(\xi_1^*, 2r) \cap [\Gamma_2] = \emptyset$, $\mathbf{C}(\xi_2^*, 2r) \cap [\Gamma_1] = \emptyset$.

The next step in the proof is to show that for almost all $n$, $f(z_n)$ and $f(w_n)$ lie in different components of $\Delta - [\tau]$: the situation is illustrated in Diagram 11.5.1. According to Theorem 11.4.3, $\Delta - [\tau]$ is the disjoint union of two domains and the proof of Theorem 11.4.3, contains a description of this (unique) pair of domains. If $\Gamma_{12}$ denotes the arc of $\Gamma$ from $\xi_1^*$ through $\beta$ to $\xi_2^*$ and then the curve $\tau^-$ to $\xi_1^*$, and if $\Gamma_{21}$ denotes the corresponding curve from $\xi_1^*$ to $\xi_2^*$ along $\tau$ and then back along $\Gamma$ (through $\alpha$) to $\xi_1^*$, then

$$\Delta - [\tau] = \Delta_{12} \cup \Delta_{21}$$

where $\Delta_{ij}$ is the interior of $\Gamma_{ij}$. We omit the formal descriptions (which can be given without reference to the diagram) as they are essentially given in the proof of Theorem 11.4.3. Now $\alpha$ is not in $[\Gamma_{12}]$ and any $\mathbf{C}(\alpha, t)$, $t > 0$, meets the exterior of $\Gamma$ and hence the exterior of $\Gamma_{12}$. We deduce that $\alpha$ lies outside $\Gamma_{12}$. Thus for

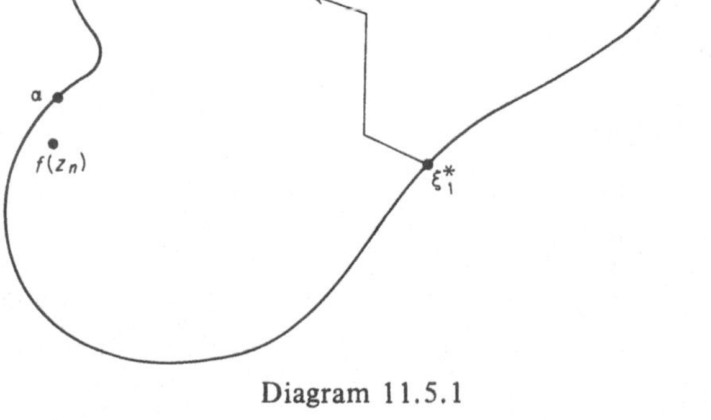

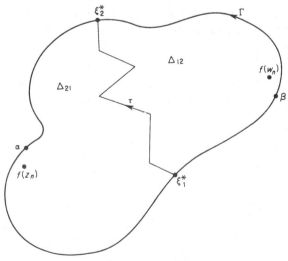

Diagram 11.5.1

sufficiently large $n, f(z_n)$ lies outside $\Gamma_{12}$ and so lies in $\Delta_{21}$. The same argument holds for the $w_n$: thus for almost all $n$,

$$f(z_n) \in \Delta_{21}, \qquad f(w_n) \in \Delta_{12}. \tag{11.5.1}$$

We shall now use Theorem 11.4.2 to obtain a sequence of curves $\sigma_n$ joining $z_n$ to $w_n$ in $D$ and with the property that as $n \to \infty$,

$$\epsilon_n = \sup \{|z - \zeta| : z \in [\sigma_n]\} \to 0.$$

We may assume that $\zeta = 0$ (this is convenient later) and then as $n \to \infty$,

$$\epsilon_n = \sup \{|z| : z \in [\sigma_n]\} \to 0. \tag{11.5.2}$$

The image curves $f \circ \sigma_n$ are denoted by $\Sigma_n$ and these join $f(z_n)$ to $f(w_n)$ in $\Delta$. Using standard arguments involving uniform continuity we may modify the $\Sigma_n$ so as to be polygonal curves and this may be done without destroying (11.5.2). We may now assume that the $\Sigma_n$ are also simple curves. Note that $|f^{-1}| \leqslant \epsilon_n$ on $[\Sigma_n]$.

It is a direct consequence of (11.5.1) that $\Sigma_n$ and $\tau$ must intersect. The fact that the curves $\sigma_n$ move uniformly to $0 \, (= \zeta)$ as $n \to \infty$ implies that the points of intersection of $\Sigma_n$ and $\tau$ must eventually be close to either $\xi_1^*$ or $\xi_2^*$. To see this, choose points $\xi_1$ and $\xi_2$ on $\tau$ such that $|\xi_j - \xi_j^*| < r$, and let $\tau_0$ be the arc of $\tau$ joining $\xi_1$ to $\xi_2$. Then $[\tau_0]$ is a compact subset of $\Delta, f^{-1}([\tau_0])$ is a compact subset of $D$ and $[\sigma_n] \cap f^{-1}([\tau_0]) = \emptyset$ for $n \geqslant n_0$, say. Thus for almost all $n$, $[\Sigma_n] \cap [\tau_0] = \emptyset$.

If all of the points of intersection of $\Sigma_n$ and $\tau$ occur near $\xi_2^*$, say, then $f^{-1}$ is analytic at $\xi_2$ and satisfies $|f^{-1}| \leqslant \epsilon_n$ on $\Sigma_n$, a curve passing near to $\xi_2$. As we shall see, this will imply that $f^{-1}(\xi_2) = 0$ and this cannot be so for $f^{-1}(\xi_2) \in D$ and $0 \, (= \zeta)$ is in $\partial D$. The general case is not so clear, for the $\Sigma_n$ may intersect $\tau$ both near $\xi_1^*$ and near $\xi_2^*$, and we then have to make a careful selection of one of these two points.

Let $\alpha_n$ be a point on $[\Gamma]$ nearest to $f(z_n)$ and $\beta_n$ a point on $[\Gamma]$ nearest to $f(w_n)$, and extend $\Sigma_n$ to a polygonal curve $\Sigma_n^*$ by the addition of the segments $[\alpha_n, f(z_n)]$ and $[\beta_n, f(w_n)]$. Again, we may assume that $\Sigma_n^*$ is a simple polygonal curve, and so by Theorem 11.4.3, $\Delta - [\Sigma_n^*]$ is the disjoint union of two domains, say $\Delta_n$ and $\Delta_n'$. As $\Sigma_n$ does not meet $[\tau_0]$ and as the segments added to $\Sigma_n$ to form $\Sigma_n^*$ lie in $\mathbf{C}(\alpha, 2r)$ and $\mathbf{C}(\beta, 2r)$ respectively (for almost all $n$) we see from (b) that $[\Sigma_n^*]$ does not meet $[\tau_0]$. As $[\tau_0]$ is a connected set in $\Delta_n \cup \Delta_n'$, either $\xi_1$ and $\xi_2$ are both in $\Delta_n$ or they are both in $\Delta_n'$, and we may assume that they are both in $\Delta_n$. A typical situation is illustrated in Diagram 11.5.2.

The boundary of $\Delta_n$ consists of $[\Sigma_n^*]$ together with one of the two components of $[\Gamma] - \{\alpha_n, \beta_n\}$. Moreover, $\alpha_n \to \alpha$ and $\beta_n \to \beta$ as, for example,

$$\begin{aligned} |\alpha_n - \alpha| &\leqslant |\alpha_n - f(z_n)| + |f(z_n) - \alpha| \\ &\leqslant 2\,|f(z_n) - \alpha|. \end{aligned}$$

Now let $\Gamma_n$ be the arc of $\Gamma$ which, together with $\Sigma_n^*$, forms the Jordan curve bounding $\Delta_n$. We assert that either $\xi_1^*$ or $\xi_2^*$ (and we assume that it is $\xi_2^*$) is not on $\Gamma_n$ and also, that for almost all $n$,

$$[\Gamma_n] \subset [\Gamma_1] \cup \mathbf{C}(\alpha, 2r) \cup \mathbf{C}(\beta, 2r).$$

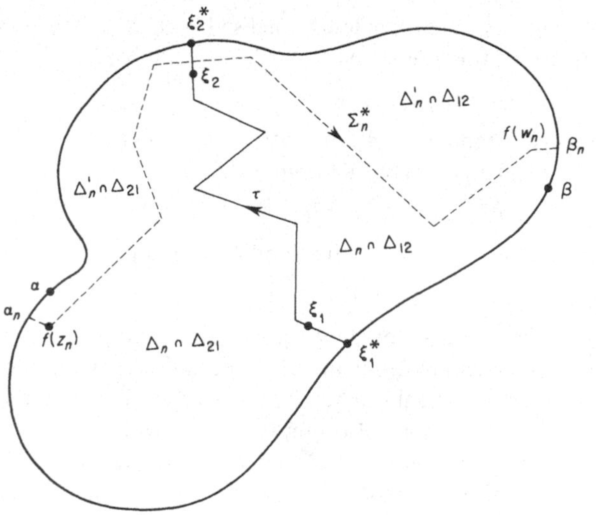


Diagram 11.5.2

Indeed, these results are easily proved when $\Gamma$ is a circle, and as $[\Gamma]$ is homeomorphic to a circle, these deductions remain valid in the general case.

Because $\xi_2^*$ is not on $\Gamma_n$, $\xi_2^*$ lies outside $\Delta_n$. Also, the above inclusion together with the inclusion

$$[\alpha_n, f(z_n)] \subset \mathbf{C}(\alpha, 2r)$$

(and similarly for $\beta$, $\beta_n$ and $w_n$) shows that

$$\partial\Delta_n \cap \mathbf{C}(\xi_2^*, 2r) \subset [\Sigma_n]$$

We are now almost finished. The crucial properties enjoyed by $\xi_2$ are

(a) $\xi_2$ lies in the domain $\Delta_n$;
(b) $\mathbf{C}(\xi_2, r)$ contains a disc $Q$ of positive radius lying outside $\Gamma$ and hence not meeting $\Delta_n$ for any $n$;
(c) $\partial\Delta_n \cap \mathbf{C}(\xi_2, r) \subset [\Sigma_n]$

and

$$|f^{-1}(w)| \leq \epsilon_n \text{ when } w \in [\Sigma_n].$$

These properties are illustrated in Diagram 11.5.3.

The disc $Q$ given in (b) subtends an angle of at least $2\pi/k$ (some positive integer $k$) at $\xi_2$ and we denote by $\Delta_n^p$ the domain obtained by rotating $\Delta_n$ by an angle $2\pi p/k$ about $\xi_2$. The open set

$$\Delta_n^0 \cap \Delta_n^1 \cap \cdots \cap \Delta_n^{k-1}$$

contains $\xi_2$ and the component of this set, say $\Delta^*$, which contains $\xi_2$ lies in $\mathbf{C}(\xi_2, r)$ because $Q$ and its images under the rotations separate $\xi_2$ from the circle $|z - \xi_2| = r$. Write $\omega = \exp(2\pi i/k)$ and without loss of generality, take $\xi_2 = 0$.

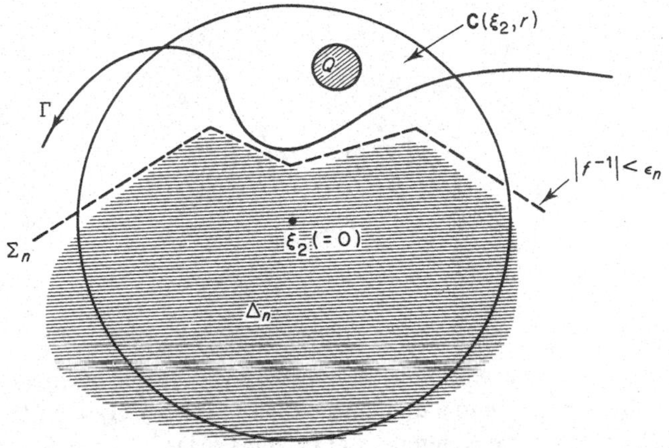


Diagram 11.5.3

Then the function

$$h(z) = f^{-1}(z) f^{-1}(\omega z) \cdots f^{-1}(\omega^{k-1} z)$$

is analytic in $\Delta^*$ and continuous in its closure. At each boundary point of $\Delta^*$, one term in this product is (in modulus) at most $\epsilon_n$, while all other terms are at most $M$, say, where $D \subset \mathbf{C}(0, M)$. The Maximum Modulus Theorem implies that

$$| f^{-1}(0) |^k = | h(0) | \leqslant M^{k-1} \epsilon_n,$$

and as $Q$ and hence $k$ are independent of $n$, we find (by letting $n \to \infty$) that $f^{-1}(0) = 0$. This is the same as $f^{-1}(\xi_2) = \zeta$ and this is false as these two points are in $D$ and $\partial D$ respectively. The proof is completed.

This final part of the proof may be considered as a substantial generalization of the result in Exercise 8.4.6(b).

**Exercise 11.5**

1. Given $z_0, z_1, \ldots, z_n$, show that there is a simple curve joining $z_0$ to $z_n$ in

$$[z_0, z_1] \cup [z_1, z_2] \cup \cdots \cup [z_{n-1}, z_n].$$

[*Hint:* proceed from $z_n$ along the last segment towards $z_{n-1}$ and use induction on $n$.]

2. Prove that $f(\bar{D}) = \bar{\Delta}$ in Theorem 11.5.1 by considering any point $w$ on $\Gamma$ and any sequence $w_n$ in $\Delta$ with $w_n \to w$.
3. Let $f: D \to \Delta$ be a homeomorphism, let $\sigma : [a, b] \to D$ be any curve in $D$ and let $\Sigma = f \circ \sigma$. Prove that given any positive $\epsilon$ there are curves $\sigma^*$ and $\Sigma^*$ $(= f \circ \sigma^*)$ such that $\Sigma^*$ is a polygonal curve in $\Delta$ and for all $t$,

$$| \sigma(t) - \sigma^*(t) | < \epsilon, \qquad | \Sigma(t) - \Sigma^*(t) | < \epsilon.$$

4. Verify (carefully) that in the proof of Proposition 11.5.1,

$$| h(0) | \leqslant M^{k-1} \epsilon_n.$$

# *Appendix*

The sole purpose of this section is to describe a proof, free of integration, of the fact that if $f$ is differentiable in a domain $D$, then $f$ is also analytic in $D$. For a fuller account of this material, see References [17] and [25].

We begin with a function $f$ differentiable in a domain $D$: thus if $\zeta \in D$, then $[f(z) - f(\zeta)]/(z - \zeta)$ is continuous in $D$ but differentiable only in $D - \{\zeta\}$. Because we shall wish later to work with such quotients, we must, at the outset, assume only that $f$ is differentiable in, say $D - \{\zeta_1, \ldots, \zeta_s\}$ and continuous in $D$.

Our first task is to establish the Maximum Modulus Theorem for such functions. If $f^{(1)}$ exists and is non-zero throughout $D$, this follows as for polynomials (see §8.1). In general, $f^{(1)}$ may not exist or may exist and equal zero at certain points.

We select any circle, say $\gamma(t) = z_0 + re^{it}$, $0 \leqslant t \leqslant 2\pi$, which, together with its interior $D_0$, lies in $D$. The image curve $\Gamma$ $(= f \circ \gamma)$ divides the plane into the domains $\Delta_\alpha$ : that is, $\Delta_\alpha$, $\alpha \in A$, are the components of $\mathbf{C} - [\Gamma]$. Further, let $n_\alpha = n(\Gamma, w)$ when $w \in \Delta_\alpha$. It is easy to see that if $n_\alpha \neq 0$ then $\Delta_\alpha \subset f(D_0)$. Indeed, if $w \notin f(\bar{D}_0)$, then we may shrink $\gamma$ to $z_0$ (by decreasing $r$ to 0) without changing $n(\Gamma, w)$ : hence

$$n(\Gamma, w) = n(z_0, w) = 0.$$

The next step is to show that if $\Delta_\alpha \subset f(D_0)$, then $n_\alpha > 0$: this is motivated by the Argument Principle. Observe that once this has been established we do have a weak form of the Argument Principle: if $w$ is not on $\Gamma$ then $n(\Gamma, w) \geqslant 0$ and is positive if and only if $f(z) = w$ has a solution in $D_0$. Of course, we cannot yet count the number of solutions.

To prove the result in the previous paragraph, let us begin by calling $z$ in $D$ a *regular point* if $f^{(1)}(z)$ exists and is non-zero and an *exceptional point* if either $f^{(1)}(z)$ exists and is zero or if $z$ is some $\zeta_j$. If $w$ is the image of *only* regular points in $D_0$, then there can only be a finite number of these, say $z_1, \ldots, z_m$ in the compact set $\bar{D}_0$ (otherwise they accumulate at a point $z'$ and as $f(z_n) = w$, $z'$ is necessarily exceptional). The differentiability of $f$ at the $z_j$ implies that we can write

$$f(z) = w + (z - z_1) \cdots (z - z_m) g(z),$$

where $g$ is continuous and non-zero in $D$. This yields

$$\begin{aligned} n(\Gamma, w) &= \sum_{j=1}^{m} n(\gamma, z_j) + n(g \circ \gamma, 0) \\ &= m \\ &> 0. \end{aligned} \tag{1}$$

We have now proved that if $w$ is in $f(D_0) - [\Gamma]$ and if $n(\Gamma, w) \leqslant 0$, when $w$ is the image of *some* exceptional point. Let us suppose that such a $w$ is also the image of some regular point $z^*$. Then by taking a small circle $\sigma$ about $z^*$ we easily find that

$$n(f \circ \sigma, w) = n(f \circ \sigma, f(z^*)) = 1.$$

This means that all points near $w$ are images of points inside $\sigma$ and hence in $D_0$. But all points $w'$ near $w$ satisfy

$$n(\Gamma, w') = n(\Gamma, w) \leqslant 0$$

and so are necessarily images of exceptional points. Thus if $E$ is the set of exceptional points in $\bar{D}_0$ and if there is a point $w$ which is the image of both regular and exceptional points, then $f(E)$ contains a disc of positive radius.

As $E \subset \bar{D}_0$, $E$ has area at most $\pi r^2$ and (except for a finite set) $f^{(1)}(z) = 0$ when $z \in E$. One can now prove that for any given positive $\delta$, $f(E)$ can have area of at most $\delta^2 \pi r^2$ (this is intuitively clear because near $E$, $f$ shrinks distances by a factor of at most $\delta$): thus $f(E)$ has area zero. This part of the argument can be made quite elementary (in terms of rectangles) and does not depend on any deeper measure theory. We can now conclude that if $w \in f(D_0) - [\Gamma]$ and if $n(\Gamma, w) \leqslant 0$, then $f^{-1}\{w\}$ consists *only* of exceptional points. In fact, we shall see that no such $w$ can exist.

Now choose a point $z^*$ in $D_0$ with $f(z^*) = w$ and $n(\Gamma, w) \leqslant 0$ (if such a $z^*$ exists). By continuity, there is an open disc $Q$ with centre $z^*$ which maps to points close to $w$: for $z$ in $Q$, $n(\Gamma, f(z)) \leqslant 0$ and so $Q$ contains only exceptional points. We deduce that $f^{(1)}$ is zero throughout $Q$ and so $f$ is constant on $Q$. This shows that $Q \subset f^{-1}\{w\}$ and so $f^{-1}\{w\}$ is an open set. It is also closed (by continuity) and, as $D$ is connected, $f^{-1}\{w\} = D$. In other words, $f$ is constant.

To summarize, if $f$ is differentiable on $D - \{\zeta_1, \ldots, \zeta_s\}$ and is continuous but not constant on $D$, and if $w$ is not on $\Gamma$, then $n(\Gamma, w) \geqslant 0$ and is positive if and only if $w \in f(D_0)$. Exactly as for polynomials (§8.1) we conclude that such functions satisfy the Maximum Modulus Theorem.

It is now easy to see that if the functions $g_n$ $(n = 1, 2, \ldots)$ are differentiable in $D$ and if $g_n \to g$ locally uniformly in $D$, then $g$ is also differentiable in $D$. For any $\bar{C}(z, r)$, $r > 0$, contained in $D$ define

$$G_n(h) = \begin{cases} \dfrac{g_n(z+h) - g_n(z)}{h} & \text{if } 0 < |h| \leqslant r, \\ g_n^{(1)}(z) & \text{if } h = 0. \end{cases}$$

The Maximum Modulus Theorem is applicable to $G_n - G_m$ and so

$$\begin{aligned} &|G_n(h) - G_m(h)| \\ &\quad \leqslant r^{-1} \sup \{ | [g_n(z+h) - g_n(z)] - [g_m(z+h) - g_m(z)] | : |h| = r \}. \end{aligned}$$

This shows that as the $g_n$ converge uniformly on $\bar{C}(z, r)$ so too do the $G_n$. Thus $G_n \to G$, say, uniformly on $\bar{C}(z, r)$ and so $G$ is continuous there. We conclude that $G(h) \to G(0)$ as $h \to 0$: in particular,

$$\lim_{h \to 0} \left\{ \lim_{n \to \infty} G_n(h) \right\}$$

or, equivalently,

$$\lim_{h \to 0} \left\{ \frac{g(z+h) - g(z)}{h} \right\}$$

exists. Thus $g$ is differentiable in $D$.

Finally, we show that if $f$ is differentiable in $D$, then

$$F_n(z) = \frac{f(z + 1/n) - f(z)}{1/n} \to f^{(1)}(z)$$

locally uniformly in $D$. As each $F_n$ is differentiable in $D$ the previous result shows that $f^{(1)}$ is differentiable in $D$. Thus derivatives of $f$ of all orders exist in $D$ and we can obtain Cauchy's inequalities and the Taylor expansion of $f$ as before.

First select any $\bar{C}(\zeta, r)$ contained in $D$ and any positive $\epsilon$. Now define

$$\varphi(z, w) = \begin{cases} \dfrac{f(w) - f(z)}{w - z} & \text{if } w, z \in D,\ w \neq z, \\ \\ f^{(1)}(z) & \text{if } w = z. \end{cases}$$

The Maximum Modulus Theorem (applied to functions of $w$) shows that if $z \in C(\zeta, r)$, then

$$|\varphi(z+h, z) - \varphi(z, z)| \leqslant \sup \{ |\varphi(z+h, w) - \varphi(z, w)| : |w - \zeta| = r \}$$

and this upper bound is at most $\epsilon$ if, say, $|z - \zeta| < \delta$ and $|h| < \delta$ (this is an easy straightforward estimate).

We conclude that given any $\zeta$ in $D$ and any positive $\epsilon$, there is a positive $\delta$ such that (with $h = 1/n$) $|F_n(z) - f^{(1)}(z)| < \epsilon$ when $z \in C(\zeta, \delta)$ and $n \geqslant n_0$, say. This shows that $F_n \to f^{(1)}$ uniformly on compact subsets of $D$.

# *Bibliography*


[1] Ahlfors, L. A. (1966). *Complex Analysis* (2nd ed.) McGraw-Hill, New York.

[2] Cartan, H. (1963). *Elementary Theory of Analytic Functions of One or Several Complex Variables*, Addison-Wesley, Reading, Mass. and Hermann, Paris.

[3] Chinn, W. G., and Steenrod, N. E. (1966). *First Concepts of Topology*, Random House, New York.

[4] Depree, J. D., and Oehring, C. C. (1969). *Elements of Complex Analysis*, Addison-Wesley, Reading, Mass.

[5] Dieudonné, J. (1960). *Foundations of Modern Analysis*, Academic Press, New York.

[6] Eggleston, H. G., and Ursel, H. D. (1952). On the lightness and strong interiority of analytic functions, *J. Lond. Math. Soc.*, **27**, 260–271.

[7] Heins, M. (1962). *Selected Topics in the Classical Theory of Functions of a Complex Variable*, Holt, Rinehart and Winston, New York.

[8] Hille, E. (1962). *Analytic Function Theory*, Vol. II, Ginn, New York.

[9] Jameson, G. J. O. (1970). *A First Course on Complex Functions*, Chapman and Hall, London.

[10] Kuratowski, K. (1961). *Introduction to Set Theory and Topology*, Pergamon, Oxford, and Polish Scientific Publishers, Warsaw.

[11] Lang, S. (1977). *Complex Analysis*, Addison-Wesley, Reading, Mass.

[12] Morse, M. (1947). *Topological Methods in the Theory of Functions of a Complex Variable*, Princeton University Press, Princeton.

[13] Nevanlinna, R. and Paatero, V. (1969). *Introduction to Complex Analysis*, Addison-Wesley, Reading, Mass.

[14] Newman, M. H. A. (1964). *Elements of the Topology of Plane Sets of Points*, Cambridge University Press, Cambridge.

[15] Ohtsuka, M. (1970). *Dirichlet Problem, Extremal Length and Prime Ends*, Van Nostrand Reinhold, New York.

[16] Porcelli, P. and Connell, E. H. (1961). A proof of the power series expansion without Cauchy's formula, *Bull. Am. Math. Soc.*, **67**, 177–181.

[17] Read, A. H. (1961). Higher derivatives of analytic functions from the standpoint of topological analysis, *J. Lond. Math. Soc.*, **36**, 345–352.

[18] Redheffer, R. (1969). The homotopy theorems of function theory, *Am. Math. Monthly*, **76**, 778–787.

[19] Rudin, W. (1966). *Real and Complex Analysis*, McGraw-Hill, New York.

[20] Saks, S. and Zygmund, A. (1965). *Analytic Functions* (2nd ed.), Polish Scientific Publishers, Warsaw.

[21] Stoïlow, S. (1938). *Leçons sur les Principes Topologiques de la Théorie des Fonctions Analytiques*, Gauthier-Villars, Paris.

[22] Veech, W. A. (1967). *A Second Course in Complex Analysis*, Benjamin, New York.

[23] Wall, C. T. C. (1972). *A Geometric Introduction to Topology*, Addison-Wesley, Reading, Mass.
[24] Whyburn, G. T. (1950). Open mappings on locally compact spaces, *Mem. Am. Math. Soc.*, No. 1, American Math. Soc., New York.
[25] Whyburn, G. T. (1964). *Topological Analysis*, University Press, Princeton.
[26] Whyburn, G. T. (1968). What is a curve?, Studies in modern topology, *Math. Assoc. America Studies in Mathematics*, **5**, Math. Assoc. America.

## CLASSIFICATION OF REFERENCES BY SUBJECT MATTER

*General reading*: [1], [2], [4], [9], [11], [13], [18], [20].

*Chapters 7, 8 and the Appendix*: [2], [3], [6], [11], [12], [16], [17], [21], [24], [25].

*Sections 11.1, 11.2 and 11.5*: [1], [4], [7], [8], [11], [13], [15], [19], [22].

*Sections 11.3 and 11.4*: [4], [7], [10], [13], [14], [21], [23], [25], [26].

# Index